SAGE Criminology & Criminal Justice: Our Story

Believing passionately in the POWER OF EDUCATION to transform the criminal justice system, **SAGE Criminology & Criminal Justice** offers arresting print and digital content that UNLOCKS THE POTENTIAL of students and instructors. With an extensive list written by renowned scholars and practitioners, we are a RELIABLE PARTNER in helping you bring an innovative approach to the classroom. Our focus on CRITICAL THINKING AND APPLICATION across the curriculum will help you prepare the next generation of criminal justice professionals.

Criminal Investigations Today

Criminal Investigations Today

The Essentials

Richard M. Hough
University of West Florida

Los Angeles | London | New Delhi
Singapore | Washington DC | Melbourne

FOR INFORMATION:

SAGE Publications, Inc.
2455 Teller Road
Thousand Oaks, California 91320
E-mail: order@sagepub.com

SAGE Publications Ltd.
1 Oliver's Yard
55 City Road
London EC1Y 1SP
United Kingdom

SAGE Publications India Pvt. Ltd.
B 1/I 1 Mohan Cooperative Industrial Area
Mathura Road, New Delhi 110 044
India

SAGE Publications Asia-Pacific Pte. Ltd.
18 Cross Street #10-10/11/12
China Square Central
Singapore 048423

Acquisitions Editor: Jessica Miller
Content Development Editor: Adeline Grout
Editorial Assistant: Sarah Manheim
Marketing Manager: Jillian Ragusa
Production Editor: Sarah Downing
Copy Editor: Talia Greenberg
Typesetter: C&M Digitals (P) Ltd.
Proofreader: Jeff Bryant
Indexer: Amy Murphy
Cover Designer: Dally Verghese

Copyright © 2021 by SAGE Publications, Inc.

All rights reserved. Except as permitted by U.S. copyright law, no part of this work may be reproduced or distributed in any form or by any means, or stored in a database or retrieval system, without permission in writing from the publisher.

All third party trademarks referenced or depicted herein are included solely for the purpose of illustration and are the property of their respective owners. Reference to these trademarks in no way indicates any relationship with, or endorsement by, the trademark owner.

Printed in Canada

Library of Congress Cataloging-in-Publication Data

Names: Hough, Richard M., author.

Title: Criminal investigations today: the essentials / Richard M. Hough, University of West Florida.

Description: Los Angeles: SAGE, [2021] | Includes bibliographical references and index.

Identifiers: LCCN 2019030657 | ISBN 9781544308005 (paperback) | ISBN 9781544396590 (epub) | ISBN 9781544396606 (epub) | ISBN 9781544396583 (pdf)

Subjects: LCSH: Criminal investigation—United States. | Police—United States. | Criminal justice, Administration of—United States.

Classification: LCC HV8073 .H828 2021 | DDC 363.250973—dc23
LC record available at https://lccn.loc.gov/2019030657

This book is printed on acid-free paper.

20 21 22 23 24 10 9 8 7 6 5 4 3 2 1

Brief Contents

Preface	xix
Digital Resources	xxiii
Acknowledgments	xxv
About the Author	xxvii

Chapter 1	Criminal Investigation Then and Now	1
Chapter 2	Forensic Evidence: The Crime Scene and the Laboratory	23
Chapter 3	People in the Process: Victims, Suspects, Witnesses, and Detectives	49
Chapter 4	Managing the Criminal Investigation	73
Chapter 5	Searches, Seizures, and Statements	97
Chapter 6	Homicide, Death, and Cold Case Investigations	117
Chapter 7	Assault, Battery, Robbery, and Sex Crimes	143
Chapter 8	Familial Crimes	165
Chapter 9	Burglary, Theft, White-Collar Crime, and Cybercrime	185
Chapter 10	Drug Crimes, Organized Crime, and Gangs	207
Chapter 11	Terrorism and Homeland Security	227
Chapter 12	Criminal Investigation in Court	245

Glossary	265
References	271
Index	279

Detailed Contents

Preface	xix
Digital Resources	xxiii
Acknowledgments	xxv
About the Author	xxvii

Chapter 1 • Criminal Investigation Then and Now   1

Running Case: Possible Homicide at the Floridan	**1**
Introduction: Criminal Investigation Then and Now	3
The History of Criminal Investigation in London	3
Pinkerton's National Detective Agency	4
First Criminal Investigation in the United States	4
Development of State and Federal Law Enforcement Agencies	4
Development of the FBI	5
The Goals of Investigation	6
What Makes a Successful Investigation?	7
Four Measures of a Successful Investigation	8
Types of Criminal Investigations	8
Which Agencies Investigate Which Crimes?	9
Changing Times Equals Changing Crimes and a Need for Law	
Enforcement to Adapt	10
Four Stages of Criminal Investigation	10
Two Phases of Criminal Investigation	11
Preliminary Investigation	11
How It's Done: First Responding Officers	**12**
Follow-Up Investigation	13
Organizing the Investigative Unit	14
The Investigator—General Assignment and Specialized	14
Task Force Investigations	15
Contemporary Law Enforcement and the Role of Criminal Investigation	16
Community-Oriented Policing	16
Problem-Oriented Policing	17
Intelligence-Led or Evidence-Based Policing	18
Science and Technology in Criminal Investigation	19
Explore This	19
Summary	20
Key Terms	21
Discussion Questions	21

Chapter 2 • Forensic Evidence: The Crime Scene and the Laboratory — 23

Running Case: Crime Scene Analysis of Evidence at the Floridan — 24
Introduction — 25
How It's Done: Field Tests — 26
The Developing Science of Criminal Investigation — 27
    The Prescientific Era — 27
    The Scientific Era — 27
    The Technological Era — 28
Methods of Collecting and Processing Evidence — 29
    Crime Scene Searches: Inside, Outside, All Around — 29
How It's Done: Where Would You Seek Outside Assistance? — 31
Types of Physical Evidence and Its Processing — 32
    Advanced Scientific Methods and Sophisticated Technology — 32
    Collecting and Preserving DNA Evidence — 33
    Firearms, Explosives, and Arson — 33
    Expert Witnesses — 35
Equipment Utilized in Investigations — 36
    Basic Crime Scene Equipment — 36
    Advanced Crime Scene Equipment — 37
    Personnel Safety — 37
The Contemporary Forensic Laboratory — 37
Personnel — 38
    Crime Scene Technicians at the Initial Scene — 38
    Laboratory Technicians — 39
Documenting the Crime Scene and Actions — 40
    Case Management Software — 40
    Photography — 41
How It's Done: Crime Scene Photography — 42
    Diagramming — 42
    Report Writing and Notetaking — 43
The CSI Effect — 44
Explore This — 45
Summary — 45
Key Terms — 46
Discussion Questions — 46

Chapter 3 • People in the Process: Victims, Suspects, Witnesses, and Detectives — 49

Running Case: The Detectives Interview Two Witnesses From the Floridan — 49
Introduction — 52
Victims — 52
    Victimology — 54
How It's Done: Victims' Rights — 55

Special Victims	55
Victim Advocate	55
Suspects	56
Investigators	57
Investigator Characteristics	58
Investigator Training	59
Witnesses and Informants	60
Eyewitness Identification and the Nonsmoking Gun	60
Informants	62
Interviews	62
Setting the Tone	63
Guiding the Interview	63
Interrogations	64
Planning the Interrogation	65
Interrogation Methods	65
How It's Done: Good Cop–Bad Cop	**67**
Technology	67
Detection of Deception Devices	67
Explore This	69
Summary	69
Key Terms	69
Discussion Questions	70

Chapter 4 • Managing the Criminal Investigation 73

Running Case: The Detectives Interview Frank Denney, Suspect in Homicide at the Floridan	**73**
Introduction	76
Investigation as a Process	77
Crime Analysis	78
Preliminary Investigation	79
Case Assignment and Solvability Factors	80
How It's Done: Solvability Factors	**81**
Follow-Up Investigation	81
Sources of Information	82
Coordination With the Prosecutor	83
Supervising the Investigative Unit	83
Case Management Software	84
Pitfalls, Fallacies, and Sloppiness in Investigations	86
Cognitive or Confirmation Bias	86
Insufficient Time	87
Framing and Groupthink	87
Probability Errors	87
Strategies to Avoid Criminal Investigative Failures	88
Developing an Ongoing Relationship With the Media	89
Improving Productivity	89
Civilianization	90
Increased Investigative Duties Performed by Patrol Officers	90

Training	91
Adapting to New Technology and the Changing Nature of Crime	92
Benefits of Greater Productivity	92
Explore This	93
Summary	93
Key Terms	94
Discussion Questions	94

Chapter 5 • Searches, Seizures, and Statements   97

Running Case: The Detectives Begin Documenting the Floridan Case and Applying for Search Warrants	**97**
Introduction	99
Fourth Amendment	99
Search Warrants	100
Search Warrant Exceptions	102
How It's Done: Most Common Search Warrant Exceptions	**102**
Voluntary Consent	103
Exigent Circumstances and Emergency Searches	103
Stop and Frisk and Plain View	104
Open Fields	105
Abandoned Property	105
Protective Sweeps	105
The Carroll Doctrine	106
The Exclusionary Rule	106
Making the Arrest	107
Fifth Amendment	108
Suspect Statements	108
Lineups and Showups	110
How It's Done: Live Lineup	**112**
Sixth Amendment	112
Explore This	113
Summary	113
Key Terms	114
Discussion Questions	114

Chapter 6 • Homicide, Death, and Cold Case Investigations   117

Running Case: The Detectives Attend Crime Scene Investigation in the Morgue on Floridan Victim and Begin Preparing for Courtroom Trial	**118**
Introduction	119
Scope of the Problem	120
Definitions and Legal Classifications of Cases	122
Medical Death Classifications	123
Preliminary Investigation and First Officers on the Scene	124

First Officers Responsibilities	124
Managing Suspects	125
Securing the Crime Scene	125
Managing Witnesses and Completing Neighborhood Canvass	126
Identity	126
Arrival of the Detectives	127
Medical Examiners, Coroners, and the Autopsy	127
Medical Examiner or Coroner (ME/C)	128
Autopsy	128
Time of Death	128
Other Specialists	130
Criminal Homicide Methods	130
Firearm, Edged Weapon, and Blunt Force Wounds	130
Asphyxiation	131
Poisoning	132
Suicide	133
Challenges in Death Investigations	134
Equivocal Death	134
Clearing Homicide Cases	135
Working With the Media and the Public	135
Death Notifications and Survivors	136
Cold Cases	137
Cold Case Investigations	137
Serial and Mass Murder	138
Serial Murder	138
Mass Murder	138
Explore This	139
Summary	139
Key Terms	139
Discussion Questions	140

Chapter 7 • Assault, Battery, Robbery, and Sex Crimes — 143

Running Case: Robbery at the 7–10 Split Bowling Alley	**143**
Introduction	145
Assault and Battery	146
Aggravated Assault	146
Documentation	148
How It's Done: Documenting the Crime Scene	**149**
Confrontational Violence	150
Robbery	151
Documentation	152
Robbers' Methods of Operation	152
Robbery Statistics	154
Victimology	154
Sex Crimes	154
Megan's Law	155
Rape and Sexual Assault Typologies	155

How It's Done: National Protocol for Sexual Assault Medical Forensic Examinations — **156**
    Investigating Sexual Battery/Rape — 157
    Sexual Abuse and College Rape — 159
    Criminal Deviant Sex Acts — 159
How It's Done: Sexual Assault Examinations — **160**
Explore This — 161
Summary — 161
Key Terms — 162
Discussion Questions — 163

Chapter 8 • Familial Crimes — 165

Running Case Study: Intimate Partner Violence and Child Abuse — **165**
Introduction — 167
Domestic Violence — 167
    Investigating Domestic Violence — 168
Intimate Partner Violence (IPV) — 169
How It's Done: Determining Level of Risk for Intimate Partner Violence — **170**
    Changing Views on Intimate Partner Violence — 170
    Correlation Between Intimate Partner Violence and Child Abuse — 171
    Restraining Orders and Arrest of Intimate Partner Violence Offenders — 171
    Responding Officers' Safety — 173
    Stalking — 174
How It's Done: Investigating Suspected Stalking Crimes — **175**
How It's Done: Law Enforcement Stalking Protocol — **176**
    LGBTQ Victims — 176
Child Abuse — 177
    Incidence of Child Abuse — 177
    Risk Factors for Child Abuse — 178
    Child Abuse Legislation — 178
    Investigating Child Abuse — 179
    Sexual Abuse of Minors — 180
    Infrequent Situations of Sudden Death or Mistreatment — 180
Elder Abuse — 181
Explore This — 182
Summary — 182
Key Terms — 183
Discussion Questions — 183

Chapter 9 • Burglary, Theft, White-Collar Crime, and Cybercrime — 185

Running Case Study: Multiple-Car Burglary on Halloween Night — **185**
Introduction — 187
Burglary — 189

Types of Burglaries	189
Preliminary Investigation of Burglary	190
Types of Burglars	191
How It's Done: Burglary Investigation	**192**
Theft	192
Preliminary Investigation of Theft	192
How It's Done: Organized Retail Crime (ORC)	**194**
Employee Theft	194
Motor Vehicle Theft	195
Investigating Motor Vehicle Theft	196
White-Collar Crime	196
Fraud	197
Embezzlement	198
Identity Theft	199
Warning Signs of Identity Theft	199
Reporting Identity Theft	200
Cybercrime	200
Types of Cybercrime	201
Investigating Cybercrime	201
Cybercrime Targeting in the United States	202
Explore This	203
Summary	203
Key Terms	203
Discussion Questions	204

Chapter 10 • Drug Crimes, Organized Crime, and Gangs　207

Running Case Study: Undercover Work—Regional Gang Surveillance and Investigating a Potential Grow House	**207**
Introduction	209
Drug Crimes	210
History of the U.S. Illegal Drug Problem	210
Defining and Classifying Controlled Substances	211
Scope of the Problem	213
Fentanyl: A Synthetic Opioid	213
Investigating Drug Crimes	214
How It's Done: Handling Narcotics Evidence	**215**
Undercover Operations	215
How It's Done: Undercover Investigations	**216**
Grow Houses and Labs	217
Knock and Talk	218
Asset Forfeiture	218
Organized Crime	219
Extortion	219
Investigating Organized Crime	220
Gangs	220

Identifying Gangs	221
Combating Gang Crime	222
Public Response	222
Explore This	223
Summary	223
Key Terms	223
Discussion Questions	224

Chapter 11 • Terrorism and Homeland Security — 227

Running Case Study: Hate Crime–Motivated Mass Shooting Planned at Sullivan's Spa	**227**
Introduction	229
Terrorism	230
Background	231
Domestic Terrorism	232
Recruiting Homegrown Violent Extremists	233
Left- and Right-Wing Groups	233
Militia Groups	234
International Terrorism	234
Suicide Bombings	225
How It's Done: Bomb Threat Checklist	**236**
Lone Wolf	236
Hate Crimes	237
Role of the Media	237
Homeland Security	239
State and Local Level	239
Federal Level	239
Fusion Centers	240
Explore This	241
Summary	241
Key Terms	242
Discussion Questions	242

Chapter 12 • Criminal Investigation in Court — 245

Running Case Study: Possible Homicide at the Floridan—Courtroom Trial	**245**
Introduction	247
The Process: Screening and Filing	248
Stages	249
The Pretrial Stage	249
The Arraignment Stage	249
Juror Selection	250
The Trial	250
Building the Case and Charging Decisions	251
How It's Done: Overarching Principles for Prosecutors	**252**

Evidentiary Issues	254
Defense Attorney Functions and Tactics	254
Defenses	256
Failure of Proof Defenses	256
Alibi	257
Justification Defenses	257
Excuse Defenses	257
Ethical Issues	258
Plea Bargaining	259
Sentencing	261
Explore This	261
Summary	262
Key Terms	262
Discussion Questions	263

Glossary	265
References	271
Index	279

Preface

Criminal investigations, when they are studied, should be interesting. They may not always be exciting, but the subject matter and the importance of the function in society render them inherently interesting. Given that fact, an instructor needs to properly assemble and lay out in front of the student the various interesting aspects of criminal investigations as well as a clear and sequential explanation of the steps in that process. An important aim of the text is to cover the essentials of criminal investigation, not write an encyclopedic document that hinders rather than helps instructors and students examine major topics and themes effectively in a variety of semester or session lengths and formats.

The text provides a brief look back at the development of criminal investigations and continues through observations about contemporary investigative methods and the importance of the investigative function in law enforcement agencies within a democratic society. Next, the book explains the general parameters of a criminal investigation, with comments about the different actors along the way. In addition to understanding the law as it applies to investigations and the tools used, students need to understand the scope as well as the limitations of the methods and tools currently available to investigators.

Criminal Investigations Today: The Essentials discusses the issues surrounding the all-important search. This includes law and rules about when and how law enforcement may conduct a search as well as guidelines on how such searches are physically carried out. Once a search reveals an item of physical evidence, that evidence must be gathered and delivered either for storage or further examination and scientific analysis. These procedures are also discussed, including the professionals who carry out these tasks in small, midsized, and large agencies. The book goes on in Chapter 3, "People in the Process," to give students a deeper understanding of victims, suspects, witnesses, informants, and detectives. In Chapter 5, "Searches, Seizures, and Statements," I discuss an appreciation for the interviews and interrogations that seek to illuminate what each of these participants in the criminal investigation process may know.

A criminal investigation textbook or a course about criminal investigation rarely comments on the causes of criminality. This is perhaps an oversight, because effective investigators try whenever possible to understand *why* a crime was committed to gain insight into *who* may have done it. *Criminal Investigations Today* discusses at some length the frequently examined theoretical approaches for various crimes, with commentary on how such knowledge can aid investigators.

Another unique aspect of the book is within the chapter addressing management of criminal investigations: a section discussing some of the pitfalls, fallacies, and most frequent areas of mistakes in criminal investigations. It is often said that we learn more from our mistakes than from our successes. With the stakes so high in a criminal investigation, it behooves investigators to flatten that learning curve by examining frequent errors made by detectives who came before.

A number of the most common crimes committed and examined are covered in chapters in the second part of the book. Topics are combined with respect to their frequency or traditional placement with other topics. These include homicide and cold cases, robbery, crimes against children, intimate partner violence, sex crimes, burglary and theft, drug crimes, and arson and explosives. Next, the book turns to the critically important area of information technology and cybercrime. Not only are these topics of vital importance in our increasingly technological society, they are areas of great interest for students and present potential job opportunities as well. Terrorism and homeland security have not been far from the collective conscious of society for two decades. The book examines criminal investigative efforts in this area, and how federal, state, and local law enforcement agencies, as well as private security, work together to address these challenges.

The text contains a chapter on managing the criminal investigation. This topic is important for students to understand. Investigations presented in books, movies, and television shows generally focus on one crime and one investigator. The reality is that for investigations to be successful and for agencies to effectively pursue the causes and perpetrators of crime, the overall process of investigations must be guided by policies, procedures, and monitored and supervised through human and technological resources. Individual investigators exist within the network of their agency and the criminal justice system. It is important for students to explore this perspective and reality. Finally, the book includes a chapter that addresses criminal investigations in court. Much crime is never discovered. A good bit of crime is never reported. Some reported crime is not adequately investigated. Crime that is investigated may not yield a suspect, an arrest, or a conviction. But what happens to cases that do make their way from commission through discovery to suspect development to arrest and then prosecution? This chapter explains the dynamics and components of cases that make it all the way.

We have seen the increasing role the social sciences have played in the construction of policies and programs in criminal justice in U.S. society. I did not want to set considerations of policy as a stand-alone chapter, nor pretend such things are not implicated in learning about how to investigate, so appropriate policy impacts are noted to some extent. I hope that instructors will incorporate discussion of how policies come about. I always ask students to bring in current news articles related to criminal investigations that might serve as relevant discussion points. I hope current users of the book will do this as well.

The chapter features reinforce student learning of concepts through examples, critical thinking questions, a series of running cases to open each chapter, and material under the titles of "How It's Done," "Explore This," and chapter summaries to suggest further relevant learning activities. Learning objectives are included at the beginning of each chapter. We include on the ancillaries website a series of career profiles of professionals involved in investigations and topics in the book. The book includes all of the requisite features and ancillaries to allow an instructor to effectively teach a course on criminal investigation. The combination of materials in one book with PowerPoints®, chapter outlines, etc., provides every criminal justice or criminology program the ability to offer a course that is always in demand. The ability for a faculty member to have all the materials necessary to teach this popular course is a tremendous advantage. The material is turn-key. Even someone who specializes in the topic of investigations often prefers to have ancillaries to edit or tailor to their own approach. The ancillaries cover important concepts and points. A pure academic orientation often results in an unbalanced emphasis on only theory. This book provides more of what interests our students while also requiring them to think critically and apply what they have learned.

I bring a perspective to this textbook as a "pracademic." While I have taught college and university criminal justice courses for more than 30 years, I have taught law enforcement recruits, officers, and investigators for even longer. This is combined with my own law enforcement career and a significant part of my time being involved as a subject matter expert and expert witness in police practices and investigations. I am active in both academic and law enforcement groups, and I hope that all of this adds up to an approach that students find helpful.

Digital Resources

edge.sagepub.com/houghci

Instructor Resources

SAGE edge for instructors supports your teaching by making it easy to integrate quality content and create a rich learning environment for students with:

- a **password-protected site** for complete and protected access to all text-specific instructor resources;
- **test banks** that provide a diverse range of ready-to-use options that save you time. You can also easily edit any question and/or insert your own personalized questions;
- **editable, chapter-specific PowerPoint® slides** that offer complete flexibility for creating a multimedia presentation for your course;
- **lecture notes** that summarize key concepts by chapter to help you prepare for lectures and class discussions;
- **career profiles** from practitioners in the field that offer helpful career insights to students;
- carefully selected **video and multimedia links** that enhance classroom-based exploration of key topics; and
- **a course cartridge** for easy LMS integration.

Student Resources

SAGE edge for students enhances learning, it's easy to use, and offers:

- an **open-access site** that makes it easy for students to maximize their study time, anywhere, anytime;
- **eFlashcards** that strengthen understanding of key terms and concepts;
- **eQuizzes** that allow students to practice and assess how much they've learned and where they need to focus their attention; and
- carefully selected **video and multimedia links** that enhance classroom-based exploration of key topics.

Acknowledgments

My acknowledgments begin with the person whose insight and intellect I seek out on every project, Dr. Kimberly McCorkle. Kimberly and I cowrote the successful SAGE textbook *American Homicide* together, but her role as the Vice-Provost of the University of West Florida kept her from signing on again to toil through this project. Nonetheless, Dr. McCorkle provided invaluable feedback on every chapter and helped discuss the original concept to provide instructors with a more manageable text to form the foundation for an effective criminal investigations course. I am grateful to the reviewers who gave of their valuable time and experience to review the initial proposal as well as groups of chapters along the way. Measured and professional reviews intended to help one another are gifts—thank you.

The reviewers were:

Jeffrey J. Ahn, University of Phoenix

Samantha Carlo, Miami Dade College

Scott Duncan, Bloomsburg University of Pennsylvania

Seth A. Dupuis, Springfield Technical Community College

Gregg W. Etter Sr., University of Central Missouri

Patrick J. Faiella, Massasoit Community College

James E. Guffey, National University

Daniel Hebert, Springfield Technical Community College

Michael Herbert, Bemidji State University

Daniel Holstein, University of Nevada, Las Vegas

Richard J. Mangan, Florida Atlantic University

Melissa A. Matuszak, Riverside City College

Tiffany L. Morey, Southern Oregon University

William James Morgan Jr., SUNY Erie

Michael J. O'Connor, Upper Iowa University

Ralph Peters, Police Chief Natchitoches Police Dept. (Retired)

Kelly Sue Roth, Bloomsburg University

David M. Scott, University of Texas at Tyler

Chernoh M. Wurie, Virginia Commonwealth University

I thank Jerry Westby, former publisher at SAGE, for initially approaching me to write this book. Jessica Miller, acquisitions editor for Criminology & Criminal Justice at SAGE, picked up the ball, and she ran with it! I am grateful for the meetings and conversation with Jessica along the way. Adeline Grout, associate content development editor for Criminology & Criminal Justice at SAGE, shepherded the final phase through to production, and her responsiveness is greatly appreciated. Solid suggestions and astute observations combined with an unfailingly pleasant and supportive approach made Adeline a joy to work with. Along the way, we gained the help and talents of Louise Bierig, freelance writer and development editor. Louise prodded and polished the manuscript into better shape, but her work really shone in her conceptualization of our chapter-opening running cases. And thanks as well to Talia Greenberg, copy editor, for much of that important final polishing during copyediting. And finally, Sarah Downing, production editor for SAGE, is due high praise for chasing, pushing, and pulling the book across the finish line.

My thanks to you who are reading and using the book and, hopefully, sending me constructive feedback for future enhancements. Finally, my gratitude to all of those students in college and police academy classrooms, and colleagues in criminal justice, who took part in the knowledge construction over the last 40 years that resulted in how I think about the world of policing.

Richard M. Hough

About the Author

Dr. Richard M. Hough has taught criminal justice and criminology courses in the college and university setting for 30 years and has taught investigative methods and other law enforcement and corrections topics in the academy and in-service law enforcement setting for 35 years. This has added to the author's extensive professional experience, including as a criminal investigator. Dr. Hough has held increasingly responsible positions in law enforcement and has served as director of law enforcement and corrections in sheriff's offices, as well as superintendent of detention centers for the Florida Department of Juvenile Justice. Dr. Hough has served as both a law enforcement and corrections academy director. He is currently a faculty member in the Department of Criminology and Criminal Justice, as well as the Department of Administration and Law, at the University of West Florida.

Dr. Hough has conducted more than 100 training seminars, conference presentations, and international briefings on criminal justice issues, and he is the primary instructor for contemporary policing practices, homicide investigation, and gangs and hate groups at the regional law enforcement academy in Pensacola, Florida. Dr. Hough is a member and vice president of the Homicide Research Working Group (HRWG), the Police Executive Research Forum (PERF), the International Homicide Investigators Association (IHIA), the International Association of Chiefs of Police (IACP), the Criminal Investigation Resource Network (CIRN), the Southern Criminal Justice Association (SCJA), and the Academy of Criminal Justice Sciences (ACJS). Dr. Hough is one of only two American International Ambassadors to the British Society of Criminology. He has been interviewed by local, regional, and national news media, and has appeared on radio and television speaking on criminal justice issues. He has published journal articles and book chapters. His coauthored SAGE text *American Homicide,* with Dr. Kimberly McCorkle, is currently in its second edition. Dr. Hough actively consults as an expert witness on police and correctional practices, criminal investigations, and the use of force.

William A. Pinkerton with railroad special agents.
PPOC, Library of Congress via Wikimedia Creative Commons

1 Criminal Investigation Then and Now

RUNNING CASE: POSSIBLE HOMICIDE AT THE FLORIDAN

Introduction: Criminal Investigation Then and Now
  The History of Criminal Investigation in London
  Pinkerton's National Detective Agency
  First Criminal Investigation in the United States
    Development of State and Federal Law Enforcement Agencies
    Development of the FBI

The Goals of Investigation
  What Makes a Successful Investigation?
    Four Measures of a Successful Investigation

Types of Criminal Investigations
  Which Agencies Investigate Which Crimes?
  Changing Times Equals Changing Crimes and a Need for Law Enforcement to Adapt

Four Stages of Criminal Investigation
  Two Phases of Criminal Investigation
    Preliminary Investigation

How It's Done: First Responding Officers
    Follow-Up Investigation

Organizing the Investigative Unit
  The Investigator—General Assignment and Specialized

Task Force Investigations

Contemporary Law Enforcement and the Role of Criminal Investigation
  Community-Oriented Policing
  Problem-Oriented Policing
  Intelligence-Led or Evidence-Based Policing

Science and Technology in Criminal Investigation

Explore This

Summary

Key Terms

Discussion Questions

LEARNING OBJECTIVES

1.1 Summarize the history of criminal investigation in the United States and abroad.

1.2 Discuss the goals of criminal investigation.

1.3 Describe the different types of criminal investigations.

1.4 List and explain the role and tasks of patrol officers during the preliminary and follow-up investigations.

1.5 Discuss the reasons why the investigative function is organized in different agencies.

1.6 List the benefits of utilizing a task force approach to investigations.

1.7 Compare and contrast the policing models discussed in the chapter.

1.8 Explain the impact of science and technology on criminal investigation.

Running Case: Possible Homicide at the Floridan

Detective Bradley Macon was racing down Third Street in his unmarked cruiser when the buzzing of his smartphone cut into his dream. He untangled himself from the bedclothes and grabbed for the phone on the nightstand, knocking it to the floor. As he finally grasped the phone, its glaring light showed him it was 3:45 a.m. At that hour, he knew the call would be from the communications center, and because he was one of the on-call major crimes detectives for the week, he expected that the call would be for a possible homicide (Signal 5).

The dispatcher, Rayna, stated just the facts, with no sympathy for waking him up. "Up and at 'em, Bradley. You've got a Signal 5 near the Floridan off of Ninth Street."

Bradley Macon smiled when he heard the crime location. The Floridan was an old club with an equally old reputation for trouble.

Rayna continued with the facts of the case. "The patrol units are parked in a narrow alleyway called Milton Way, which runs behind the bar. One of the bar employees left after closing. As he reached the mouth of Milton Way, he heard yelling and then several gunshots. He ducked away from the alley and called 9-1-1. While on the phone with the dispatcher, he saw someone run the opposite direction on Milton Way. It looked like the person fleeing was running away from someone on the ground."

With all the facts in mind, Bradley Macon signed off, dressed, and walked out of his apartment. As soon as he opened the exterior door and stepped into his driveway, he groaned. How could it be so hot and sticky at four in the morning? At least it wasn't raining: Rain had a way of affecting crime scene evidence.

He climbed in his assigned unmarked car and switched on the police radio, then asked the dispatcher to have the patrol supervisor on-scene meet with him on one of the tactical radio channels. These were used for operations or chatter that did not need to interfere with normal radio traffic by other officers responding to calls. Sergeant Kevin Lloyd came up on the radio channel and let Brad know that several of his officers had secured the scene. One was talking to the bar employee, who would be treated as a witness. The homicide sergeant, Mike Joseph, was also en route and had called for crime scene investigators to respond. Detective Bradley Macon asked Kevin Lloyd to have officers begin canvassing the immediate area to determine if any other businesses were open and to locate anyone who may have seen or heard anything that could be connected to the possible murder.

Bradley Macon was satisfied the scene was secured. A number of people were going to be writing a number of reports this morning, he thought. The preliminary investigation was underway. For now, all he needed to do was drive to the scene.

As his cruiser skimmed through the silent streets, Bradley Macon realized he still felt excited each time he was assigned as the lead detective on a serious case. After 6 years serving as a patrol officer, with the last year as a field training officer (FTO),

Bradley Macon had been assigned to the major crimes squad 2 years ago. The exciting part, in a homicide case, was that everything he had learned in his previous 8 years could come into play. He needed to coordinate all the moving parts and make sure the whole group of investigators worked as a team. He also felt grateful that he worked at an agency where investigators received the additional support they needed to handle complicated investigations. He knew officers from smaller departments in the area who often mentioned how tough it was to properly handle large cases without outside help.

As Detective Bradley Macon pulled his car to a halt across the street from the entrance to Milton Way, he was running through a series of tasks the team would have to complete as the preliminary investigation gave way to the follow-up stage.

How do you imagine detectives and other personnel mentally prepare for a homicide investigation? Even at this point, how do you think the follow-up investigation will differ from the preliminary one described?

Introduction: Criminal Investigation Then and Now

The History of Criminal Investigation in London

As with most practices in American jurisprudence, our contemporary model of investigative efforts has its origins in England and Western Europe. In the 1750s, local writer turned magistrate James Fielding worked hard to curtail rampant crime in the Westminster area around London by appointing a night watch and utilizing men known as "thief-takers" to investigate various crimes and bring the accused before the magistrate to answer for their crimes. Fielding's group became known as the Bow Street Runners because of its base of operations on Bow Street, which became the location of the Magistrate's Court. The thief-takers were ostensibly a voluntary group, but they were compensated on commission from private citizens seeking return of stolen goods, or by securing convictions at court. The arrangement resulted in many thief-takers coming from the ranks of London's criminals and sometimes working in concert with confederates to initiate crimes and then "solve" them to receive a reward for returning the property.

Eventually replacing the Bow Street Runners and other groups in London was the London Metropolitan Police Department, founded by Home Secretary Sir Robert Peel in 1829. The Met, as it became known, was a primarily uniformed department, and plainclothes investigators were not widely used in its early years because of the unease many Londoners had with groups such as the thief-takers. The Met eventually formed a group of detectives at their headquarters in the area of London known as Scotland Yard. The name of the area quickly became the way of referring to the investigative division, which has been called Scotland Yard since.

Pinkerton's National Detective Agency

In 1850, Scottish-born Allan Pinkerton (1819–1884) formed the first private detective agency in the United States. Based in Chicago, where Pinkerton had been the city's first police detective, the Pinkertons operated alongside government agencies to provide both private services for clients and quasi-governmental investigation and apprehension of a wide range of criminals.

With the company's distinctive symbol of the unblinking eye and the motto "We never sleep," Pinkerton detectives soon became known as "private eyes." The Pinkerton agency had responsibility for the protection of President Abraham Lincoln during much of his time in office, and the Pinkerton agents tracked down and apprehended many notorious outlaws. Pinkerton today is a company that operates worldwide risk management, investigative, protective, and intelligence services to private and corporate clients.

First Criminal Investigation in the United States

As the United States, Great Britain, and other countries began experiencing all of the benefits and challenges of the Industrial Revolution, economic and workforce shifts spelled a new era of crime and an increased need for law enforcement efforts. In the United States, the transition from a predominantly agrarian society to one of industry drew significant numbers of people to the swiftly growing cities. The unprecedented pace of workers from the countryside and new immigrants thronging to the places of industry created an environment and conditions that allowed criminals to flourish as well. In addition to the inadequate infrastructure, education resources, health protections, and other aspects of developing cities, the accelerated urbanization brought many people of varied means into close contact and gave opportunities for crime and circumstances for conflict.

The first police departments in America arose in response to the dynamics of growth in larger cities, modelling themselves somewhat on Peel's British model with uniformed forces patrolling the streets in an effort mainly to address crimes of disorder. Beginning in Boston (1837), New York (1844), and Philadelphia (1854), most major cities established municipal police forces in the span of a few decades. The violence and vice of dense urban areas were compounded by rioting between and among various groups. Lawlessness in the western expansion of the country also brought the need for additional law enforcement mechanisms; the Americanized "shire-reeve" of England became the county sheriff, as states and the federal government began to form agencies to deal with specific crime problems. As with policing in England, investigations by police were not common, only "peace-keeping."

At the beginning of organized American law enforcement, policing consisted almost entirely of patrolling, rarely becoming involved in the investigation of a crime that had already occurred. Private citizens might employ individuals to recover items that had been stolen. The manner in which "recovery" occurred could be violent or involve a scam to deceive the victim into paying for recovered goods that the purported investigator had originally arranged to have stolen.

DEVELOPMENT OF STATE AND FEDERAL LAW ENFORCEMENT AGENCIES

The Texas Rangers were established before the state of Texas was even formally admitted to the Union. With duties ranging from military service to tracking train robbers, the storied Texas Rangers became the template for many other state agencies that followed. After the creation by Congress of the Department of Justice in 1870, several small (by today's standards) enforcement organizations were formed. The U.S. Secret Service began its task of investigating counterfeit currency at the end of the Civil War. The Secret Service was a part of the U.S. Department of the Treasury (now in the Department of Homeland Security), and in 1901 the agency took on the duties of presidential and, eventually, foreign dignitary protection. In the 1890s the U.S. Postal Service took on criminal investigations involving the mail, and the Bureau of Immigration was formed with the Treasury Department to regulate immigration and conduct investigations as needed. At the end of the 1900s the challenges of unregulated prescribing of drugs such as cocaine and morphine for pain set the stage for conflict of state and federal power, as there were no national laws to confront what had become a significant social problem. The Harrison Narcotics Act of 1914 enabled the Department of the Treasury to begin regulatory efforts on certain drugs, and the Narcotic Division of the Internal Revenue Bureau was formed to investigate related crimes. Unlike local police officials, who responded to the needs of individual citizens, federal law enforcement agencies focused on a narrow range of crimes that violated federal law or crossed state boundaries.

DEVELOPMENT OF THE FBI

Perhaps the best known federal law enforcement agency in the United States, and focusing largely on investigative functions, is the Federal Bureau of Investigation (FBI). First formed as the Bureau of Investigation (1908) within the Justice Department, the agency became the FBI in 1924 under the avowed anticorruption and apolitical J. Edgar Hoover. While the FBI under Hoover's leadership did not keep complete faith with the public through actions such as maintaining secret files on political opponents of the Bureau, there is little doubt that worldwide the FBI became one of the preeminent and most professional investigative agencies in the world. Today, the FBI investigative priorities are protecting the United States

- from terrorist attack
- against foreign intelligence operations and espionage
- against cyber-based attacks and high-technology crimes
- by combatting public corruption at all levels. (FBI.gov)

Today, local, state, and federal agencies perform a wide array of investigative functions, sometimes working together to prevent, detect, and interdict crime and arrest those responsible.

Criminal investigation in the modern era involves the determination of a violation of a criminal statute, determining who is responsible for that violation, compiling *evidence* that is relevant and legally admissible, and then working with prosecutors to effectively present the case to the court. This definition seems straightforward, but accomplishing this list is often quite challenging. In the United States and other democratic nations the process of the criminal investigation is very much concerned with ensuring that the rights of accused persons are protected at critical stages of the case. When it comes time for a criminal trial, the evidence presented will persuade the jury about most matters. The American Bar Association (ABA) describes the two types of evidence:

Direct evidence
Evidence such as personal knowledge or information that may prove or disprove a fact. Direct evidence does not require inference.

Circumstantial evidence
Evidence that alone may infer a variety of facts. The trier of fact will reach his or her own conclusion about what the presence of a piece of evidence ultimately means in a case.

- **Direct evidence** usually is that which speaks for itself: eyewitness accounts, a confession, or a weapon.
- **Circumstantial evidence** usually is that which suggests a fact by implication or inference: the appearance of the scene of a crime, testimony that suggests a connection or link with a crime, physical evidence that suggests criminal activity.

Both kinds of evidence are a part of most trials, with circumstantial evidence probably being used more often than direct. Either kind of evidence can be offered in oral testimony of witnesses or physical exhibits, including fingerprints, test results, and documents. Neither kind of evidence is more valuable than the other.

Strict rules govern the kinds of evidence that may be admitted into a trial, and the presentation of evidence is governed by formal rules. (ABA, *How Courts Work*)

Searching for physical evidence must follow established rules, and taking testimonial evidence likewise has guidelines to protect against law enforcement from inappropriately compelling suspects to speak against their own interest. An investigation can amass quite a bit of information, but not all of it may be allowed to be considered in determining the guilt or innocence of an accused person.

The Goals of Investigation

The average citizen would likely say that the number one goal of a criminal investigation is to figure out who committed the crime and arrest that person. What law enforcement officers and investigators learn once they are on the job is that the work is not quite that straightforward. Often, an initial challenge exists in determining whether a crime was actually committed. People will report to agencies a variety of circumstances that seem compelling or suspicious and that necessitate agency members looking into whether a crime can be shown to have occurred. Not all reported crime is investigated. While a patrol officer may take an initial report from a victim or complainant, if there are no clues or evidence, a case may simply be recorded and await the possibility that the victim may discover more information or an unrelated crime discovery may bring to light a connection to the perpetrator. Sometimes a situation is misunderstood by the citizen and is subsequently determined not to be a matter that involves police action. Some incidents reported as crime are disagreements between people that must be resolved through the civil court system. Some reports, of course, are false allegations and as such the reporting of false information to law enforcement may turn out to be the crime.

If a law enforcement officer determines that a crime has been committed within his jurisdiction, he works with victims and witnesses and other criminal justice professionals to legally locate evidence, compile reports documenting the evidence, and then coordinate with prosecutors to formally charge (and usually arrest) the responsible party. If the crime is one that involves a loss of property, then the investigator will also have a duty to try and recover what has been taken and return it to the victim, or connect the victim with other agencies that may be able to secure crime compensation funds for the victim. After a suspect has been arrested, and working closely with the prosecutor, officers will participate in assembling a case for trial and testifying as needed to various aspects of the investigation.

What Makes a Successful Investigation?

The most commonly accepted measure of an effective or successful investigation is one in which a suspect is arrested and charged. Through movies, television, and novels, the public generally understands the desired outcome of an investigation to be "catching the bad guy." In fact, most crime in the United States goes unsolved. There is a general sense by many people that if adequate time and resources are devoted to the investigation of virtually any crime that it can be solved. This simply is not the case. Some crimes are never discovered. Some that are discovered are never reported. Some of those reported are not reported promptly and thereby suffer from deteriorated physical evidence or the absence of witnesses or others who can give information. Some reported crimes lack any easily discovered physical evidence and without a witness or suspect coming forward with information will go unresolved. According to FBI and Bureau of Justice Statistics data (2017) from the 2015 completed data, just under half of violent crime is reported and 35% of property crime. Of that *reported* crime only 46% of violent crime was cleared and 19% of the property crimes (see Figure 1.1). Some crime investigations falter when fresh leads run out, and these are typically considered cold cases if they were of a serious nature to begin with, such as homicide or sexual battery. The term **cold case** is not frequently used if someone's bicycle was taken from the front yard and no thief was readily discovered.

Cold case Generally, a crime that is no longer being actively investigated, usually due to a lack of leads or evidence.

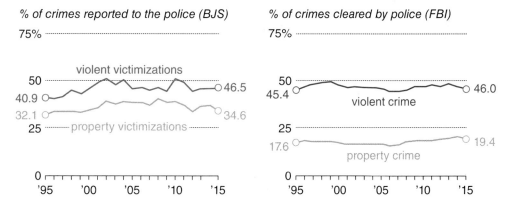

FIGURE 1.1 Reported and Cleared Crimes

Note: BJS and FBI crime definitions differ for some offenses. 2006 BJS estimates are not comparable with other years due to methodological changes. FBI figures reflect percentage of crimes cleared through arrest or "exceptional means," including cases in which a suspect dies or a victim declines to cooperate with a prosecution.

Source: "Most violent and property crimes in the U.S. go unsolved." Pew Research Center, Washington, D.C. (March 1, 2017). https://www.pewresearch.org/fact-tank/2017/03/01/most-violent-and-property-crimes-in-the-u-s-go-unsolved/

FOUR MEASURES OF A SUCCESSFUL INVESTIGATION

Aside from the arrest of the person accountable, other measures of investigative effectiveness have been considered. Referring to homicide, Brookman and Innes (2013) proposed four different ways that such investigations might be considered successful:

Why is the arrest and conviction of a suspect such an important measure of the success of an investigation?

©iStockphoto.com/RichLegg

1. The first measure is the most commonly accepted one of identifying and arresting a suspect who is subsequently convicted.

2. A second success may be seen as "procedural." That is that officers and detectives have performed a competent and thorough investigation and exhausted every avenue of information and legal means of obtaining evidence without identifying or being able to arrest a suspect.

3. The third conceptualization involves reassuring the public that crime is proactively investigated and addressed by the community's law enforcement agencies.

4. And finally, success may be viewed as coming through the "prediction, prevention and preemption" (p. 293) of crimes based on the work of investigators and agency personnel. While these four themes were applied to homicide investigations by Brookman and Innes, the ideas can extend to much if not all criminal investigation to broaden our view of the work that is being done by law enforcement agencies and investigators.

Types of Criminal Investigations

Persons crimes A common way of referring to crimes committed against the body of a victim, such as an assault, battery, robbery, sexual battery, or homicide.

Property crimes Generally refer to those offenses that affect the property of a person or business rather than the actual person. Some examples include theft, vandalism, burglary, or arson.

Vice crimes Generally, those crimes made unlawful as the result of a community's moral stance on issues. Examples include gambling, prostitution, and pornography.

Most local level law enforcement agencies separate crime investigations into a few categories. One of the most frequent divisions of crime types are referred to as *persons* crimes and *property* crimes. **Persons crimes** are typically those involving one person physically victimizing another. These would include simple and aggravated assault, simple and aggravated battery, sexual battery or rape, robbery, and homicide. This list is not all-inclusive but serves to illustrate the nature of directly harming another person. In the **property crimes** category are acts such as theft, burglary, trespassing, and criminal mischief, what some refer to as vandalism. Again, the illustration here is that physical property is taken, damaged, or destroyed. Based on the number of investigators an agency has, an added category of **vice crimes** may exist. Vice crimes are offenses that are generally viewed as not having a specific victim but rather are considered illegal because they offend the morals of society, which has labeled them in this way. This category typically includes drug use crimes, prostitution, and illegal gambling as examples.

In addition to these rather broad categories of crime are specific offenses that an agency may or may not include in this separation. Examples might include crimes against children, though these may be classified as persons crimes; white-collar crimes, though these can be included within property crimes; or crimes requiring specialized investigative resources, such as arson or illegal use or possession of explosives. Some agencies use a designation of a **major case squad** for crimes that are violent or significant such as kidnapping, or serialized crimes such as burglary, arson, or rape. Some agencies will also initiate proactive investigations that may have started with information from the public or from information arising in an unrelated case. Crimes such as money laundering, corruption, and planned terrorist acts are some examples.

Major case squad Unit tasked with investigating crimes that require significant resources. Such squads in larger agencies may handle, for example, homicides.

Which Agencies Investigate Which Crimes?

Because of the different duties and responsibilities of agencies from state to state most local departments that a citizen might telephone 9-1-1 to reach would include municipal police departments, county sheriff's offices, and some special district agencies such as those in national parks, at schools or universities and colleges, or agencies that enforce laws on the waterways. Most states have a highway patrol that is largely responsible for enforcing traffic laws on state highways. Some states also assign general police duties to their state agencies, especially if there are portions of the state that have few or no local law enforcement officers to patrol unincorporated parts of counties. Each of these local or state agencies may conduct all of the mentioned investigations and many more. Dozens of federal law enforcement agencies investigate specific violations of federal law. While people often think of the FBI, the DEA, U.S. Customs and Border Protection, and others, there are lesser-known agencies such as the Federal Air Marshal Service, the United States Park Police, the United States Capitol Police, and many others.

State-level agencies are most often tasked with traffic enforcement duties, such as a highway patrol, or investigations. A few states have agencies referred to as state police that will perform general police functions in unincorporated areas of the state where municipal police or county sheriff's deputies are unavailable. Investigations will include many that span the boundaries of counties or perhaps more than one state. The investigation of criminal happenings in multiple states will often bring in the interest or efforts of one of the many federal law enforcement agencies. Some of these include the Federal Bureau of Investigation (FBI); the Drug Enforcement Administration (DEA); the Bureau of Alcohol, Tobacco, Firearms and Explosives (ATF); and the United States Marshal Service (USMS).

Criminal investigation of law violations occurs constantly at each of the local, state, and federal levels. To give you just a sense of the variety of investigations at the federal level, here are several recent actual press release headlines from the United States Department of Justice:

- 3,800 Gang Members Charged in Operation Spanning United States and Central America
- Alabama Man Arrested on Production of Child Pornography Charges
- Two Real Estate Investors Plead Guilty to Bid Rigging in Northern California
- Alaska Department of Health and Social Services to Pay Nearly $2.5 Million to Resolve Alleged False Claims for Supplemental Nutrition Assistance Program (SNAP) Funds
- Former Utah CEO Pleads Guilty to Tax Evasion

The U.S. Justice Department also engages in the investigation of actions alleged to have violated the civil rights of individuals. These are a few examples:

- Department of Justice Settles Employment Discrimination Claim on Behalf of U.S. Army Reservist
- Justice Department Settles Employment Discrimination Lawsuit Against the State of Rhode Island and the Rhode Island Department of Corrections

Changing Times Equals Changing Crimes and a Need for Law Enforcement to Adapt

It is an understatement to say that as society has changed so too have the methods of criminal investigation. With the advent of the motor car came the ability for some criminals to travel to and from the place of their crime. And so was also created the crime of auto theft! Increased capacity and lethality of firearms led to their use as an instrument to commit crimes, while police often lagged behind in the area of weapons when confronting well- or heavily armed criminals. Communications technology and eventually the computer facilitated old crimes in new ways and allowed for innovative ways to victimize others. Once again, law enforcement has generally been behind the curve based on the relatively small size of most agencies and the subsequent lack of sufficient resources, including equipment and advanced training.

Four Stages of Criminal Investigation

Criminal investigation is a process with several identifiable but overlapping stages:

1. The first stage involves both the detection of what is believed to be a crime and the reporting of that crime. As we have already noted, many crimes that occur are either not discovered or not reported, and this is often in the hands of the potential victim.

2. The second stage comes after law enforcement is aware of an apparent criminal act, and typically at the local level a uniformed patrol officer will respond to conduct a preliminary investigation. If there is little evidence to work with or follow-up on at this point, many crime reports become inactive, still recorded by the agency and available to reopen if further information comes to light.

3. The third stage will be reached if there is actionable information to warrant a follow-up. In this third stage a detective will pick up the case and move it forward. If the personnel assigned to a case develop enough information that a prosecutor decides to enter a formal charge against a suspect, then it will move forward to a formal charge.

4. The fourth stage is preparing the case for potential trial in working with the prosecution to see it through to conclusion.

Each stage of the investigative process can contain multiple tasks and involve many different individuals in a coordinated team effort. The fictional portrayal of the insightful and dogged detective is not how contemporary crimes are solved, if it ever was. An investigation is also fluid, as information continues to come to the attention of investigators, filling in gaps and additional pieces of a puzzle that the investigator can see taking shape before completing the entire picture. These are decidedly qualitative-sounding and hardly concrete.

Two Phases of Criminal Investigation

The investigation into most crimes has a preliminary and a follow-up phase, though the former often blends and overlaps with the latter, as discussed in the stages above.

PRELIMINARY INVESTIGATION

During the **preliminary investigation**, a patrol officer is the first responder in most cases. This critical phase determines whether an investigation gets off on the right foot, or stumbles and perhaps falls down completely. The RAND study conducted in 1975 by Greenwood and Petersilia revealed that "information taken by the patrol officer first on the scene was the most important factor in determining whether an arrest was made" (Thistlewaite & Wooldredge, 2010, p. 73). This is an important reminder to aspiring police academy recruits who generally picture themselves in the glamorous role of homicide detective, not realizing that as a patrol officer they begin day one on the job as an investigator.

Preliminary investigation The initial inquiry into a reported crime, usually conducted by a patrol officer.

When a patrol officer arrives on the scene of a reported crime his initial multiple tasks include being cautious for any danger, determining if anyone is in need of medical care or protection, learning the basic facts of what has happened, and securing the crime scene if there is one until determining what steps should be taken next. As mentioned, many agencies do not have formally assigned investigators and therefore the patrol officer may conduct the complete investigation of a majority of reported crimes. Well-trained officers in agencies of any size are aware of the risks of failing to secure a crime scene from contamination or unauthorized intrusion. At the same time, insufficient personnel either in terms of training or adequate assistance can lead to damaging, obscuring, or failing to identify evidence.

After the first officer determines that there is no active danger at the scene or persons in need of care, he will establish a perimeter broad enough to protect potential items of evidence and hold back onlookers from interfering with personnel or evidence at the scene. While officers or investigators can shrink the crime scene area as appropriate, it is very difficult to expand a perimeter and not face potential of contamination based on people walking through an area without knowing to be careful. Many if not most officers carry crime scene tape in their squad cars to mark these boundaries, though media representatives and onlookers have been known to disregard even obvious barriers at a crime scene. Excessive criminal justice personnel at the scene of a crime can also result in inadvertent scene contamination, as the officers believe themselves exempt from what the crime scene tape signals. An important task for officers is to faithfully record the movement of every person into and out of the crime scene.

How It's Done
FIRST RESPONDING OFFICERS

First responding officers to the report of a crime have a number of concerns and responsibilities. The International Association of Chiefs of Police uses the acronym PRELIMINARY to outline most of those duties.

In addition to carrying out the duties listed, the patrol officer must notify the police communications center concerning injured parties and any dangerous conditions that are present and request any additional support personnel as necessary. He or she should also determine if there actually was a crime committed that is definable by statute. Once the scene has been controlled, a crime scene perimeter should be established and protected from entry by unauthorized persons.

Source: Training Key #558, Criminal Investigations.

P Proceed to the scene promptly and safely
R Render assistance to the injured
E Effect the arrest of the individual
L Locate and identify witnesses
I Interview the complainant and the witnesses
M Maintain the crime scene and protect the evidence
I Interrogate the suspect
N Note the crime scene and protect the evidence
A Arrange for collection of evidence
R Report the incident fully and accurately
Y Yield the responsibility to the follow-up investigator

At a microscopic level, each person has the potential of introducing substances into a crime scene or carrying away evidence from a crime scene that has adhered to the person's shoes, clothing, or equipment. In 1910, in Lyon, France, Dr. Edmond Locard created the first official police laboratory to examine and categorize evidence and to show the connection between the evidence and a suspect's presence at the scene of a crime. The idea that every contact between two items leaves a trace is generally referred to as Locard's Exchange Principle. Law enforcement officers and crime scene technicians must always keep the implications of this theory of transfer in mind as they arrive at and move around the potential crime scene.

The initial officers on the scene of a crime are also in a unique position to identify witnesses who may have knowledge regarding what happened or the people involved. This is an important duty of patrol officers, since it may occur to many people that they do not wish to remain and be questioned or identified by law enforcement. The ubiquity of recording devices is an aid to documenting who comes and goes at the scene of a crime as well as the condition of the scene when officers arrive. If forensic technicians arrive to complete the processing of a suspected crime scene, they will likely use recording technology to aid in the documentation and later reconstruction of the scene or events. It is important that the officers learn what they can from anyone at the scene, as well as obtain contact information so that investigators can follow-up with the individuals if they think it's necessary. It is the responsibility of the initial officer at the scene to notify his supervisor or the communications center to contact an investigator and brief him about the crime and the condition of the scene. Finally, any officers at the scene of the crime should complete either the primary incident report (if the officer initially assigned to respond) or a supplemental report if they were assigned to assist the primary officer.

Midsized and larger agencies will often have a dedicated crime scene unit (CSU) staffed by sworn officers, civilian technicians, or a mix of the two. Due to the relatively small size of a large percentage of U.S. law enforcement agencies, full-time crime scene technicians may not be available, leaving the patrol officer or detective to identify and collect evidence or to seek the assistance of a larger local agency or in some cases a state agency to assist. These types of arrangements are common and are important to the proper conduct of the investigation. The activities of crime scene personnel are integrated with but distinct from the testing conducted on evidence in a laboratory or the contributions made by a medical examiner, coroner, or other medical professional in relevant cases. As part of the preliminary investigation the crime scene technician is generally concerned with identifying and collecting evidence, as well as responsible for transferring that evidence to a local, regional, or federal laboratory for further analysis.

For serious crimes such as homicide, sexual battery, or robbery, an investigator may respond to the initial crime scene or become involved shortly after the crime is reported. He will take over coordination of additional efforts beyond the initial work of patrol officers or personnel on the scene. This involvement helps guide the evidence identification and collection activities, and the detective will generally not only do an initial walk-through of the crime scene but, along with the lead evidence technician, conduct a final walk-through of the scene as well.

FOLLOW-UP INVESTIGATION

Depending on the seriousness of the crime and other **solvability factors**, a detective will be notified and possibly respond immediately or be assigned the case later to follow-up whatever leads have been identified. Eck (1983) categorized cases as (1) unsolvable; (2) having a known offender who must simply be caught; and (3) needing the trained efforts of an investigator. Again, not all reported crimes are investigated, and one of the reasons is a lack of leads or evidence. Most agencies have a procedure in place to reactivate and assign a case to an investigator if evidence comes to light. Either at the outset of a reported crime or as a follow-up at some later point, a detective may be assigned primary responsibility for investigating the case.

The detective assigned to the **follow-up investigation** will review all of the information available up to the point she takes the case. Then she will begin planning the steps necessary to move toward a case resolution. The plan will vary based on the reported crime but gathering information from victims, witnesses, suspects, and records including the initial incident report and any supplementary reports, and is a significant and time-intensive component of the follow-up. The investigator will ensure any evidence needing further analysis has been routed to the appropriate facilities or specialists and, if sufficient progress is made in the investigation, she will confer with a prosecutor.

During the follow-up investigative process, most work that will be accomplished by laboratory technicians, medical doctors, and other specialists will occur. It is also in the follow-up phase that investigators will have the most interaction with prosecutors as they coordinate to further the investigation and assemble a case for court. The prosecutor's office is critical in deciding if a case will move forward, but it is also important to assist with search

Solvability factors Those pieces of information or evidence that provide adequate information to move an investigation from the preliminary to the follow-up stage. Examples include pictures of a suspect or the tag number from a car involved in a crime, a viable way to identify stolen property.

Follow-up investigation Additional resources or an extended investigation may occur in serious cases or those where sufficient evidence exists.

or arrest warrants and other critical tasks including lineups of suspects or using the power of the subpoena to gain access to information people are not willing to voluntarily supply. The prosecutor may also provide input into interviews and interrogations to ensure as much information as possible is gathered at critical phases. This progression seems logical and succinct, but like the hydra, each new bit of information can spawn multiple leads and case considerations that must be followed and documented.

The management of an assigned criminal investigation follows a protocol developed by the agency and will often involve periodic reviews of case status with a supervisor to help ensure all tasks have been identified and acted upon, to provide an opportunity to get a second perspective or seek additional resources, and to maintain accountability of the assigned detective. If at some point during the investigation sufficient progress has stalled, a decision will be made as to whether to inactivate the case. If the case has failed to progress after a predetermined period of time, perhaps 6 months or 1 year, the case may be considered *cold*. A cold case may remain on the detective's caseload and be reviewed by him periodically or when new information is reported or has been developed through a source. Crime analysis may suggest a link between or among crimes spanning a period of time. This type of computer-aided investigation may also include crime mapping and geographical information that reveals connections that might otherwise be missed.

Organizing the Investigative Unit

In addition to the categories of crime listed above, agencies organize the investigative function based on the volume and type of crimes they have reported. Some departments may have only one full-time investigator or none at all and rely on patrol officers to conduct all aspects of simple investigations and seek assistance from larger local agencies or state organizations for serious or complex matters. Midsized and larger departments may have investigators working a variety of shifts and sometimes partnered with other investigators for some cases. According to the Bureau of Justice Statistics approximately 85% of the nearly 16,000 local level law enforcement agencies in the country have fewer than 50 sworn personnel (2015). Nearly half employed just 10 or fewer officers. While many people tend to think of large agencies such as the New York or Los Angeles police departments, only 5% of local police departments employ 100 or more sworn officers (2015).

The Investigator—General Assignment and Specialized

Within the different configurations of the investigative function in agencies some detectives fill the role of a generalist and are the lead investigator on any reported crime that the agency intends to examine beyond an initial report and the efforts of the patrol personnel who respond to the incident. The investigator may be referred to as a general assignment detective. The categorization of a generalist does not imply a lack of competence in performing the investigative function. The protocols and procedures established by individual agencies and augmented by established investigative practices and external resources provide a sound basis for generalist investigators to capably handle most crimes. Other departments, either because they have many more people assigned to investigations or because of the complexity of certain types of cases, will assign a specialist investigator as the lead on certain cases. Examples include homicide, mid- and upper-level narcotics investigations, white-collar crime, and increasingly cybercrime.

The challenge, as we have noted, is that having sufficient resources, most notably investigators, to explore each and every possible aspect of a reported crime is simply not within the realm of what is possible given public budgets devoted to the criminal justice system. For investigations of a serious nature that do exceed the capabilities of one agency, there is the potential to seek assistance from other departments. Additionally, some crimes involve more than one jurisdiction, either in their commission or in seeking a suspect who is within the local area or who has left the jurisdiction or region altogether. For these reasons departments that are smaller or have fewer resources will often seek assistance from larger local agencies, state or federal agencies for expertise and assistance.

Task Force Investigations

The nature of some crimes or the limited resources of individual agencies can lead to creation of task forces to better address investigative challenges. The nature of many crimes leads to the crossing of jurisdictional boundaries. The distribution of illegal drugs, human trafficking, child pornography, and arson cases are all examples of crimes that are more effectively investigated and sometimes legally necessary by involving multiple agencies and personnel with varied expertise. Such groups are sometimes referred to as major case task forces. Agencies will generally have an inter-local agreement in place that outlines the responsibilities and jurisdictional limitations for each participating agency. Sometimes local departments will involve state or federal agencies that have a focus or expertise in a particular area that is critical to the local or regional task force. Examples of these would include the drug enforcement administration assisting with a narcotics distribution enterprise that is discovered to be more widespread than initially thought, or a domestic terrorist group investigation calling upon the Bureau of Alcohol, Tobacco and Firearms (ATF) to lend expertise in the examination of explosive devices.

What role does surveillance play in an investigation, and how have advances in technology impacted surveillance?

©iStockphoto.com/ EvgeniyShkolenko

Such multijurisdictional task forces require ongoing communications procedures regarding personnel and resources needed, assignment of responsibility for various tasks, evidence protocols, and consistent reporting requirements. The budgetary limitations of any one agency may be partly mitigated by the synergy of cooperation and leveraging the resources of larger agencies or combining efforts with multiple agencies. In the same way that detectives can access resources not immediately available to patrol officers and can adapt their movement and schedules to meet the demands of an investigation, a combination of investigative efforts of multiple agencies exceeds and expands beyond the ability of a single investigator or investigating agency. Over time agencies develop effective and relatively seamless methods for meshing the expertise and resources into effective groups to address the challenges and complexities of mobile and geographically dispersed criminals.

The Violent Criminal Apprehension Program (ViCAP), created by the Federal Bureau of Investigation (FBI) in the 1980s, focuses on researching multijurisdictional and specifically interstate crimes and assists agencies in identifying and helping to track or link serial offenders. The Joint Terrorism Task Force (JTTF) that arose after 9/11 is a regional effort led by the FBI to coordinate information and resources to combat terrorism. The Internal Revenue Service (IRS) leads task forces' efforts into major fraud and tax evasion schemes. The Office of Justice Programs, through its Office for Victims of Crime Training and Technical Assistance Center, provides guidance for task force development in the investigation of human trafficking. And the U.S. Department of Justice sponsored an evaluation of multijurisdictional task forces (Rhodes, Dyous, Chapman, Shively, Hunt, & Wheeler, 2009).

An example of one of the most notorious crime challenges that led to the use of a large multijurisdictional task force was the D.C. Sniper shootings in October 2002. Two individuals terrorized the area around Washington, D.C., and Central Virginia as they indiscriminately shot people over a 23-day period. The Police Executive Research Forum (PERF) distilled many of the lessons from this effort in a lengthy document published in 2004 and discussed the Sniper case itself in the context of high-profile investigations, leadership, managing investigations, information management, local law enforcement operations, media relations, community issues, and summarizing considerations for jurisdictions facing these types of events (Murphy & Wexler, 2004). One important difference from non–task force investigations that can significantly impact the conduct of such investigations is the "top-down" nature of the task force. Chief executives may have difficulty accepting a subordinate role to other agency heads, and coordination and information-sharing may suffer as the result. The review by PERF provided lessons learned about large-scale investigations involving multiple jurisdictions. The overall categories, which contain many important considerations, were: investigators, managers, controlling and coordinating investigative resources, securing and processing the crime scene, managing leads, and integrating the role of prosecutors (PERF, 2004).

Contemporary Law Enforcement and the Role of Criminal Investigation

Contemporary law enforcement agencies in Western democracies operate under laws and policies that seek to balance the due process rights of individuals suspected or accused of crimes with the crime control imperative that instills the confidence of the public that their law enforcement agencies are doing all that they can to protect the community. The tools by which crime is prevented, and committed crimes are detected and solved, are not all in the hands of law enforcement. To the extent that police agencies are able to influence crime prevention, reduction, and successfully solve cases, they do so in cooperation with others. To accomplish the tasks set to them, law enforcement agencies continue to evolve strategies and methods to prevent what crime they can and to effectively partner with the community and other organizations to retroactively and effectively investigate crimes that have occurred. Three important conceptual approaches to modern policing are the community-oriented, problem-oriented policing, and intelligence-led models.

Community-Oriented Policing

The origins of policing both in the United Kingdom and in the United States involve the concept of community members sharing the responsibility for community safety. This involved voluntary participation in watching the village overnight to protect against criminals or even the risk of a fire getting out of control and spreading through the wooden structures of the day, to eventually organizing groups of citizens on a more regional basis to seek outlaws or perform other protective functions as a group. As American policing passed through the political era and into the reform era (Kelling & Moore, 1988), it adopted training of officers and embracing of professionalism as a way to progress as a field and to be accepted as legitimate by the public. A component of reform and professionalism was donning the mantle of expertise in all things crime related. Police (with good intentions) told the public: "Step back and we'll handle this." As competent crime-fighting through claimed deterrence and retroactive investigation became the hallmark of reform policing, the role of the individual citizen once again receded into the background.

When crime in the 1960s and the 1970s rose significantly and law enforcement leaders took stock and realized that not only were they under-resourced but that no amount of resources could adequately address the crime problem in the United States, they turned once again to the public and **community-oriented policing (COP)** was introduced. Most crime occurs indoors and out of the sight of police or the general public. This reality calls for the involvement of more than just law enforcement in discovering, investigating, and resolving much of the crime in society that does get solved. The U.S. Department of Justice created the Community Oriented Policing Services (COPS) to help support efforts in communities around the United States in adopting a model of community policing.

Much discussion surrounding community-oriented policing efforts involves the concept of partnering. Law enforcement agencies follow an intentional approach of partnering with other governmental entities as well as private sector, civic, and citizen groups. This synergistic approach allows for two-way communication of important information and the perspectives of stakeholders in the business of public safety and quality of life. "In 2013, about 7 in 10 local police departments, including about 9 in 10 departments serving a population of 25,000 or more, had a mission statement that included a community policing component" (Reaves, 2015, p. 8). Many people think of the neighborhood watch program as one example of how citizens and local law enforcement work together. Another example would include law enforcement agencies working with the code enforcement department of local government to repair or remove dilapidated buildings. This relates to the "broken windows" concept (Wilson & Kelling, 1982) of addressing disorder issues within communities to help reduce crime that gravitates to uncontrolled areas that may also exhibit high turnover of residents and the flight of small businesses in the absence of civic organizations.

Community-oriented policing (COP)
An approach that relies on partnering with public, private, and volunteer organizations as well as individual outreach to neighborhoods. Officers are often assigned to work in the same area of a community for extended periods of time so they can become thoroughly familiar with the residents and businesses, and so those businesses and residents may come to know and trust the officers.

Problem-Oriented Policing

Problem-oriented policing (POP) has sometimes been seen as a component of community-oriented policing or a standalone concept or approach. Regardless, the idea is to work with the community to identify specific problems and work to address them. If an area is plagued with graffiti and vandalism, directed patrol may address or displace some of the activity and an added component of involving local faith or educational organizations or other groups to interact with those defacing property may provide a more thorough response. Similarly, open-air drug dealing in a rundown neighborhood may be partly reduced through the code enforcement efforts mentioned above in conjunction with efforts at turning some residential through streets into cul-de-sacs, which is a strategy within the crime prevention through environmental design (CPTED) movement.

Communities and agencies realize the expanded benefit that usually accrues from improving one area of the community, as this often has a ripple effect into other parts

Problem-oriented policing (POP)
A strategy built on the idea of identifying the underlying causes of a crime or other police problem in devising a strategy to combat that problem rather than the symptoms.

of a community. In a Bureau of Justice Statistics (BJS) report based on the 2013 Law Enforcement Management and Administrative Statistics (LEMAS) survey sponsored by the BJS, "A majority of departments serving 25,000 or more residents maintained problem-solving partnerships or agreements with local organizations" (Reaves, 2015, p. 1). Approaches such as COP and POP are intended to improve cooperation with the community, and these methods are critical to improving criminal investigations, not only interactions with uniformed officers.

Intelligence-Led or Evidence-Based Policing

Research in policing, while a fairly recent endeavor, is nonetheless quite active and addresses many different aspects of the police role in society. One of the first recognized studies to impact policing in the United States was the Kansas City Preventive Patrol Experiment of the early 1970s (Kelling, 1974). As one of the first large-scale experiments to examine policing methods, it had limitations perhaps not foreseen by the researchers. The study did, however, draw attention to the fact that patrol officers spent a significant portion of their shift in nondirected and possibly unproductive patrol with the public relatively oblivious to their presence or deterred from committing most crimes. In the early years that researchers looked at policing with an eye for what would improve the police or their methods they used, the personalities and types of people who gravitate to policing were examined (Skolnick, 1966; Van Maanen, 1973; Wilson, 1968), attributes inherent in the job such as the use of discretion (Piliavin & Blair, 1964), or the tools that seemed to make a difference such as the power of arrest (Sherman & Berk, 1984), how quickly police arrived at a call (Pate, Ferrara, Bowers, & Lorence, 1976), whether the arriving officers were one or two to a car (Boydston, Sherry, & Moelter, 1977), and whether the officer's gender made a difference (Bloch & Anderson, 1974).

In the decades since this awakening of policing research, studies have expanded to examine more and more practices and, importantly, the impact and use of research in formulating agency policy and statutory law by elected legislatures.

Evidence-based should not be used interchangeably with the term *best practices*. Frequently, the term *best practices* may be applied more appropriately to methods that are "frequent practices." Early efforts at policy may be deemed evidence-based but they are often not tried or tested when the term is applied. Pawson (2006) said that, "Evidence based policy is much like all trysts, in which hope springs eternal and often outweighs expectancy, and for which the future is uncertain as we wait to know whether the partnership will flower or pass as an infatuation" (p. 1).

One of the few early comprehensive examinations of the investigative function in law enforcement agencies was conducted by the RAND Corporation more than 40 years ago (Greenwood & Petersilia, 1975). This often-cited early examination of the work of law enforcement was designed to look at the process of criminal investigation in major metropolitan police departments. In addition to describing how investigations were organized and managed in the 1970s in America, the researchers attempted to assess "the contribution of various investigation activities to overall police effectiveness" (Greenwood, 1979, p. 1). Twenty-five departments representing jurisdictions of more than 100,000 population or an agency with at least 150 sworn officers were selected for the study.

Although academic research interest has increased into some aspects of criminal investigation, the status of the field as art and not science, but *involving* science, has likely contributed to the dearth of theoretical studies leading to any consensus on foundational principles or practices. To be sure, various books have been written on the subject of criminal investigations, mainly by practitioners for practitioners to recount methods and techniques of information-gathering including the interview of witnesses and victims and the interrogation of suspects. Yet law enforcement agencies have shown themselves receptive to research in general in the last several decades and they recognize that while experience serves them in many ways, participating in and learning from research can improve effectiveness (Telep & Lum, 2014).

Science and Technology in Criminal Investigation

No discussion of contemporary criminal investigation would be complete without talking about the contributions of science and technology to the process of criminal investigation. Technological developments in crime detection and analysis have accelerated, as they have in other fields. While an expanded discussion will take place in the next chapter, it is important to note that people have trained the lens of science onto criminal acts for a very long time. In the last 200 years, as the ability to test substances and determine their presence has advanced, the application of these methods has proven their value in the investigation of crime. Saferstein provides a workable definition of forensic science as "the application of science to those criminal and civil laws that are enforced by police agencies in a criminal justice system" (2007, p. 5). That science provides the mantle of credibility to physical objects and test results that tend to prove or disprove certain facts, which benefit the justice system while also opening to debate the views of experts at odds over what an established physical fact indicates in the context of an investigation. The importance of science has also been underscored with infrequent yet consequential problems in some crime labs. The need for laboratory certification and accountable management remains a priority.

The scientific process applied to questions of crime works in tandem with advancements in the technology that allow the discovery, identification, collection, and testing of various items of physical evidence in increasingly smaller amounts and from locations beyond the previous reach of human eyes or hands. Identifying and recording the presence of impressions including fingerprints have progressed from application of fine powders that adhere to the oily ridge characteristics on a flat nonporous surface, to the use of laser and alternate light sources and scanning technologies to capture such impressions from many more surfaces and locations. Ballistics testing, gas chromatography, even devices to measure with sensitivity the voice quality and bodily responses of suspects during interview or interrogation—all provide opportunities to learn more completely the circumstances surrounding the crime.

EXPLORE THIS

Visit the website of the U.S. Department of Justice at www.justice.gov. Navigate to the Resources section of the website and then go to Publications. Scroll down to the link for the National Institute of Justice (NIJ). Go there and explore a number of the publications that the NIJ has created covering a wide variety of policing and law enforcement topics. Look at the ongoing efforts of the NIJ with detailed research and discussions of topics benefitting the law enforcement community.

SUMMARY

The protection of community members from fellow citizens is a role assigned to government actors throughout much of history in most developed cultures. Retroactive investigation of acts labeled as crimes within a given culture has only enjoyed the status of a distinct police function in Western society for perhaps 150 years. Individuals within police forces were assigned to follow-up initial reports of crime with efforts to uncover the persons responsible, locate evidence to prove their guilt in court, as well as to recover stolen property. The process of criminal investigation begins with determining if a crime has actually been committed and then proceeds with legally finding evidence of a suspect's guilt, arresting the person based on the probable cause established by the evidence, and working with the prosecutor to assemble and coordinate an effective case for court.

The arrest and prosecution of an offender is the most commonly understood measure of effective criminal investigation, but other outcomes worthy of note include thorough and competent investigation following all laws and procedures even if a suspect is neither identified nor convicted; signaling to the community through aggressive investigations that the law enforcement agency is working to protect citizens; and, potentially, some deterrent effect from the pursuit of investigations.

Among the many types of investigations are crimes specifically committed against another person such as assault, battery, rape, robbery, and homicide. Another broad category of crimes are those that target property. These include such acts as criminal mischief, theft, burglary, and white-collar or financial crimes. There are many investigative agencies at the local, state, and federal levels that look into crimes in a wide variety of areas. Most people think in terms of crimes committed locally, in which you reach your local police department or sheriff's office by calling 9-1-1. A patrol officer will typically conduct the preliminary investigation and, if there are sufficient leads or evidence, a detective may be assigned to conduct a follow-up investigation. Many others, including criminal justice students, see opportunity in federal law enforcement agencies.

Based on the size of a law enforcement agency and the available resources, detectives may fill the role of a generalist or a specialist. The generalist investigator will handle investigations for virtually any reported crime that is to be followed up. In the case of the specialist, personnel are assigned to the more complex matters that benefit from lengthy experience or advanced training such as homicide, child crimes, arson, and the like. In some instances the complexity of cases or the knowledge or belief that a criminal is operating across more than one jurisdiction calls for the involvement of multiple agencies. This may be a matter of seeking the assistance of a larger department, an adjacent agency where crimes of similar nature are occurring, a state or federal organization with expanded resources, or the development or involvement of a task force that draws upon multidisciplinary and multijurisdictional resources and strengths.

Most law enforcement agencies in the United States follow a mode of policing that seeks to partner with the public and various groups and agencies to more effectively prevent or solve crimes. Most agencies use some version of community-oriented policing as a model to accomplish this partnering. Often incorporated is the problem-solving policing model that has officers and agencies using data and crime mapping to search for the causes of crime underlying the calls for service in the hope of eliminating those causes and thus reducing the commission of crime. The use of data and analysis is also an important component of what are often referred to as intelligence-led or evidence-based policing practices, which incorporate such data as well as ongoing and valid research into crime and criminal justice practices.

KEY TERMS

Circumstantial evidence 6
Cold case 7
Community-oriented policing (COP) 17
Direct evidence 6
Follow-up investigation 13
Major case squad 9
Persons crimes 8
Preliminary investigation 11
Problem-oriented policing (POP) 17
Property crimes 8
Solvability factors 13
Vice crimes 8

DISCUSSION QUESTIONS

1. What are the goals of criminal investigation?
2. List and discuss the different types of criminal investigation.
3. How does the work of patrol officers affect the outcome of a criminal investigation?
4. What are the factors that affect how the investigative function is organized in different agencies?
5. How do community-oriented and problem-oriented policing relate to one another?
6. Do advances in science and technology benefit the work of officers investigating crime? If so, how?

- Get the tools you need to sharpen your study skills. SAGE edge offers a robust online environment featuring an impressive array of free tools and resources.
- Access practice quizzes, eFlashcards, video, and multimedia at **edge.sagepub.com/houghci**

AP Photo/Joseph Kaczmarek

2 Forensic Evidence
The Crime Scene and the Laboratory

RUNNING CASE: CRIME SCENE ANALYSIS OF EVIDENCE AT THE FLORIDAN

Introduction

How It's Done: Field Tests

The Developing Science of Criminal Investigation
- The Prescientific Era
- The Scientific Era
- The Technological Era

Methods of Collecting and Processing Evidence
- Crime Scene Searches: Inside, Outside, All Around

How It's Done: Where Would You Seek Outside Assistance?

Types of Physical Evidence and Its Processing
- Advanced Scientific Methods and Sophisticated Technology
- Collecting and Preserving DNA Evidence
- Firearms, Explosives, and Arson
- Expert Witnesses

Equipment Utilized in Investigations
- Basic Crime Scene Equipment
- Advanced Crime Scene Equipment
- Personnel Safety

The Contemporary Forensic Laboratory

Personnel
- Crime Scene Technicians at the Initial Scene
- Crime Scene Technicians in the Laboratory

Documenting the Crime Scene and Actions
- Case Management Software
- Photography

How It's Done: Crime Scene Photography
- Diagramming
- Report Writing and Notetaking

The CSI Effect

Explore This

Summary

Key Terms

Discussion Questions

LEARNING OBJECTIVES

2.1 Contrast the scientific periods of criminal investigation.

2.2 Compare and contrast indoor and outdoor crime scene searches.

2.3 Explain the challenges in collecting physical evidence.

2.4 List the most common equipment used at a crime scene.

2.5 List and describe major functional areas of the contemporary forensic lab.

2.6 Contrast the roles of crime scene technicians and criminalists.

2.7 Explain the importance of documenting activities at a crime scene.

2.8 Explain the CSI Effect.

Running Case: Crime Scene Analysis of Evidence at the Floridan

For the next 2 hours, Detective Bradley Macon watched Charla Lynne, the crime scene supervisor, oversee the investigation of the obvious homicide. Two crime scene investigators directed the scene search, documented evidence, and handled items. These items would either be taken for processing and lab analysis or be logged in to the property room and held in the case.

Charla Lynne immediately asked the bar owner for closed-circuit television (CCTV) footage from inside the Floridan. Watching this footage, it seemed that the victim and another man may have met and possibly argued. Bradley Macon knew that argument-based, or confrontational, homicide was the largest category of criminal homicide. Usually it involved a young man who had perceived some insult from another (usually) man. Often there had been alcohol or drug influence on the thought processes of the men, and perhaps even an "audience" of others who did nothing to break up an escalating argument, or actively encouraged it, probably never thinking it could end in someone being killed. In this case, the video showed two men on the dance floor engage in a standoff with aggressive body language. When the two men began shoving one another, it was obvious they were in a heated argument. Shortly after the fighting, they both fled the dance floor and exited the CCTV footage. Charla Lynne assigned another investigator to follow the footage to see if the men reentered it.

FTO Wes Thompson and his trainee, Officer Carl Jayden, two of the patrol officers who initially responded to the shooting, found a 9 mm semi-automatic pistol lying on top of cardboard boxes in a dumpster close to the north end of Milton Way, in the path of the person who fled. Charla Lynne photographed the weapon, gathered the pistol and "made it safe" by first noting the bullets in the gun and their position, and then removing them, then safely packaging the gun to be transported to the ballistics section of the local lab. "We'll get this test-fired. Try to find a match with the bullet from the victim," she said to another technician. "Order gunshot residue (GSR) testing."

Charla Lynne and the second technician used paper to bag the victim's hands and torso area. It would be important to rule out that the victim had not also held and fired a gun.

A uniformed officer stood outside the bar and controlled and noted who entered the scene. Milton Way was short enough that there were no intersecting streets. A half-dozen businesses had backdoors opening into the alley, and there was a streetlight at either end, though the one where Milton Way intersected Ninth had a burnt-out bulb. One of the crime scene technicians documented this with a photograph and a written notation.

Because this was a public street, the investigators did not have to obtain a search warrant, although they did assign someone to contact each business owner on either side of the alley to ask if they had security cameras or any further information. Police department administrators had learned that a community-oriented policing approach meant keeping the business people up to date about the homicide was good policing.

In addition to the two potentially valuable pieces of physical evidence, Detective Richard Ashley was interviewing the bar employee, Bubba Paul. This quick on-scene interview would be followed by a complete one at headquarters.

By now it was 7:30 a.m. and the sun was already turning the night's dew to steam on the pavement. Satisfied that all evidence had been documented, Detective Bradley Macon prepared to leave the scene. Working with professionals, he did not need to remind anyone that reports and preliminary observations had to be completed before anyone went off-duty. By 10 a.m. that day, Sergeant Mike Joseph would be conducting the first of several formal reviews to ensure all steps were being taken. The case management software, or CMS, the department now used improved the comprehensive approach by applying a relational database and automated checklists to help avoid missing an important task or case detail.

By 8 a.m. the crime scene specialists released the victim's body to be transported to the morgue. Detective Bradley Macon left the crime scene and drove to headquarters.

Why is it important that each action taken by personnel is fully documented? Why did officers canvass the businesses in the area?

Introduction

Science and technology as well as the training and education of those who wield it cannot be overemphasized as we consider identifying and collecting physical **evidence** as well as criminal investigation overall. Because various scientific methods of the last two centuries have been brought to bear to assist the investigation of crime, tremendous improvements have been made in the process and the outcomes of the criminal justice system. Few people in contemporary society have not absorbed information about science and technology specifically related to criminal justice and a fair few have succumbed to the fanciful efforts of Hollywood in believing that a criminal investigation is not complete without whiz-bang technology and the employment of science. That such a version of the intersection of crime and technology exists in society is not shocking, but it does require perception and expectation management.

These fanciful stories of crime, criminals, and the detectives-cum-scientists who solve the mysteries are tempered somewhat by reality shows such as A&E's *First 48*. There is a concern by many in the public that along with popular television and movie depictions, as well as documentaries about crime and solving crime, people are being taught how to cover up crimes and effectively evade law enforcement efforts. This is not an unreasonable concern, though there is little evidence those criminals who were not already overly cautious have grown more sophisticated in covering their tracks. Crimes such as homicide often occur

Evidence Facts supporting the proof of a thing.

rather quickly and lack the element of premeditation or planning. In such circumstances the perpetrator is hard-pressed to perform the professional "cleaning" that appears with some frequency in fictional portrayals. Even if some degree of knowledge gives the criminal a potential edge, the ever-increasing expertise of criminal justice personnel using increasingly sophisticated equipment likely mitigates any effect. Murderers may on the one hand try to obscure any evidence of their presence at a crime scene. On the other hand is the admonition by many professionals with a great depth of experience in death investigations (and other crimes) to stick to established protocols and remain persistent in the human process aspects of the investigation. This is captured well by Adcock and Chancellor when they point out, "If modern detectives have an Achilles heel or weak point, it is in the overreliance and dependence on forensic evidence" (2013, p. 21). As we discuss gathering and processing evidence, it is important to consider the role of officers and investigators to understand the continuity needed in matters of physical evidence.

Entire books are written on the subject of crime scene work, laboratory analyses of different types of evidence, and crime scene reconstruction. Our aim in the book is to review some of the aspects of crime scenes, some services provided by forensic laboratories, and the personnel who devote their talents and expertise to both. Just as investigations are generally categorized as either preliminary or follow-up, so too can the use of technology be divided. Tasks that patrol officers or crime scene technicians perform at an initial scene are intended to document and preserve potentially perishable evidence, whereas follow-up tests in the laboratory generally are confirmatory of what has been gathered at the crime scene. The criminalists who work to test evidence in the laboratory setting coordinate closely with detectives during the follow-up phase of an investigation.

Some of the initial questions at the scene of a crime may be about what a particular substance might be, for example, is a stain blood, or is there gunshot residue on a suspect's clothing? So-called **presumptive tests** can provide an initial answer to such questions. Presumptive tests are also to determine if certain circumstances are drugs or explosives and if liquids at an arson may be flammable fluids or vapor residue. When certain substances are exposed to a chemical reagent they can provide a positive or a negative reaction that suggests a need for further laboratory testing for confirmation of the makeup of a substance.

Presumptive test Analyzes a sample and determines that it is either not a particular substance or that it probably is a particular substance.

How It's Done
FIELD TESTS

Field or *presumptive* tests are used to determine the probable nature of a substance found. These can include tests for human blood, semen, drugs, gun-shot residue, and more.

The initial testing done at the crime scene, if it shows a positive result, will likely need to be sent along to a laboratory for verification.

The use of field tests can assist investigators in understanding the incident and can even provide probable cause for an arrest. Importantly, though, the use of such presumptive tests can help eliminate a person from consideration as a suspect so that officers can move on to focus on more viable suspects and clues to follow.

The Developing Science of Criminal Investigation

Isolated efforts at applying the science of the day to a criminal matter have been noted throughout recorded history. Observations by physicians about the cause of death have been made since the time of Julius Caesar. Various Eastern and Middle Eastern cultures utilized the uniqueness of fingerprints in business dealings. The identification of traces of poisons in the body has similarly been developed in various countries over a long period of time. These scientific periods in the evolution of U.S. and Western criminal investigation generally act as an overlay to the human and governmental efforts at progress in the investigative field. To aid you in considering the advances of science related to crime detection, the following periods illustrate how we have gotten to where we are today and why that matters in considering contemporary use of science and technology. These periods of time do not have specific beginning and ending dates; rather, they coincide with society developing, as have scientific advances.

The Prescientific Era

The prescientific era describes the time when little or no use was made of recognized scientific methods or technological devices. The period was marked by physically harsh interrogation techniques and the questionable use of information by informers accusing and convicting with little hindrance of physical facts. We noted previously the early efforts in the 1800s at organizing the police function itself, first with Sir Robert Peel in London, and later as formal agencies were created in the larger cities in the United States before spreading west to catch up with the pioneering migration. Crime as a topic in literature has existed for a long time, yet the genre of crime fiction only began to develop in England during the 18th century. Intrepid individuals observed clues in the behavior of people to deduce guilt for the entertainment of the reader. With the slowly developing institution of policing as it moved from addressing crimes of disorder in the streets to an additional role of retroactive investigation, popular stories followed. It would be some time before the plot device of the scientific method would feature prominently in the fictional reflection of the societal endeavor of determining "whodunit."

The Scientific Era

By the time that Sherlock Holmes was becoming popular in the late 1800s as Sir Arthur Conan Doyle's deductive genius, applications of the science of the day were being increasingly seen. The fictional Sherlock Holmes was patterned after real-life Frenchman Eugene François Vidocq, who, like most of the Bow Street Runners, began as a criminal himself before creating a private detective organization that evolved into France's national police force. Vidocq (1775–1857) was first to use plaster cast shoe impressions, early ballistics, and undercover agents that grew out of the model of informers who had close ties to the criminal underworld.

Also in France, Alphonse Bertillon (1853–1914) introduced the concept of identifying people by their unique body structure and features. The system, known as anthropometry, also included scars, marks, and tattoos now standard shorthand in the lexicon of law enforcement and embraced by U.S. and UK police in the 1880s. While fingerprints were increasingly recognized as unique from person to person during the late 1800s, Sir Francis

Galton (1822–1911) systematized the shapes and patterns of the prints that allowed an organized classification method of using fingerprints for identification and published his findings in the book *Fingerprints* in 1892.

In addition to work in the area of **biometrics** and measuring individual human features, others examined substances by applying the scientific method and increasingly technological devices. Mathieu Orfila (1787–1853), a French chemist, is credited with a good deal of work involving the effects of poison on the human body and specifically in death cases. Orfila is credited with founding what we still refer to as toxicology, science that has aided in numerous death investigations. Calvin Goddard performed possibly the most recognized early work in forensic ballistics, writing an article of the same name in 1925. Goddard researched and wrote on ballistics and firearms identification and did much to move the fledgling field forward and creating the first private crime laboratory in the United States. Goddard also popularized the use of the comparison microscope and performing ballistics comparisons, which spread back across the Atlantic to the UK. With the ever-increasing use of firearms in American crime, the development of this particular subfield of the **forensic sciences** was inevitably important.

Edmond Locard began the first police laboratory in France in 1910. August Vollmer, then chief of the Los Angeles police department, formed a similar laboratory in 1923 using the services of an officer who received his science training as a dentist. As the 20th century unfolded, the FBI and many larger cities formed crime laboratories of their own. This brought the need for the training of criminalists that was somewhat standardized by an expectation of applicants possessing a bachelor's degree in one of the hard sciences such as biology or chemistry. Hans Gross, an Austrian lawyer, published his 1893 book *Criminal Investigation,* which further spurred the development of the field as a serious endeavor. The use of the scientific method gained more widespread support as applied to forensic science throughout the 20th century as technological advancements aided the gathering and examination of evidence to be tested by science.

The Technological Era

The capabilities brought about through scientific examination of evidence required increasingly sophisticated and sensitive equipment to detect and collect. Cameras became common for examination and collection duties at crime as portability increased and cost decreased. Various powders and tools used to collect fingerprint evidence, magnifying equipment to observe trace evidence, and specialized tools to gather the evidence have continued to develop and to become more effective.

The technological period can be seen as beginning in the 1960s as computer databases allowed for the storage and retrieval of large amounts of information about people, vehicles, weapons, stolen property, and more. The FBI's National Crime Information Center (NCIC) was created in 1967 to aid local and state law enforcement in their efforts to identify and locate people and the various items. The resources of the federal government later led to the development of the Integrated Automatic Fingerprint Identification System (IAFIS), a repository for the fingerprint records of millions of suspects and offenders, as well as prints obtained from crime scenes and deceased victims. While many people consider the use of **deoxyribonucleic acid (DNA)** as the "rock star" in the identification of suspect presence or involvement in a crime, it is interesting to note that the method only came to prominence just over 30 years ago. But undoubtedly this application by Sir Alec Jeffries in England to

Biometrics Measurements of the human characteristics that serve to identify individuals.

Forensic science The application of science to criminal and civil matters of law.

Deoxyribonucleic acid (DNA) Two coiled chains of molecules that carry genetic functions of an organism.

assist in a famous case there was a landmark in the technological period. Once again, the FBI provided the needed backbone to create a database to leverage this developing technology with the **Combined DNA Index System (CODIS)** in 1990.

Combined DNA Index System (CODIS) Searchable database of stored DNA profiles.

Methods of Collecting and Processing Evidence

Whether a patrol officer is looking for fingerprints at a burglary scene or crime scene technicians and investigators are combing through the scene of a violent homicide, a search for evidence is combined with the methods for collecting and processing that evidence. The size, complexity, or perhaps even the notoriety of those involved may dictate the volume of resources devoted to a particular crime scene search and forensic effort. In the case of homicide, for example, research has tied the number of investigators initially assigned to a case to clearance rates in some studies. Yet it is not a simple example of how many detectives assigned results in a case closure. If a large number of officers and support personnel responded to a crime scene but there is little direction or coordination of the activities that go on, a case may actually be damaged more than it is helped.

After law enforcement officers respond to a reported crime and a decision is made to search for or collect evidence the perimeter of the crime scene must first be established. It is generally best to set the perimeter at the outermost distance that seems likely to contain evidence, since investigators can contract the scene as needed but if they have not set the scene big enough evidence may be damaged or lost before realizing the scene should expand. The division of labor at a crime scene investigation is an important aspect of managing the case. Patrol officers secure a scene and determine if suspects or witnesses are present when they arrive. Crime scene technicians will photograph, test for, and collect potential evidence at the scene. And detectives will coordinate with both to formulate further plans for carrying an investigation forward. The patrol officer is crucial at the outset of an investigation to ensure the greatest likelihood of success. He does this by the protection of the scene, the gathering of information on witnesses or others who can provide important information, and by properly notifying and briefing others who become involved in a case. Crime scene technicians and detectives are each specialists who perform specific functions and contribute to the construction of a case.

Crime Scene Searches: Inside, Outside, All Around

Searching even a relatively simple crime scene requires a plan. Ideally, patrol officers will have closed off the access to an indoor crime scene if they are awaiting other personnel to respond to conduct the processing. Outdoor crime scenes can be challenging to secure if they are in a public place or somewhere of high traffic volume. An additional challenge of the outdoor scene is weather. Evidence can be moved or washed away in a storm, covered by snow, or simply degraded due to heat or exposure to sunlight for a lengthy period of time. Securing either type of scene is a function intended to protect potential evidence from inadvertent or intentional damage as well as to limit access to witnesses who need to be interviewed and to reduce the distractions for official personnel working at the scene of a crime.

The initial crime scene walk-through allows investigators and technicians to develop the plan and approach for locating and collecting evidence. For a large or complex crime scene there may be multiple individuals performing both functions. The planning phase

What are some challenges these technicians may face when processing this outdoor crime scene?

Ben Torres/Stringer/Getty Images

limits confusion or redundancy as well as ensuring a comprehensive scene search. By mapping out the crime scene, personnel can determine what sequence areas will be searched in and who will be responsible for marking locations, photographing or recording images, and collecting and documenting actions taken by everyone at the scene as well as the cataloging and transportation of any located evidence utilizing a chain of evidence that lists the hand-off of items from one person or place to the next. Some crime scenes are large enough to challenge the available resources of an agency and may require assistance from neighboring departments or volunteers who can maintain a perimeter until a search can be conducted. Some volunteer efforts involve the search of large areas such as open fields or words trying to locate victims, weapons, or other items of evidence. Searchers are given a description of what is sought and clear instructions not to touch the evidence but to signal search leaders to examine what has been found and determine whether it is to be collected for processing.

Search efforts often follow one of several patterns (Figure 2.1) that segment an area to allow searching smaller sections one at a time. This can be done by having a number of people side-by-side walking in a *lane search* or *strip* pattern to try and locate what is directly in their path, or followed with an additional lane search proceeding perpendicular to the first path resulting in a *grid* pattern to increase the possibility of finding an item missed on the first path done in a different direction. A search may also be done in a *spiral* pattern that starts at a central location and works its way outward in an expanding circle shape.

Some crimes may involve multiple scenes at which one or more criminal actions occurred within the same reportable incident. Imagine an individual accosted on the street where he is pulled into a vehicle, driven to another location where he is further harmed, put into a different vehicle where he subsequently dies en route to yet another location, where his body is dumped. Many crime scenes, one criminal incident. Each scene, if it can be located, must be processed with the intention of tying suspects and evidence together in a

■ **FIGURE 2.1** Search Patterns

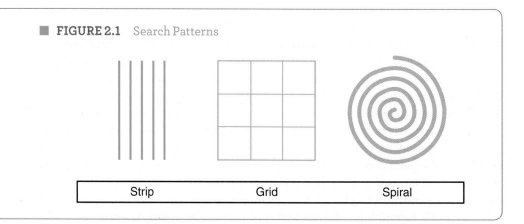

meaningful way. The early and thorough processing of crime scenes is critical, as they generally represent the one best opportunity to identify and collect physical evidence that may be essential in reconstructing a criminal act and identifying those responsible.

Photography and sketching are two methods of documenting a crime scene. Both are important. Patrol officers may sketch a simple layout of a crime scene to assist them in later writing their incident report. Crime scene technicians can perform extensive photography, videography, and computer mapping to create elaborate and precise drawings or animations of a scene. A cautionary note here is that some people including jurors after the fact may expect a sophisticated level of crime scene reconstruction that is not necessarily needed to establish what happened and who was responsible. This is one aspect of the CSI Effect discussed later in this chapter.

How It's Done
WHERE WOULD YOU SEEK OUTSIDE ASSISTANCE?

Many local agencies are too small to have specialized personnel to process crime scenes. Departments will often rely on cooperative agreements with state or other local agencies to provide crime scene processing (Hough, McCorkle, & Harper, 2019). Often, the evidence will also be stored at another agency, and laboratory processing is typically handled by regional or private crime laboratories with only the largest departments housing their own labs. Ballistics examination, fingerprint analysis, and documents, let alone DNA processing, are just some of the specialized analyses requiring well-trained criminalists and laboratory scientists.

In the majority of cases access to the suspected scene of a crime is not at issue. This is when an authorized person gives voluntary consent to officers for the search of a building, a car, or other location. In some circumstances, an emergency may be believed to exist that permits an officer access to a place without consent. If officers can articulate probable cause that someone may be destroying evidence or preparing to flee with it they may be able to enter and search long enough to secure the evidence before gaining a search warrant. Officers may also have a reasonable suspicion that someone's safety or life is in jeopardy, and this too would allow them access to a location. However, in some instances investigators must seek a search warrant from the court to allow access to a location. If there is no imminent threat to life or safety or evidence, or consent from an authorized individual, officers will likely need to present probable cause to the court to obtain a search warrant to seek evidence. The Fourth Amendment to the U.S. Constitution is implicated in these searches, and courts have provided guidance about the circumstances when a warrant is not required.

Whether the officer or technician is searching a building, a private residence in the area closely surrounding it (curtilage), a vehicle, or even a garbage can, the same organized method should be used in each case. Such an approach allows for a thorough examination, as personnel systematically examine one area at a time. As with searching live suspects or dead victims, a structured search method will reduce or eliminate distractions or a lapse in searching. For live suspects this also has implications for officer safety.

Types of Physical Evidence and Its Processing

Almost any object could be physical evidence, and each type of physical evidence involves specific methods of collecting and subsequently processing. Physical evidence that most people are familiar with includes blood and other fluids, fingerprints, hair and fibers, marks or impressions from tolls or weapons on or in the surface of an object, and the list goes on. Physical evidence is also referred to as "material" or "real" evidence and is often presented in court for the judge or jury's consideration. Once evidence has been recognized as such and documented as to location, officers or crime scene technicians must collect the evidence and arrange for it to be transported to the agency's property room, laboratory facility, or outside laboratory. This identification, documenting, and collecting, taken all together, is referred to as processing the crime scene. The physical evidence, once identified, may be compared to a **control specimen** to see if they came from the same source. Fiber and paint evidence may be compared in this way to determine if their presence on the suspect or his car, for example, are similar to the same type evidence found on a victim or at a crime scene. One type of analysis that the forensic laboratory can provide are the characteristics of an item of evidence. **Class characteristics** are those that put an item in a category of like items. An example would be width of a blade impression indicating a tool mark to be made by a screwdriver. **Individual characteristics** distinguish an item from others within a category. An example might be a gouge within the screwdriver's blade that other screwdrivers do not possess.

Control specimen Allows comparison between a known sample and an unknown item of evidence.

Class characteristics Those attributes of an item that place it within a broader group of objects.

Individual characteristics Narrow down the class characteristic that put an item into a group of objects to one emanating from a single source.

Advanced Scientific Methods and Sophisticated Technology

Some of the technology that makes its way into movies and shows does exist. The availability of such technology to all agencies is not consistent and often has a waiting time that may render the technology all but irrelevant. Consider DNA samples waiting 6 months or more to be tested. An investigation may have concluded based on other factors or a suspect may flee because sufficient probable cause does not exist without the test results. However, the availability of DNA evidence in, for example, cold cases allows the matter to be reopened when little other physical evidence appeared to link a suspect to a crime (Kirsch, 2006). The idea of DNA comparison results in less than a day is not realistic. A television show may use a computer that automatically scans fingerprints to find a match. This device is the **Automated Fingerprint Identification System**, or AFIS. In real life, once a "hit" or match is located, a trained technician must still examine and compare the prints to determine a match.

Automated Fingerprint Identification System (AFIS) Database that allows computer matching of unknown to known fingerprints.

Without question, technology has benefited society in general and the investigation of crimes in particular. Many have noted that technological advancement requires careful reflection about the morality of its usage as well as evolving issues in the law. While the public at large is often unaware of or only somewhat concerned with the seizure, for example, of DNA samples from arrestees, there are others who point out that such broad gathering of individual data may have implications for individual rights and carries a possible unjustified stigma that predicts future offending. Some jurisdictions gather DNA samples from virtually everyone arrested, which may have unintended consequences (Cole, 2007). The criminal justice system for a variety of reasons does not always function with the exactitude that everyone would expect or hope for. As Fourth Amendment issues continue to be considered by the courts the laws governing the gathering of DNA samples may certainly change (Maclin, 2006). At the same time, we should note that DNA analysis has also brought the exoneration of people who were wrongly convicted.

Collecting and Preserving DNA Evidence

While the variety of evidence gathered at a crime scene is almost boundless, DNA has received a great deal of attention because of the ability to match a sample to a specific person or, in some cases, a group of related individuals. DNA evidence and other items examined through scientific methods are given sometimes inordinate consideration by jurors. The profiles arrived at through DNA analysis, for example, are probabilities. As with any of the physical sciences, the probabilities can be quite high—almost to the point of statistical preclusion—and still be called into some question. And while the DNA matching process within the Combined DNA Index System (CODIS) is not exact, the 1 in 7 billion chances of incorrect identification are compelling to most jurors. DNA has been used to exonerate individuals inappropriately convicted of crimes and eliminate those suspected of some crimes. Crime scene technicians are trained and aware of the challenges of gathering DNA evidence, specifically how it can be contaminated during collection or analysis (Aronson & Cole, 2009). Because of the need to always search for additional and more effective methods of analysis, research to improve and supplement CODIS is ongoing (Widyanto, Soedarsono, Katayama, & Nakao, 2010).

Aside from physical challenges to gathering such evidence, DNA gathering from people arrested has raised questions by some people on ethical grounds. The balance of individual rights against the public's crime control mandate sets up the tensions. Some argue that widespread gathering of DNA samples from even those charged with minor crimes can compromise protections from the 4th and 14th Amendments.

CODIS and other local databases aid in solving crimes, even though the volume of requests often result in the criminal justice system awaiting results on backlogged tests (Gabriel, Boland, & Holt, 2010).

Given the challenges that DNA can present during analysis and subsequently in court, technicians and agencies must take every precaution to gather DNA evidence correctly. Still, evidence may not be viable, through no fault of technicians or law enforcement. The National Institute of Justice (NIJ) provides guidelines for the collection and preservation of DNA evidence (NIJ, 2000). NIJ recommends the following precautions:

- Wear gloves. Change them often.
- Use disposable instruments or clean them thoroughly before and after handling each sample.
- Avoid touching the area where you believe DNA may exist.
- Avoid talking, sneezing, and coughing over evidence.
- Avoid touching your face, nose, and mouth when collecting and packaging evidence.
- Air-dry evidence thoroughly before packaging.
- Put evidence into new paper bags or envelopes, not into plastic bags. Do not use staples.

Firearms, Explosives, and Arson

The use of a firearm in a murder presents a number of opportunities for evidence. Some of the well-known aspects of firearms evidence are the spent projectile and bullet casing and

Gunshot residue (GSR) Particulates left on the skin or clothing of someone discharging a firearm or in close proximity to one that has been fired.

attendant markings from the weapon used to fire them, the possibility for **gunshot residue (GSR)**, fingerprints on the firearm, and—if the weapon is recovered—the potential for determining ownership or possession. In reconstructing the circumstances of a death case involving a firearm, the statements of suspects and witnesses may be corroborated by the firearms evidence. Distance between the firearm and the victim may be able to be determined with a reasonable degree of accuracy within certain threshold distances.

According to Fisher, "Firearms identification . . . refers to the study of firearms and includes the operation of firearms, cartridges, gunshot residue analysis, bullet and cartridge case comparisons, powder pattern determination, and the like" (2003, p. 329). This should not be confused with the term *ballistics*, which Fisher defines as generally referring "to the trajectory taken by a projectile and assumes an understanding of physics" (2003, p. 329). Determining the trajectory of a projectile is also useful in reconstructing the crime. Young and Ortmeier (2011) note that it may be possible to obtain:

- Path of the bullet,
- Location of the shooter,
- Sequence of shots fired,
- Which bullet holes or impressions are entry, exit, or ricochet.

The markings imparted by the lands and grooves of a firearm's barrel to the expended bullet and cartridge casing are unique. If the same type of cartridge is fired from the same gun, it is possible to use a comparison microscope to compare the marks (striations) on both spent bullets as well as both cartridges. Even without the firearm, class characteristics such as a bullet's caliber can often be obtained from recovered projectiles and cartridge casings. While some in the legal community have challenged the validity of tool mark and firearms identification, the scientific community generally accepts and supports the rigor and validity of the discipline (Nichols, 2007).

As with CODIS for DNA evidence and AFIS for fingerprints, databases were created for projectile and casing information. While the FBI and the ATF both created such information databases, the two merged in 1999 to create the **National Integrated Ballistics Information Network (NIBIN)**. If a "hit" is obtained potentially matching two or more entered records, a firearms examiner compares the items. This protocol is the same as with the other automated systems for fingerprints (AFIS) and DNA (CODIS).

National Integrated Ballistics Information Network (NIBIN) Computerized network available to law enforcement for the comparison of ballistic evidence.

The striations imparted to a projectile from a firearm or the extractor marks left on a casing from a firing pin are two examples of *tool marks*. More generally, marks made by tools or other items used as tools are found at various scenes. Photographs of marks at the scene are collected and then, hopefully, the item that has such marks. Marks can show general size and shape of a tool and, possibly, unique abnormalities of the object that left a mark. If the tool or object is recovered it will be sent on to the crime lab for comparison to the marks left on other evidence as well as to a standard.

Gunshot residue (GSR) may also be discovered at a scene or on the clothing or person of a suspect or victim. When a firearm is discharged, a cloud of residue issues from the muzzle of the weapon's barrel and possibly from around the chamber (of a revolver). Part of the residue can travel several feet as the projectile moves forward. Some of the residue may settle on the clothes or skin of the shooter and some of the residue may land on other items in close proximity, including the victim or his clothing. This GSR will be tested at the laboratory using a scanning electron microscope (SEM). As some changes in ammunition have occurred, science and technology must keep pace with new analysis methods sensitive to rounds that may not contain lead as in traditional bullet projectiles.

While firearms may leave tool mark impressions on projectiles, employing explosives or accelerants and igniting devices may leave some amount of residues or other physical evidence. Technology has the potential more than ever before to detect, collect, and identify such substances. While criminal homicide in the United States is not frequently accomplished by arson or explosives, arson may be used to attempt to cover a homicide. Most people are familiar with the tremendous explosive destruction and death caused by Timothy McVeigh and Terry Nichols in the bombing of the Alfred P. Murrah federal building in Oklahoma City in April 1995. This homicidal attack killed 168 people and injured more than 680. Forensic testing showed traces of residue in and on McVeigh's clothing that corresponded with the ammonium nitrate and fuel oil used in the bomb. The use of explosives to commit murder in the United States is not frequent, despite the extensive media coverage of this infamous case.

Fires occur where fuel of some type and heat come together in an environment with oxygen. Firefighters and arson investigators are trained to look for artificially introduced oversupply of any of these three elements required for a fire. When an abnormal amount of fuel, heat, or oxygen is present at a scene it can be an indicator of intentional fire-setting. In arson cases there is frequently the presence of an **accelerant**, a liquid fuel source such as gasoline or kerosene intended to ensure the fire builds and spreads. For the laboratory analysts such presence of a liquid accelerant brings into play the gas chromatograph. The gas chromatograph is used to analyze liquids to determine their ingredients. Matching liquid residue possessed by a suspect to a similar residue at the scene can provide important evidence. A spectrophotometer may be used to examine trace evidence found at the scene. This instrument analyzes the waves of energy that make up the characteristic colors given off when a substance is burned. When the various colors are exposed on film they allow for identification of the burned material.

Accelerant Any substance used to increase the speed that a fire burns or spreads. The substance or mixture may be liquid or solid.

An important consideration in arson cases is the cooperation needed with other agencies. Typically, an arson investigation involves one (or more) fire department and a law enforcement agency. Such cases can, however, involve a task force or the use of state or federal agencies with the expertise and resources to assist. Insurance companies often also have a stake in the outcome of a fire investigation. This fact can lead to assistance from an insurance company, or occasionally interference based on the insurance company's investigators pursuing their own agenda at a scene or in follow-up investigation. Whether it is gun-shot residue, tool marks, or trace accelerant, the discovery and analysis of physical evidence may be the key to solving a case.

Expert Witnesses

In conjunction with the use of advanced scientific methods and sophisticated technology comes a need for **expert witnesses** to conduct testing and subsequently testify about their methods and results. This expertise is obviously necessary, but there can also be a tendency among some jurors to defer to expert testimony if the jurors simply do not grasp the science (Singer, Miller, & Adya, 2007). The expert opinion can be critical to proper understanding of a case, but as an opinion, the trial judge has to determine if the expert will be allowed to testify based on her credentials and the reliability and relevance of the evidence

Expert witness A person whose experience, training, and education provide the basis for acceptance by a court to testify to his or her opinion on matters before the court.

to be given. This is required based on what is called the *Daubert* standard (*Daubert v. Merrell Dow Pharmaceuticals,* 1993). Jurors must still *weigh* the testimony of experts. Experts must use the accepted methods of analysis in their particular field to evaluate information and arrive at the opinions they offer to the court. Jurors are instructed that though the analysis and information provided are intended to give them a better understanding of a particular process or topic, it is an opinion nonetheless and each juror will determine what weight to give such testimony. The *Daubert* standard was extended to apply to the testimony of non-scientist experts as well (*Kumho Tire Co. v. Carmichael,* 1999).

Equipment Utilized in Investigations

As we mentioned earlier, the tools used in the location, collection, identification, and comparison of physical evidence are varied—some simple, some quite complex. Just as with the preliminary and follow-up investigations, the equipment utilized in these two phases of the investigation can generally be distinguished. Initial recording of a scene involved one or more photographic and video recordings, measurements and sketches, in-depth descriptions of the scene, and narrative accounts of what personnel did at the scene. The narrative would also include what contamination or human incursions were observed or known.

Basic Crime Scene Equipment

Latent fingerprint Visible print after powder is applied to the residue from oils of the hand.

Plastic print Impression of a print left in a soft surface.

Patent print Transfer of material from a print to the surface leaving an impression.

Alternate light source (ALS) Used to discover various types of evidence; ultraviolet light, for example, illuminates natural substances such as semen or saliva.

Reference An item of evidence from a known source that allows comparison from a suspect source.

Patrol officers commonly have access to basic crime scene equipment that allows them to perform rudimentary examination of the scene and perhaps dust for fingerprints and take limited photography. Processing and photography by patrol officers are limited by the equipment the officers have available. Various powders, chemicals, and the applicators to apply them are used for **latent fingerprint** development of oils left behind and visible after applying fingerprint powder, while those impressed in a soft surface, or **plastic print**, collection may be achieved through photography and possibly the use of a casting substance. A bloody fingerprint is an example of a **patent print** where a substance is transferred from someone's fingertips to a surface resulting in the print image being visible. In what is called sequential processing, each print may be processed in more than one way, but in every instance a photographic record should be made before the attempt to process the print in case it is smudged or otherwise degraded. A technician may use an **alternate light source (ALS)** to better view evidence before photography or collection. Basic equipment for the officer will also include lifting tape and print cards to collect latent prints revealed through the use of powders.

An officer may have casting material for impression evidence. Processing this type of physical evidence would also follow photographic documentation, which would include the use of a ruler, scale, or other **reference** in the picture to indicate relative size. The picture is also taken without the reference to ensure adjacent evidence is not covered by whatever item is used for size comparison. Crime scene technicians and specialists also use a variety of basic equipment including measuring devices, plastic forceps for gathering evidence that may otherwise show pressure impressions such as spent bullets, or lighting equipment to facilitate their work. Officers or technicians may carry a variety of presumptive test kits for different types of drugs. Cotton swabs and swatches may also be used to gather fluid or wet evidence, which may be allowed to air dry before transportation and placed into paper containers that can breathe as opposed to plastic or glass, which would seal moisture and often result in degrading or destruction of a wet sample through microorganism growth.

An important piece of basic equipment is often a roll of crime scene tape for the officer to use to erect a visual barrier around the area until it can be properly searched. For crime scene technicians, basic equipment would also include a variety of bags, both paper and plastic, and other containers suited to gathering various types of physical evidence. In addition to items commonly thought of to gather evidence are important pieces of **personal protective equipment (PPE)** . These include filter masks or re-breathing equipment if necessary, impermeable gloves to prevent transmission of chemicals or liquid-borne contaminants to the technician, and clothing covers of various types to limit exposure to harmful agents or the transfer of evidence from a scene to the clothing of an investigator or technician.

Personal protective equipment (PPE) Items such as goggles, helmets, and protective clothing that shield an individual from various hazards or injuries.

Advanced Crime Scene Equipment

More advanced equipment, even at the crime scene, may now include devices to scan for fingerprint evidence, chemicals present at the scene, trace materials of many sorts, and measurement and photographic capabilities of great sensitivity and accuracy. Back at the laboratory, science and technology combined have reached new heights in identification of substances in the linking of such substances to people and to places. While tweezers may be viewed as basic equipment, they are used to gather extremely small or fragile pieces of evidence such as hair and fibers that can then be examined through more advanced devices in the laboratory. It is not uncommon to use vacuum devices to attempt to gather small evidence at some crime scenes.

Investigators use orange powder that fluoresces under ultraviolet light. What role might these fingerprints play in an investigation?

U.S. Air Force photo/Airman 1st Class Micaiah Anthony/ RELEASED

Personnel Safety

The safety of all personnel who assist in processing a crime scene is paramount. From the patrol officer who first arrives at a scene with often unknown people, substances, and evidence, to the crime scene technician tasked with identifying and collecting all manner of items, professionals should be furnished with (and use) personal protective equipment (PPE). Various scenes may contain chemical hazards such as those associated with clandestine drug labs, biological hazards from blood-borne pathogens, and physical hazards from sharp and broken objects or even loaded firearms. Each agency should have established safety protocols to be followed by all personnel. Assessing risk at a crime scene is essential and should be done while also planning the processing of a scene. Various types of protective apparel and equipment exist to safeguard the hands, eyes, respiratory system, and more of employees.

The Contemporary Forensic Laboratory

As discussed earlier, the current technological era has built upon the scientific era to provide modern investigators with tools and testing capable of great sensitivity and accuracy for the examination of many types of physical evidence. Some of the capabilities found in various laboratories include trace evidence including fibers and residues, fingerprint comparison, firearm ballistics and tool marks examination, questioned documents (e.g., forgery), DNA testing, other chemical testing including blood, semen, saliva, drugs, and explosives components, to list just a few. Due to the cost of equipping and maintaining such laboratories, as well as acquiring highly trained technicians, criminalists, and scientists, many are run by the federal government or state governments. Some local or regional labs may handle only certain types of analyses, while sending on other types of evidence to larger or private laboratories for lengthier or less common analyses.

While there is no centralized requirement for laboratories to become accredited, many in the United States voluntarily participate in the accreditation offered by the American Society of Lab Directors. As with many accreditations in various fields, it signifies that the facility or organization adheres to accepted best practices and meets the standards set within the profession.

Personnel

The story of technology and forensics is also the story of the professionals who conduct investigations and examinations. Markedly different from the "triple-threat" fictional characters from television shows such as *CSI*, these real-life professionals do not simultaneously work crime scenes, conduct laboratory analyses, and interrogate suspects. They also do not generally wear high heels or tight suits while visiting crime scenes. We hope you are not surprised by this. Many undergraduate college students enter course or degree programs believing that this type of omni-skilled, high-fashion career awaits them. Each role in forensics is fascinating and important, and requires a singular focus on the requisite skill sets and functions.

Crime Scene Technicians at the Initial Scene

At the initial scene examination, crime scene technicians locate, document, and collect evidence at the various scene locations. Patrol officers frequently fill this role as well. The job of crime scene technician has gradually transitioned from one occupied by certified law enforcement officer to one of highly trained civilian employees of an agency. This follows the civilianization trend of other positions including communications specialist (dispatchers and call-takers). This evolution recognizes that the higher salary cost of certified officers is not needed for crime scene work and that by having personnel who only conduct scene investigations, the level of specialization increases and results in higher quality work at scenes.

The variance in qualifications for the job is based on factors such as departmental budget, crime rate and volume, top management philosophy, or available interagency assistance. Undergraduate students with various related degrees have had success in acquiring crime scene technician positions. Some retired officers also seek out the job of technician. Hands-on training after joining an agency can be the primary source of training in the various functions of the job.

Crime scene technicians may perform virtually all documentation and evidence collection functions, or there may be role separation resulting in crime scene photographers distinct from technicians who locate and collect evidence as well as perform limited presumptive testing. The photographer may also perform digital mapping of a crime scene or sketching as appropriate. Once again, this varies with available resources and personnel.

Laboratory Technicians

Beginning in the early 1900s crime laboratories began to be established, though these were available to relatively few agencies. The FBI established a laboratory in 1932 to make forensic science available to law enforcement agencies around the country. While this marked a step forward in the use of science in criminal investigations, the general lack of sophistication of medium- and small-sized agencies in applying science meant that fully exploiting new and developing capabilities was still some time off.

Personnel who analyze evidence in the lab are generally referred to as **criminalists**. Some crime scene technicians choose to eventually transition to laboratory work. The qualifications for criminalist typically include at least a bachelor's degree in biology, chemistry, biochemistry, pharmacology, or a related core forensic discipline. In some cases a person with a degree in criminal justice who completes a diverse selection of courses in the hard sciences may be accepted to laboratory training programs.

Criminalist An individual (usually degreed) who applies scientific knowledge to analyze evidence.

In the laboratory setting criminalists may work in any of a number of sciences, including anatomy, anthropology, bacteriology, biology, chemistry, entomology, pharmacology, and even psychology. Different labs provide different services around the country. Just as with other key areas of public safety resources, the federal government has been a leader in establishing and funding laboratory services.

A lab technician examines an article of clothing for residue. What types of evidence might be obtained from this shirt?

AP Photo/Steve Helber

Crime laboratory growth has been significant, but there is much concern that insufficient national standards exist to ensure the thoroughness of qualifications for personnel and solid integrity of testing protocols for the labs. Various court decisions over the second half of the 20th century required law enforcement agencies to rely more on physical evidence and science in building cases against suspects. This too led to the proliferation of laboratories in the United States. No doubt that the tremendous volume of drug cases inundates labs with testing, as well as the rapid increase in the use of DNA in forensic identification since the mid-1980s.

Technology and forensic capabilities for testing may or may not be readily available to various-size agencies. Large local agencies routinely assist smaller agencies. State law enforcement agencies similarly assist local-level departments in large or complex cases. Additionally, the FBI has various programs and personnel available to assist with coordination, training, and other forms of assistance. The National Center for the Analysis of Violent Crime (NCAVC) is the central entity to facilitate assistance. Within the NCAVC the Behavioral Analysis Units (BAU) are most known to laypersons. The Violent Criminal Apprehension Program, or ViCAP, works to identify linkages among crime locations and methods for homicides, sexual assaults, and unidentified or missing persons.

Documenting the Crime Scene and Actions

Documentation begins when a department is notified of a potential crime. The call-taker or communications specialist will enter some amount of information, typically, into a computer-aided dispatch (CAD) system, which may integrate to a records management system (RMS) as well as to supply information to an officer in her squad car via a mobile computer terminal (MCT). The initial officer dispatched can enter additional notes via her car computer or terminal, which may also act as a laptop that can be dismounted from the squad car to allow the officer to bring it into a crime scene or location where she is completing an incident report. Developing skills in notetaking and report writing are among the most crucial for law enforcement officers and criminalistics personnel.

Case Management Software

As the preliminary and follow-up investigations move forward, officers and technicians are documenting their observations and actions through notes and official reports. The management of all information available as well as activities assigned to and completed by various personnel have long been the focus of management efforts to ensure a comprehensive investigation. Modern case management software assists in this effort by more efficiently tracking activities and linkages among many factors present in a given case. Relational databases and other software algorithms feature search capabilities that provide investigators and supervisors a powerful tool. Crime mapping, already mentioned, can further guide investigators to examine whether a pattern exists in reported crimes.

Few would question the impact of the microchip on the world. The original mission of Microsoft to put a computer on every desk and in every home has largely been fulfilled, and this has been a boon to investigators worldwide. It would likely be difficult to locate a law enforcement agency that does not have access to and benefits from personal computers, as well as networked systems that give officers in agencies access to an overwhelming volume of information. Laptop computers are normally found in squad cars, and these devices not only allow officers to query various databases for wants, warrants, driver's license, and other information, but integrate with records management systems to upload reports and information from the investigations officers conduct. The same input capability and database functions allow supervisors to monitor investigative progress. In addition to accountability, computer programs allow various criminal justice and court personnel to upload relevant files and media to the building computer-based record of a case. Crime scene technicians utilize computer-aided drawing programs in the normal course of their duties where before such computing power would be reserved to perhaps architects and engineers.

Because officers and agencies can search and access nearly unlimited sources of information and connect with personnel and other criminal justice agencies with relative ease, intelligence efforts are far more fruitful than ever before, which in turn leads to more effective investigations and the apprehension of offenders. The enhanced effect brought about through application of computer technology also improves the efficiency and effectiveness of police or private contract laboratories processing the evidence located and collected by criminal justice personnel. The exaggerated time frames for processing evidence seen in fictional dramas do a disservice to the complexity of the testing and the high level of skill of the criminalists conducting testing on evidence. The public is generally unaware that circumstantial (also known as indirect) evidence and testimonial evidence are still far more common than the direct evidence results of esoteric laboratory analyses in securing convictions. And this point is important for prosecutors to make during a trial as jurors await sci-fi-like presentations filled with motion and animated recreations that simply do not exist—and are not necessary in the case at hand. Because resources are sometimes limited and the reality of available forensic testing does not match the fictional depiction on television, agencies often send evidence to private outside laboratories for tests to be performed. Even priority cases take time to process, let alone the backlog that exists for a great deal of evidence at crime labs across the country. And, of course, when those results finally arrive, they may not link a suspect to a crime or a scene (Roane, 2005).

Photography

As previously mentioned, photography has become an important and expected practice at serious crime scenes and in many instances involving cases that may benefit from a visual record. Improved digital photography, images, and recording have also become relatively inexpensive for agencies to acquire. This extends to the use of such photographic devices in jail facilities for booking an arrestee and helping to manage the inmates throughout their time in the correctional system. Patrol officers routinely use photography to document crime or incident scenes, graffiti or damage to public areas, and to take pictures of suspects, witnesses, or informants. For arrestees as well as suspects, witnesses, and informants, pictures of tattoos and other distinguishing marks can be useful later in an investigation or to identify individuals. Sometimes unidentified bodies have limited identifying features but might have tattoos or scars that can be recorded and noted in state or nationwide databases.

Photos and videos aid the work of detectives and technicians as well as tell the story in court of a crime and the actions taken as a result of the crime. While it is generally true that more photos are better than fewer, personnel should be deliberate in recording images and noting their relevance in the officer's report so that the picture is also authenticated as having been taken by a specific person. Photography is immediate, and though lacking multiple dimensions, can provide a faithful representation of what was seen by the photographer of the evidence or scene. The photos can help present the overview during case preparation. While we again caution about the CSI Effect, we also note that photos can be linked by software to present something of a walk-through that can help the case team, prosecutor, court, and jurors understand the setting. Capture technology allows a three-dimensional mapping of a scene, though this is rarely needed. And while the photographs of a scene can be helpful, the prosecutor must be mindful that the picture can also be cluttered with too many items not relevant to the crime or investigation.

The use of digital or video recording at the scene of some crimes provides perspective that goes beyond photos and aids court testimony by personnel. Scene photography will proceed from the general to the specific. Personnel take pictures of the approach to a scene, overall views, and then photographs that move closer to specific items or aspects of a scene. Close-up views allow examination of evidence in greater detail once the overall state of the crime scene is understood. Poor camera work that wanders or creates distractions is a downside to some use of recording. Photos and videos are also useful in the training of crime scene technicians, lawyers, and obviously, law enforcement officers. **Body-worn cameras (BWC)** for officers continue to grow in usage across the United States, and images are useful in many investigations. Other uses of photography include static and dynamic surveillance, mug shots and lineups, and scene management by crime scene technicians. For detectives

Body-worn camera (BWC) Camera devices typically attached to the uniform of a police officer to gather evidence, enhance officer safety, and improve public relations through accountability of officer actions.

and crime scene technicians, the photography allows them to inventory the scene and identify the locations of key pieces of evidence. Technicians may render a drawn-to-scale sketch using computer software generated by the measurements taken at the scene. Case-management software and criminal justice system databases available to court personnel can also make use of uploaded photography and video.

How It's Done
CRIME SCENE PHOTOGRAPHY

Crime scene photography is a blend of personnel and equipment. Crime scene technicians follow established methods of photography that include long-range photos to provide perspective on overall layout of a scene that could include how someone may have gotten into or exited from a location, medium-range pictures that give views of specific areas around a crime scene to focus attention, and close-range images of items of evidence. Personnel use a variety of still and video recording and have available film or digital imaging to accomplish documentation of the scene.

While the presence of recording devices in the public is ubiquitous, it is still necessary to explain the context and overall place that a given picture or piece of video has in a full understanding of a scene or event. Dashboard-mounted cameras in police cars, body-worn cameras (BWC) used by many agencies, and closed-circuit or other pictures and recordings in public areas or in private homes and businesses can all aid investigators.

In general:

- Photograph the exterior of a scene or location first.
- Take photos to provide a sequential depiction of the crime scene narrative.
- Take close-up shots of evidence as well as pictures that provide the broader scene to situate the item of evidence.
- For small or unique pieces of evidence, use a ruler to show scale of the item.
- Use evidence placards to ensure later clarity in referring to individual items of evidence.

Diagramming

Computer-aided drawing (CAD), or computer-assisted drafting and design (CADD) Software used to facilitate design.

Whether preliminary handmade estimates of a scene on the notepad of a patrol officer, or the finished quality of **computer-aided drawing (CAD)**, or computer-assisted drafting and design (CADD) renderings of a crime scene technician, the "drawing" of a scene can be a useful if not critical component of documenting an investigation. The crime scene sketch has value by telling quite a bit about a scene without the embellishment of words that may unintentionally misdirect or place too much emphasis on one item or another. The available reports, and certainly the statements or testimony of the person completing the sketch, can add narrative to the graphic depiction. The sketch is not a replacement for the reports of investigating personnel by any means.

Sketches give perspective to others—and from different angles, depending on the manner in which the sketch was completed. The sketch can aid a witness in recall, refresh officers and investigators as they review evidence or testify, and help jurors understand the interconnections between pieces of evidence and the actions of those who were involved.

A rough sketch, much like initial notes, can be completed later in detail. The accuracy of the measurements are important so that the final sketch or drawing will be a scaled representation of the scene so the context of witness, offender, and victim actions are understood as well as possible. Multiple sketches may illustrate the larger area outside the crime scene itself, as well as smaller areas of the scene blown up to more clearly understand small items of evidence or their relationship to objects or people.

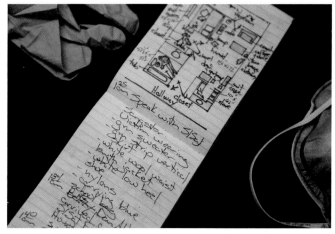

How might these investigator's notes, complete with an initial sketch of the crime scene, be used later in the course of the investigation?

Creative Touch Imaging Ltd./Alamy Stock Photo

As with other aspects of a criminal investigation, a local department may not have speciality personnel or equipment to produce high-quality sketches. A larger agency may have the resources to assist, and this cooperative agreement should be established against the eventuality of needing large-scale or complex drawings and sketches. Graphic designers can also fill this need for a department. For the purpose of understanding the common tasks of a crime scene, let's briefly look at two of the most common measurement methods for sketching: rectangular coordinates and baseline, and triangulation.

The rectangular coordinate method is perhaps the most simple and involves measuring the position of an object from two other fixed points. This is often done by drawing a *baseline* between the two fixed points, working from left to right, and then noting the distance along and away from the baseline. Triangulation is an overhead perspective sketch that locates the item in question via straight-line measurement from two or more available points of reference.

Report Writing and Notetaking

Report writing begins with notes: the more comprehensive the better. The initial notes taken allow personnel to re-create the sequence of actions they took including notification of others and assisting other personnel in various tasks. Patrol officers, crime scene technicians, and investigators must all complete thorough reports documenting what they found at the scene of a crime and incorporating information from as many of their senses as apply. It is not unusual for experienced firefighters and arson investigators to note specific smells at fire scenes; officers familiar with the odor of burning marijuana will recognize it on a traffic stop or at the scene of a crime (which also may provide probable cause for certain searches). Each of these personnel may also create sketches and record various photographic or video records.

Officers also document all the information they take in about evidence, suspects, and other persons involved in a crime. Some agencies provide a template or structure to the narrative portion of reports to ensure major aspects of a case are covered. Description of the crime scene, a brief reconstruction of the crime, all investigative actions taken by the reporting officer, and synopses of victim and witness statements provide the basis of a report. The notetaking begins almost immediately and continues as time and activity allow. The notes, while taken sequentially by the officer, will likely be piecemeal and require reordering to construct a chronologic description. Officers develop their own method of taking note, including abbreviations and other shorthand to allow quick chronicling of information or statements. Some officers may use a mix of recording notes and writing based on the circumstances.

The formal written report should be thorough and written in clear language without jargon or emotion-laden terminology. An officer or investigator states known facts and not opinions or suppositions. If a victim, witness, or informant makes a statement that is not verified by the officer, the statement should be noted as such. Once all of the information is gathered in the officer's notes, she must organize it and complete the report. No written work is complete without proofreading for spelling and grammar as well as assessing whether all required elements have been included.

The CSI Effect

The so-called CSI Effect refers to the general belief by laypeople and some in the criminal justice system that they know a good deal more about forensic science and technology than they actually do. Watching any television crime drama and many movies will give one the sense that all things evidentiary in all crimes are vulnerable to a well-dressed technician-detective-scientist who (in the space of 45 minutes and a few commercial breaks) will solve the matter. In an equally misguided alternate version, scores of personnel from various disciplines will descend upon a scene to process every particle, interview every witness, and have someone hand the lieutenant in charge a clipboard to sign every few minutes. We can assure you that neither view is particularly accurate.

The reality of technological aids to investigation is no less impressive just because they are not used as universally on all crimes and with the same emphasis on an eclectic assortment of team members including the cyber-guru ripping and pasting critical pieces of data from, well, everywhere, in time to move the team onto the next exciting scene. Prosecutors must establish juror expectations early on when a case opens at trial when it comes to the contributions of science and technology to a specific case. Lovgren (2004) points out that jurors need to be made aware that neither crimes nor investigations always provide for conclusive physical evidence. Finding trace evidence is not simple and in some cases is beyond the capabilities of the equipment and personnel available. There may also have simply been little evidence at a scene. In this event, technicians and investigators must be prepared to explain what collection efforts they did make and observations that support why potential evidence was not collected. Attorneys on both sides of a case should be up to date on technological capabilities in forensics so that they can request appropriate tests, and formulate a court strategy respecting of the realities.

Prosecutors have become acutely aware that they must educate jurors from the moment when they are selected to serve. If one or more jurors hold unrealistic expectations of what forensic evidence should be present in a case, the result may in effect be an entertainment media–based acquittal. Prosecutors always devote a significant amount of attention to jury selection. Even during this selection process, attorneys may remind prospective jurors that crime, evidence collection, and court are not like what they have seen on TV. In a strategy designed to counteract the CSI Effect, prosecutors may request more scientific evidence to sate the technology appetite of jurors (Hayes-Smith & Levett, 2011). The presentation of more physical evidence, whether it shows guilt or not, can be used to win over the jury.

Research has also shown that people who watch CSI-type programs had higher expectations than those who did not (Shelton, 2008). However, these higher expectations did not directly translate into an acquittal when scientific evidence is not presented.

The higher expectations of both the prosecutor and law enforcement is one of four ways the CSI Effect has been described as influencing people (Smith, Patry, & Stinson, 2007). When the state fails to meet the unreasonable standards of evidence brought about by media expectations, jurors tend to side with the defense. This reality brings a second way the CSI Effect is described; prosecutors have had to change how they present their cases in order to deal with *CSI*-watching jurors. Third, the CSI Effect has created a huge increase in college students who are interested in the study of forensic science. A fourth impact of the CSI Effect is how it has "educated" some criminals in how to avoid leaving physical evidence or obliterating it to the point it is beyond the capability of many agencies to gather.

EXPLORE THIS

Navigate the web to the FBI's Handbook of Forensic Services at https://www.fbi.gov/file-repository/handbook-of-forensic-services-pdf.pdf/view. In addition to sections on submitting evidence and evidence examinations, the handbook contains a section on crime scene safety. How might this information from the FBI impact the training and preparation of crime scene technicians?

SUMMARY

Technology is the basis for the application of much of forensic examination and testing. Science and technology have developed over time to provide investigators increasingly sensitive and powerful tools to discover and analyze evidence. From the comparison of tools marks to biological samples, laboratories have become quite effective in linking trace or other evidence from one person or place to another. Computers have also had application for several decades, and they will continue to power aspects of analysis, including crime scene drawings and databases that allow matching of evidence such as CODIS, AFIS, and NIBIN.

Some concern has been voiced that the universal awareness of forensic methods can lead criminals to be better at covering their tracks. However, the professionalism and expertise of criminalists and investigators, in addition to the sensitivity of equipment and testing methods, make us believe that law enforcement will stay ahead in this race. The presence of realistic and nonrealistic depictions of forensic technology on television and in the movies has led to the CSI Effect with which prosecutors and others must contend. The specialties and subspecialties in forensic personnel and laboratory examination will likely increase. In our brief treatment of technology and forensic examination we did not, for instance, discuss glass fragments, soils and minerals, documents, digital evidence, drugs, hair and fiber evidence, and many other items and corresponding analyses.

Technology plays an important role at the scene and later in the laboratory. At various scenes technology allows crime scene technicians to thoroughly search for evidence and then document and gather the evidence. Photography and videography continue to feature prominently during investigations and in preparing for and presenting cases in court. At the laboratory, criminalists employ many devices and databases to analyze and compare evidence gathered and establish links or connections between objects and suspects.

KEY TERMS

Accelerant 35
Alternate light source (ALS) 36
Automated Fingerprint Identification System (AFIS) 32
Biometrics 28
Body-worn camera (BWC) 41
Class characteristics 32
Combined DNA Index System (CODIS) 29
Computer-aided drawing (CAD), or computer-assisted drafting and design (CADD) 42
Control specimen 32
Criminalist 39
Deoxyribonucleic acid (DNA) 28
Evidence 25
Expert witness 35
Forensic science 28
Gunshot residue (GSR) 34
Individual characteristics 32
Latent fingerprint 36
National Integrated Ballistics Information Network (NIBIN) 34
Patent print 36
Personal protective equipment (PPE) 37
Plastic print 36
Presumptive test 26
Reference 36

DISCUSSION QUESTIONS

1. How have science and technology changed over the three periods of criminal investigation discussed in the chapter?
2. Compare and contrast indoor and outdoor crime scenes.
3. List and explain at least three challenges in the collection of physical evidence.
4. What are the differences between crime scene technicians and criminalists?
5. What is the CSI Effect and how does it impact the criminal justice system?
6. What types of presumptive tests are used at crime scenes? Why?

SAGE edge™

- Get the tools you need to sharpen your study skills. SAGE edge offers a robust online environment featuring an impressive array of free tools and resources.
- Access practice quizzes, eFlashcards, video, and multimedia at **edge.sagepub.com/houghci**

PRACTICE AND APPLY WHAT YOU'VE LEARNED

▶ edge.sagepub.com/houghci

SAGE edge™

WANT A BETTER GRADE ON YOUR NEXT TEST?

Head to the study site where you'll find:

- **eFlashcards** to strengthen your understanding of key terms.
- **Practice quizzes** to test your comprehension of key concepts.
- **Videos and multimedia content** to enhance your exploration of key topics.

©iStockphoto.com/SDI Productions

3

People in the Process

Victims, Suspects, Witnesses, and Detectives

RUNNING CASE: THE DETECTIVES INTERVIEW TWO WITNESSES FROM THE FLORIDAN

Introduction

Victims
    Victimology

How It's Done: Victims' Rights
    Special Victims
    Victim Advocate

Suspects

Investigators
    Investigator Characteristics
    Investigator Training

Witnesses and Informants
    Eyewitness Identification and the Nonsmoking Gun
    Informants

Interviews
    Setting the Tone
    Guiding the Interview

Interrogations
    Planning the Interrogation
    Interrogation Methods

How It's Done: Good Cop–Bad Cop

Technology
    Detection of Deception Devices

Explore This

Summary

Key Terms

Discussion Questions

LEARNING OBJECTIVES

After reading this chapter, students will be able to

3.1   Discuss the various actors in a criminal investigation.

3.2   Explain the impact a crime may have on a victim and what support is available.

3.3   Contrast what characterizes a suspect and a witness or informant.

3.4   Describe the training necessary for investigators.

3.5   Contrast what characterizes a suspect and a witness or informant.

3.6   Discuss the differences between interviews and interrogations.

3.7   Explain how detection of deception devices aid in criminal investigations.

Running Case: The Detectives Interview Two Witnesses From the Floridan

On the drive to headquarters, Sergeant Kevin Lloyd called. "The neighborhood canvass panned out. We've turned up a witness named Miley Denis."

"Good news," Detective Bradley Macon said, parking his car. He hurried into headquarters, eager to find out what Detective Richard Ashley had learned in his interview with bar employee Bubba Paul and to meet Miley.

In the investigative division offices, he found Richard Ashley had just finished his interview with Bubba Paul. The bar employee was writing out his statement.

While they waited for Miley Denis to arrive at headquarters, Detectives Macon and Ashley reviewed the digitally recorded interview. For Bradley Macon, this was a delight: Richard Ashley had been working homicide cases for 8 years, and not only did he have the skills to ask the right questions, but he knew how people thought and how to make them comfortable speaking to an officer.

Richard Ashley led with basic questions.

"How long have you been working at the Floridan?"

"Two years now," Bubba Paul said.

"What are your duties?"

"I just advanced from dishwasher to night manager. I have the keys to the place and I lock up."

"How does the job suit you?" Richard Ashley asked, smiling at Bubba Paul.

Bubba Paul shrugged. "I like managing better than washing all those glasses."

Richard Ashley smiled again.

Detective Bradley Macon admired how Richard Ashley created warm, inviting body language by keeping his feet on the floor and his arms open, resting on the chair arms. This technique allowed Richard to gain all the information and create a bond with Bubba Paul.

The interview continued with Bubba Paul relaying what he'd seen in Milton Way.

At the end of the interview, Richard Ashley said, "Thank you for all the information and your time. We appreciate your ongoing cooperation. The process can be lengthy. We will need to ask you to come in again to attend depositions and meet with the prosecutor. I know you work at night. Are there daytimes that are better for you than others?"

"I usually sleep until 8 a.m.," Bubba Paul said. "After that, my kids wake me up, and I'm up for the day. They go to school at 9 a.m. So any time after that."

Richard Ashley nodded. "We'll do our best to work around that."

"Thank you," Bubba Paul agreed.

The two men shook hands, then the video clipped off.

Just then Officer Carl Jayden knocked at the door. "Miley Denis is here."

The two detectives walked into an adjacent interview room to talk with her.

After determining that Miley Denis had gone to the Floridan around 9 p.m. to unwind after a long day working as a dental hygienist and drink a beer or two, maybe meet someone, they learned that she had met Bill Johnson, the victim. Miley Denis's hands were shaking as she relayed details of the evening.

She had sat at a table behind the bar where she witnessed Bill Johnson arguing with the suspect, whom she overheard called Frank Denney, about who scored the most goals in the local soccer club. As the men continued to drink, they argued more intensely. Miley Denis estimated that she met Bill Johnson around 10 p.m. because she was almost finished with her first beer when Bill appeared tired of the argument. He turned to her and asked her to dance.

"He was about 5' 8"," she said. "I'm 5' 6", and when we were dancing, he was a little taller than me. His dark hair was buzzed short. I saw a hook-shaped scar on the back of his head. Weighed around 200 pounds, I'm guessing, because he was athletic, like my brother. No mustache or beard."

Miley Denis was also able to provide a physical description of Frank Denney. "A blond guy with floppy, curly hair. Looked like a surfer. Probably six foot and very strong. He had one of those scruffy goatees."

This description matched what the detectives had seen on the security video.

When Miley and Bill started dancing, Frank followed them on to the dance floor.

"Come down to the field tomorrow and see for yourself," he said.

"Leave me alone," Bill said.

"You leave me alone."

Bill poked Frank in the chest.

Frank shouted, "Touch me again and see what happens."

A few minutes after 10 p.m., Miley reported leading Bill off the dance floor. Frank followed them.

How you would go about interviewing a witness? What type of questions would you ask?

Introduction

Have you ever pictured yourself as a detective? Chances are that you have, and you are certainly not alone. Fictional characters in the movies and on television built upon the growing mythology of the dogged and sometimes flawed loner private eye or police detective. This iconic figure was often set in contrast to inept or one-dimensional uniformed officers, furthering the status distinction that persists even today in police forces around the world. But there is no venue for the detective, or police at all, without the dyad of criminal and victim. In addition to the central characters, sometimes an act of crime is witnessed or various individuals possess knowledge to share with authorities—sometimes at a price.

Within the many descriptions of what a criminal investigation is, there is not a great deal of discussion of the people in the process. They are abstractions: the victim, the suspect, the investigator, the witness. Fleshing out the bare bones of these categories is not just an attempt at humanizing the process that is criminal investigation; it is a critical aspect of understanding how the criminal investigation process begins, develops, and concludes in the ways that it does.

So let us turn first to those at the beginning of the process of criminal investigation.

Victims

The history and story of victims is the story of mankind. Those who have something wanted by someone else may have it taken from them, possibly by force. Someone who is angry or perhaps in the grip of paranoia and in fear may strike out at others, resulting in an injured person. Regardless of the circumstances that brought a victim together with an offender, the experience of being victimized can range from one of annoyance, frustration, or inconvenience through terror, injury, and death. A person may be impacted for a lifetime and families, neighborhoods, and communities are all affected by crime. In modern times victims may choose to report their victimization to law enforcement with the general expectation that the appropriate agency personnel will investigate the circumstances and bring the perpetrator of the crime to justice (i.e., to court). But not all victims report the crimes they have experienced. A person who feels that he carelessly left an item unsecured outside his home, perhaps an old bicycle, may chalk it up to a life lesson and not report the crime or feel it would be a waste of the time of the police. A person may not discover that she has been the victim of a property crime for a long time and think it no use to report. A victim of sexual battery may be embarrassed and fear even further humiliation or insensitive treatment. Yet another person may have been robbed—of illegal drugs; the victim will likely not report the robbery. Yet another victim may have knowledge of who took from him and decide to handle the matter "personally."

Personal retribution Punishment for a crime or offense carried out by the victim against the one who committed the crime.

Blood feud Ongoing series of attacks typically between two families.

This last point is where most peoples of the world began the practice of **personal retribution**, or the **blood feud**. If you harm me or take from me, I will repay you in kind. If I am incapacitated, my relatives or clan will seek revenge on my behalf. At a certain point this becomes unwieldy, as family or group members thin out from excessive retribution. And so village elders, tribal chieftains, or representatives of a king or queen step in to act as arbiters of individual social wrongs to pass judgments and mete out punishments. Cultures began to codify the offensive acts and systematize the assigned sanctions, creating law and a fairly stable way to deal with much crime. As the response to a wrongdoing declared by society or the governing class transitioned from personal retribution to a central proxy authority, the *actual* victim receded into the legal background and became a spectator to the process of their own justice. The efficiency achieved by the system of law inadvertently led to a consequence of depersonalization and businesslike if not downright brusque treatment of the one who actually suffered.

Moving forward in time to the 1980s in the United States, a change in perspective and approach to the role of the victim in the process came about. Each state and the federal government passed laws during and since the 1980s to provide to victims, at a minimum, information about the process of criminal justice related to their case, information about restitution that may be available including for things such as medical bills or lost wages, what protections or notifications are due them, and what input they may expect to have in the outcome of their case. This last often takes the form of a victim impact statement given to the court to consider prior to passing judgement in the form of a sentence. Recognition of the personal impact on the victim was needed and important. The changes reflect a system-level policy response that varies in its application to individual crime incidents. The U.S. Justice Department expressed the opinion that the **Crime Victims' Rights Act (CVRA)** did not provide for any rights to victims until a prosecutor filed a criminal charge in the matter. This was roundly criticized and, observers noted that given most cases never have charges filed, would result in victims having no rights in many or most cases (Cassell, Mitchell, & Edwards, 2014). How individual criminal justice agency members interact with individual victims is often complex and changes over the time a victim is involved in the process.

Crime Victims' Rights Act (CVRA) Enacted in 2004, the act lists the rights of victims in federal crimes.

The relationship between an agency and the community can certainly be affected by how agency members interact with victims of crime. Recognizing that being a victim is an unpleasant experience to begin with and that the goal of interacting with the criminal justice system is not to make the victim "happy," *per se*, leaves the dynamic fraught for dissatisfaction or disappointment. If an offender is not identified and successfully prosecuted or stolen property is not recovered (and the majority of property crimes are not solved or cleared), victims may generalize their dissatisfaction to the law enforcement agency with which they first had contact or specifically to the investigator assigned their incident report. What is sometimes to the detective simply an assigned case on a likely full caseload is a personal and perhaps emotional matter for the specific victim. Engagement with victims is important and can be effective as a way to set and meet expectations about what the criminal investigation process can realistically achieve. Periodic communication, especially in the aftermath of more serious crimes, can help ensure that victims do not feel forgotten in the process. Once again we are talking about feelings and emotions as well as differing views about the role and objectives of criminal justice system employees. While examination of this specific point of the role of investigators to provide solace through communication to victims has received relatively little study, it perhaps mirrors a similar discussion that continues following the paradigm shift in law enforcement to the community policing model that added a focus on quality of life that had not been overtly stated previously and perhaps remains in debate at the department or officer level.

As an investigator, what are some important things to keep in mind when speaking to victims?

©iStockphoto.com/Yuri_Arcurs

Victimology

A person may be victimized on the street, at home, in a business, through online means—in many places and in a variety of ways. How the potential victim and an offender come together is the study of **victimology**. As you consider the circumstances of victimization you can imagine ways that individuals can reduce the opportunity to be a victim. At the same time, an investigator can find insight into the motives and opportunities of offenders, and agencies can, perhaps, formulate strategies to help the public in general be less vulnerable to crime of various sorts. Many of these dynamics have been studied and the variables in the process have been theorized to have various relationships. Criminological theory has a purpose to explain crime, which ideally should lead policy-makers and practitioners to design and adopt methods of preventing crime and provide detectives with insights that can be helpful in solving many crimes that have been committed.

While theories of crime are covered thoroughly in a criminology course, a few brief examples should serve to underscore the need for detectives to be familiar with the dynamics of crime creation or the motivation of certain offenders. The interaction between offender and victim must be examined and understood before the efforts and skills of officers, detective, crime scene technicians, and others come into play. Criminal justice personnel must work to understand the factors that bring offenders and their victims together to assist in re-creating the timeline of actions in a crime and build the case around this sequence.

Some of the individual explanations for deviant behavior include **biological** and **psychological theories**, though how each person's individual makeup interacts with his or her life circumstances is important to develop a more complete picture of how he or she may make decisions or take actions. **Social learning theory** can provide insights into how we model our behavior after others, especially those we want to be like. **Differential association theory** speaks to how we learn behaviors from others with whom we associate. Various **social control theories** illustrate a view that if there were no rules or constraints many people would commit acts that harm others. The **social control perspective** explains how the various controls in life, society, and in individuals constrain bad behavior. **Strain theory** reflects on how the tension we may feel at not having things other people have (whether we worked to earn them or not) can lead to various strategies to acquire those things—not all of which are socially acceptable approaches. Compelling observations about the neighborhood environment and degree of community cohesion or organization of **social disorganization theory** provide an explanation for how some people may not feel attached to an area or the people in it, and this can lead to criminal behavior.

Finally, in this very brief commentary on just a few of the many theories developed to explain crime, let us think for a moment about when and where during our day or night, and whom we are with, may lead to a higher chance of our being a victim of crime than someone else. A person's *lifestyle* can be seen as exposing her to crime. If a person not only lives in an area without the attachments addressed by *social control* and *social disorganization* theories, but "hangs out" (*differential association theory*) with individuals who are engaged in risky behavior according to a **subcultural perspective**, then this overall lifestyle puts her at greater risk than the person at home watching television. And so by the same token, lifestyle factors can serve as insulators from being a victim. Related and often considered together with lifestyle is **routine activities theory (RAT)**. As the name indicates, those things we may do as a matter of routine—travel to work, school, social places, spend time after hours at or near relatively uncontrolled drinking establishments—may bring us into contact with someone who makes us his or her victim—or becomes our victim.

Victimology The study of victims and their experiences, including interacting with offenders.

Biological theory Perspective that examines physical aspects of humans to explain psychological reasons for the behavior.

Psychological theory Describes and predicts behavior based on human thought and emotion.

Social learning theory Learning through observing and imitating others, especially admired individuals.

Differential association theory In criminal justice, perspective that interacting with others provides a vehicle to learn and adopt their values, attitudes, or motivations to commit deviant acts.

Social control theories Various and include: containment, neutralization, and self-control.

Social control perspective The use of sanctions by individuals or institutions to guide or control behavior through the informal means of socialization or the formal means of the external sanctions by government agencies.

Strain theory The concept that various societal pressures influence behavior.

Social disorganization theory Examines the circumstances of the environment where a person lives to postulate criminal involvement.

Subcultural perspective Posits the existence of alternative values and attitudes by groups or subcultures in a larger society.

Routine activities theory (RAT) The appearance at one time and in one place of a likely offender, a suitable target for that offender, and the absence of a capable guardian of the target.

How It's Done

VICTIMS' RIGHTS

The advocacy organization National Center for Victims of Crime provides information on the resources available to victims throughout the country. The organization defines a victim as "a person who has been directly harmed by a crime that was committed by another person" (victimsofcrime.org/help-for-crime-victims). In the Center's overview of victims' rights it lists the basic rights provided in most jurisdictions as:

1. Right to be treated with dignity, respect, and sensitivity
2. Right to be informed
3. Right to protection
4. Right to apply for compensation
5. Right to restitution from the offender
6. Right to prompt return of personal property
7. Right to speedy trial
8. Right to enforcement of victims' rights

Special Victims

Within the general category of people as victims are some individuals who for various reasons require additional consideration during interaction with criminal justice system personnel. Child victims of sexual assault, for example, may require the assistance of social workers or therapists to deal with trauma. And while this may be true of adult victims as well, the courts have recognized that concern for the well-being of a child may also lead to videotaping their testimony to be played at trial rather than subject them to further psychological impact. Maturity can also affect the accuracy or focus of a child's recollection or information of a crime. Similarly, individuals with developmental disabilities or certain mental health challenges can present challenges to personnel who must obtain information to aid investigations.

Victim Advocate

Working with victims after commission of a crime, **victim advocates** help provide support and information about the criminal justice process. The victim advocate can be of assistance in the immediate aftermath of a crime such as a rape, robbery, or helping the surviving family member or "co-victims" of a homicide. The needs of victims of different types of crimes can vary, but even the victim of a property crime such as larceny or burglary can be profoundly affected emotionally, and victims of all types of crime may need help completing crime compensation paperwork and the like. Child victims, **intimate partner violence (IPV)** victims, and elder victims are just a few examples of those who may have specialized needs that may best be served by a victim advocate serving a role as referral person.

The criminal investigator and victim advocate can have a symbiotic relationship in working to handle the support needs of victims and co-victims, while simultaneously providing the needed space for the detective to continue the investigation. In homicide cases, for example, co-victims and survivors have frequently felt their need for information about

> **Victim advocate** Professionals often employed in law enforcement or prosecutorial agencies to support victims throughout their interactions with the criminal justice process through direct services and providing referrals.
>
> **Intimate partner violence (IPV)** Various forms of abuse or violence committed by an individual with whom the victim is currently or had previously been in an intimate relationship.

their loved one's case is not seen as important by investigators, and that information is withheld. Victim advocates are often well positioned to provide appropriate updates to family and survivors while not comprising the integrity of the investigation. Victim advocates may accompany victims or witnesses to court during a trial, and take them to the courtroom in advance of a trial to increase the person's comfort with the setting and proceedings. Victim advocates work in law enforcement agencies, the prosecutor's office, and in various medical or social work agencies.

Suspects

Understanding why people commit crime can help investigators identify suspects where none are readily apparent as well as piece together the picture of the incident. The causes of criminality, as well as of some antisocial behavior that falls short of a criminal act, are varied and have been partly explained through various theoretical approaches discussed in the previous victimology section. It is not the aim of the chapter or book to provide a comprehensive discussion of theory, but it is helpful to point out examples of how the theoretical foundation in both criminology and criminal justice (and other fields) can aid the overall efforts at preventing crime in the first place, and in solving and resolving crimes that are committed.

The combination of causes or factors that culminate in one person committing a criminal act is not exactly the same as a *motive*, though they share a certain amount of space in the geography of intentional behavior. Motive may be broad or general, such as to acquire something through, for example, burglary, theft, or robbery. A motive may be more specific to an offender targeting a victim in crimes such as homicide, assault, or even criminal mischief (vandalism). In such crimes where an offender intends his crime against a particular person, this insight may be helpful for investigators trying to discover a suspect. What an offender's intention or state of mind was when he committed his crime is referred to by the Latin term **mens rea**, meaning "guilty mind." What the offender actually does to act on his intention is the **actus reus**, which means "guilty act." A crime occurs when both *mens rea* and *actus reus* are present.

Mens rea Knowledge or intent to commit the crime.

Actus reus Element that is the criminal act.

Identifying the suspect in a case is clearly of great importance, and there are several ways that this comes about. In certain cases the victim knows the identity of the offender or can offer suggestions as to who may have committed the crime. Witnesses can also play a key role in identifying a suspect in the technological witness of surveillance, or security video has shown its value in countless cases. Without an initial indication of a specific suspect, investigators may use aspects of the victimology discussed above to determine who may be a likely suspect. Someone who lives near the victim or who knows her through work or social interactions may emerge as suspect. Direct evidence such as a witness or video, circumstantial evidence including suspect movement or activities, or a culmination of gathered information that suggests the potential involvement of a person all may bring him to be considered under the status of suspect.

Certainly physical evidence located at the scene of a crime can also implicate or suggest a person's involvement. It is important to remember that someone who is found to have been at a location where crime occurred may have an explanation for why they were there. Think of the family member whose fingerprints will obviously be present in the home where a crime occurred or an employee whose trace DNA fingerprints or security system noting of their presence may be able to explain *why* they were at a business. Though an employee's presence may not explain the *when* of the time frame, if it was when a crime occurred.

Someone in close proximity to the commission of a crime may also be identified as a potential suspect. A **field identification** can occur if a witness or victim is taken to the location of the potential suspect near the crime. What is important in such field identifications or *showups* is the amount of time that has passed since the crime occurred. Not only is time important so that a victim or witness has fresh recollection of a suspect's features, but various jurisdictions and courts may impose time limits (usually less than 30 minutes) on the conduct of field identifications. Given that an individual is identified as a suspect in this way she must be advised of her constitutional rights, or the **Miranda warning**, prior to questioning. While the suspect does not have to respond to questions or may ask to have an attorney present before questioning, this right of representation does not extend to being passively viewed by a victim or witness in this field setting. If too much time has passed since the commission of a crime but a victim or witness is available who may be able to identify the offender, then a live lineup may be arranged or the witness may be presented with a photo array or directed to view mug shots.

If a suspect has been developed in the case he becomes a prime source of information. Even though an individual may not willingly speak with investigators or provide incriminating details about his actions, carefully examining what is known about his life, work, and acquaintances can reveal a great deal that may be helpful during an investigation. Even denials of involvement or knowledge about a particular criminal act may prove important to impeach later statements by a suspect or perhaps a hostile witness.

Field identification Viewing a suspect by a victim or witness within minutes of a crime. Also known as a showup identification.

Miranda warning Advisement of rights given by the police to a criminal suspect prior to questioning.

Investigators

The detective or investigating officer is central to our thoughts when we consider criminal investigation. The detective is who *investigates*. This verb carries all the weight of every action in furtherance of determining who did what, whether it was against the law, and how to proceed. Given the seeming importance and centrality of the role of the primary investigator it seems logical to learn about how a person becomes a criminal investigator. As we have already established, the sheer volume and varying sizes of American law enforcement agencies means that no aspect of police operations is approached in exactly the same way in every agency. The initial criterion for the selection of who is to be a detective in a municipal or county law enforcement agency generally begins with someone who has been a uniformed officer. For state enforcement agencies, it is often desirable to hire prior law enforcement from a police or sheriff's department. At the federal level, even if a former police officer is hired, a new agent will attend training at the Federal Law Enforcement Training Centers (FLETC) in Brunswick, Georgia, or elsewhere.

It is likely that absent a personal identification with one of the other actors in the criminal investigation process, most people identify with or are most interested in the role of detective, investigator, or agent. This is not a judgment of value by any means, but we view the investigator as operating with agency and intent, as in the verb *detecting*. While the empirical reality of outcome impact by investigators is grounded in a sober assessment of cases solved, they remain the linchpin of what gets done, if anything, in a case deemed suitable for follow-up. The number and volume of overall arrests by detectives for crimes reported or discovered is far less than patrol officers, but this is somewhat the nature of the protracted efforts that are more typical of the *type* of cases and arrests made by investigators.

A patrol officer drives up to some incidents such as bar fights, neighborhood arguments, and burglaries in progress—all incidents where the crime is being played out in front of the officer and therefore often enforcement action is unavoidable and probable cause to make an arrest is presented without the need for extensive investigative effort. Contrast this with the far more involved investigation activities of robberies, rapes, and murders where the offender has fled. The time consumed by extensive search for and interacting with witnesses and others, communications with prosecutors to coordinate investigative strategies and the courts to seek warrants, all are outside the purview of patrol officers—by design. Putting aside the questionable glamor of the detective's role, common attributes are the ability to exceed a standard 9-to-5 job, travel outside the officer's jurisdiction to perform investigative duties, and receive specialized training to equip him with skills exceeding patrol officers. With this seeming freedom of temporal and geographic movement comes the weighty responsibility of carrying assigned cases around inside the detective's mind on and off duty as opposed to leaving the official responsibilities of the shift at the door of the squad car or squad room by the uniformed officer.

Most people, including new officers, do not realize that even on the first day on the job as a patrol officer you assume the mantle of investigator. As time passes and the volume of calls grows and the variety expands, officers layer on experience that informs and deepens their understanding of people in different circumstances. Methods and approaches to gathering information from those people accrue as the officer becomes more effective in one of—if not the—most important skills and functions of the profession: communicating. Detectives in the modern era (and perhaps for a long time previously) are known as information managers (Dean, Fahsing, Glomseth, & Gottschalk, 2008). The variety of information systems and methods and frequency of their use is often driven by an organization's culture (Gottschalk, 2007). As information is gathered, investigators must discern or construct evidence from it (Jackall, 2000).

Investigator Characteristics

In many disciplines there have been studies to determine characteristics that indicate or are at least correlated with "successful" incumbents. A study published in 2016 examined a small group ($N = 30$) of investigators in Australia and New Zealand to get their view of what makes an "effective" detective (Westera, Kebbell, Milne, & Green, 2016). The leading factor identified was the ability to communicate with a wide variety of people in various roles and status. Following close behind communication abilities was the observation that motivation and thoroughness were also essential. Westera et al. (2016) viewed the identification of characteristics, skills, and abilities as important due in large part to the views of the public. This idea is tied to the conception of the "procedural success" case outcome discussed by Brookman and Innes (2013). Brodeur (2010) drew out the self-perception of detectives as "courtroom evidence managers," likely acknowledging the information age reincarnation of Sherlock Holmes's prodigious memory for detail that escapes mere mortals. Cook and Tattersall (2016) also give lie to the myth of the "loner" detective, saying: "Metaphorically speaking, the SIO [senior investigating officer] role can be compared to that of a musical conductor who similarly has to unify performers, set the tempo, give clear instructions, listen critically and shape the product of their ensemble" (p. 2). A team effort, indeed. Miranda (2015) described the "hybrid figure of the contemporary detective emerging," referring to "the fusion of traditional methods of criminal investigation (**hard surveillance**) with new

Hard surveillance Traditional methods of investigation.

technologies of collection and use of information (**soft surveillance**)" (p. 422) to find the modern detective trawling and reaping a harvest of deep database dives and web crawling.

Soft surveillance New collection technologies.

With the flow of information comes the need to parse and create meaning from raw data. In the same way that raw data are not "intelligence" until an analyst examines and puts them into context, a detective must piece together information using experience heuristics to integrate bits into a framework. This process involves decision-making on the part of the investigator; decisions are made as to the order of addressing clues/tips/facts/leads, and decisions are made about what specific steps are to be taken about the information so ordered. As information managers, detectives must be able to organize extensive amounts of materials and data into a coherent whole and then have the ability to communicate the various aspects of the case to prosecutors, other investigators, the media, and to the court and possibly a jury of citizens.

Building rapport with victims, witnesses, and to some extent suspects is a developed skill that effective investigators consciously use to be supportive of victims and gain information from everyone in the process. Appearance of confidence is also a hallmark of the professional investigator. How the detective carries herself, speaks, and conducts the investigation is observed by many different people and can impact how they interact with her. These soft skills may provide greater insight into the functions of investigators and key skills they must possess (Tong & Bowling, 2006). While some individuals in the investigative role seem to have a more natural style that forms the foundation for rapport building, most people selected or promoted into the role of detective are able to build and enhance this ability through training and experience. As discussed earlier in the book, there are several views about what a successful outcome is for criminal investigations, but the detective certainly bears a moral burden to gather and present information within the confines of contemporary law (Miller, 2014).

Investigator Training

Most local police criminal investigators begin their careers by spending time as a patrol officer in a law enforcement agency. In federal agencies, the path may be one of an undergraduate college degree and then on to a period of formal academy training followed by apprenticeship in the field for the department where they work. It seems that training for investigators should devote a substantial and explicit component on the role of information and its management. An interesting experiment by Britain's largest police force, the Metropolitan Police, is allowing those with a college degree to immediately enter a detective training program and bypass the traditional experience of patrol officer or "bobby" (Camber, 2017).

Most states criminal justice academy systems offer a range of specialized courses that officers may take or be assigned to as a way to expand their skill sets or prepare them for new duties. Examples of available courses whose curriculum is developed and approved by the Florida Criminal Justice Standards and Training Commission (CJSTC) include Narcotics and Dangerous Drugs Investigations, Case Preparation and Court Presentation, Injury and Death Investigations, Interviews and Interrogations, Advanced Report Writing and Review, Underwater P Gangs and Security Threat Groups, Police Science and Technology, Crimes Against the Elderly, Violent Crime Investigator Course, Adult Sex Crimes Investigations, Child Sex Crimes Investigations, Organized Crime, Elder Abuse Investigations, and many more (CJSTC, 2017).

As in uniformed patrol, most detectives and federal agents are initially paired with or under the supervision of more experienced personnel who act as trainers. This is important as the officer or federal agent trainee transitions to the more protracted mode of follow-up investigation that differs not only in length of time but complexity of tasks from the role of preliminary investigation carried out routinely by patrol officers. Many larger agencies have formalized a detective training program that includes initial classroom instruction and internship or apprenticeship by being formally partnered with a senior investigator. The veteran investigator will guide the new detective through common tasks performed and work alongside them as they handle a number of investigations. A unit supervisor or some manager within smaller agencies will typically serve this role as well. Formalized training at regional policing academies or other institutes provides ongoing training in basic and advanced skills. A hallmark of the criminal investigative process is the use of abductive inferential reasoning, contrasted with deductive and inductive (Carson, 2009), as that which develops over time as a law enforcement officer against specific training experiences.

Both education and training can also occur in the college or university setting. Criminal justice majors as well as students from various disciplines regularly sign up for courses that have investigations as a focus of study. Brickner, Mahoney, and Moore (2010) describe an interesting applied-learning exercise that can be carried out by college students that partners with the Internal Revenue Service (IRS). The students are put together in small teams and furnished with hypothetical tax fraud scenarios to solve. Working with criminal investigators from the IRS the students are guided through evidence-gathering and analysis, use of various investigative tools, interviewing, and communication skills. The research found that the applied-learning exercise improves skills that can apply to employment opportunities across a range of accounting jobs. The federal government also allows local and state law enforcement officers to attend many training courses designed using federal resources.

Witnesses and Informants

Witness A person who either sees something take place or has information about the event or people involved.

There is an occasional semantic confusion when law enforcement uses the term **witness** and when the rest of us use the word. An eyewitness accurately describes someone who *saw* something. Maybe not the offender as he committed the offense, but perhaps some other part of the lead up to the crime or some action afterward such as a car driving from the scene. Other individuals may not have watched a crime being committed but they may have information about the investigation. These individuals can be referred to as *primary* or *secondary* witnesses. The primary witness will typically possess direct information about the crime or a suspect. A secondary witness may have information relevant to an investigation or be a source of facts that can aid the investigators. The victim is also a witness and, as such, is interviewed in the same way as others, but with consideration of emotional aspects of how they are processing their status as a victim. The parlance of police and their report forms classify all of these people broadly as witnesses.

Eyewitness Identification and the Nonsmoking Gun

Eyewitness accounts have long been viewed as compelling during an investigation and carrying on into a trial. Yet such accounts do not always lead to accurate identification of a suspect. Mistakes may be inadvertent based on factors surrounding the incident or the witness, or influenced by overzealous officers leading to a false identification (Cutler & Kovera, 2010).

Contemporary recommended lineup procedures are discussed in the chapter on searches, seizures, and statement, but the need to eliminate any influence on an eyewitness is critical during a lineup. When eyewitness identification is mistaken, the remainder of the investigation process and legal safeguards of the U.S. system must be effective in protecting the rights of the accused and in presenting sufficient additional information and evidence to ensure the trier of fact can make an appropriate judgment. Defense attorneys are allowed to be present, though not intervene, at lineups involving their client. The attorney can make legal motions to the court challenging the procedures used in the lineup or show up that resulted in identifying their client as a perpetrator. The eyewitness to a crime will generally be available for cross-examination during a criminal trial, and here too the defendant's attorney will have the opportunity to pose questions about certainty and clarity in the identification. The judge also has the responsibility to make jurors aware of certain instructions during a trial, including those addressing eyewitness identification (Bornstein & Hamm, 2012).

The method used to **interview** a witness to an event can be helpful in aiding recall of details and the sequence of what happened. This need to aid victims or witnesses and recall is not at all unusual. Some witnesses can be quite confident in their recall and yet not be accurate. Memory is influenced in many ways, not the least of which is the attention paid to an event that we later try to recall. Cognitive interview technique (Fisher & Geiselman, 1992; Geiselman & Nielsen, 1986) developed as a method of eliciting greater depth of detail and recall by having a subject talk about an entire episode or even activities during the day prior to the event in question. This is largely a narrative method and achieves its success as people recall more as they fill in details and perspective about the experience they witnessed. Avoid interruption and allow a free narrative by the witness; have the person relive the event by recreating in their mind the time period, setting, and circumstances, which can evoke emotions and more vivid recollection; provide all details; after chronologic recitation, have the witness change the order and then the perspective of recall; recollect by association to familiar things or people, or prior events.

Hypnosis has featured more in fictional accounts of eyewitness recall than day-to-day activities by investigators. While the use of hypnosis and criminal investigations is not new, it is recognized in contemporary times as subject to certain weaknesses. While the method can prove helpful, caution must be taken to ensure a hypnosis-enhanced interview is conducted according to established protocols and properly documented if any results are expected to be used further in an investigation, let alone in court. Court restrictions are appropriate given the risk of suggestibility that may be introduced with memories enhanced by hypnosis. Some jurisdictions prohibit the use of such testimony, while some would examine the totality of the circumstances during a pretrial hearing before making a decision on admissibility.

There are varieties of circumstances that may impact the accuracy or reliability of people who witnessed the commission of a crime. Considerations include the physical setting and proximity to the crime as well as individual attributes or condition of the witness. If a hit-and-run accident occurs at an intersection and four different people are on each corner, not only will their viewing angle be different but imagine someone in a building on the corner on a third floor who heard the accident and rushed to the window in time to look down and see a vehicle speed off down the street. Additionally, if our hypothetical accident occurred at dawn, at dusk, at nighttime with or without streetlights, all will affect what is seen and remembered. As for the individual witness, their age may have implications for how they saw something occur and which aspects of the occurrence they were more focused on or

Interview Questioning or talking to an individual who is believed to have information regarding some incident or crime.

can recall. Picture again our hit-and-run traffic accident and now imagine that one of the witnesses was a 17-year-old young man with a keen interest in all things automotive (stereotypic, I know), which allowed him to describe with great accuracy the make, model, and attributes of the car that left the scene. Perhaps one of our witnesses wears glasses and had just taken them off to clean the lenses when the accident occurred. This person may be able to state certain pieces of information about car movement and timing but may provide limited information if their vision was temporarily impaired. The effects of alcohol were examined in one study to determine whether varying levels of intoxication affected the accuracy, confidence, or time to make or reject identification (Kneller & Harvey, 2016). Interestingly, this study in examining lineup identifications with individuals affected by alcohol intoxication found no significant difference in accuracy or decision confidence from the sober witnesses in the study. Law enforcement cannot choose who witnesses will be, or their condition or attributes, but all information should be gathered as completely as possible, as well as noting any issues that may impact the observations or statements of the witnesses.

Informants

Once again, television and movies have given many in the public an unrealistic view of the number of individuals who serve as informants, as well as the amount of knowledge they have about a given crime. Detectives do not have "go-to" informants who can quickly supply the name of a suspect or tell officers where a kidnap victim is being held. What a person classified as an informant may do is provide investigators with a wide variety of information including the general area a suspect may hang out, the people known to the informant who also know certain criminals, or the informant can sell or purchase stolen goods on behalf of law enforcement in a controlled operation that allows police to establish **probable cause** for an arrest or search. Having written agency policy for the management of informants, I can tell you that proper procedures are a must. An agency has to document the pay of such individuals, it must monitor any actions undertaken by the informant on behalf of the department, it must verify the information received from an informant, and many more actions to ensure the integrity of a case built using informant information.

Agencies generally classify a person who has given information once, and had that information verified, as a **confidential informant**, or **CI**. A person who has provided verified information multiple times is referred to as a **reliable confidential informant**, or **RCI**. Documentation of all interactions and management of the informant is critical in the use of either. Some of these methods of documentation mirror the procedures necessary when officers work undercover so that the case integrity is maintained.

Interviews

Interviews are nonaccusatory in nature and may involve anyone thought to have information pertinent to a criminal investigation. If during an interview it becomes apparent to the detective that the person thought to be merely providing information is actually a suspect, the interview must end or transition to an **interrogation**, and the investigator must read the warning of rights (Miranda warning) and gain acknowledgement of the suspect's understanding and willingness to talk.

Interviews can be brief and simply allow a patrol officer or detective to exclude someone as having relevant or helpful information, or they can suggest the need for further

---

Probable cause The level of information needed for a court to issue a warrant, for an officer to conduct certain searches or arrests, and the standard for a grand jury indictment.

Confidential informant (CI) An individual who provides verifiable information to law enforcement.

Reliable confidential informant (RCI) Sometimes distinguished from a confidential informant by virtue of having been utilized a set number of times and found to have provided accurate information.

Interrogation Questioning an individual who is suspected of committing a crime.

discussion. If this brief interview occurs at a crime scene, the officer should still obtain the personal and contact information from the person's government identification so that they can be contacted later if needed with follow-up questions as well as to ensure that they do not later appear as a defense witness with recollections that they claimed not to have had when the incident occurred. Documenting that someone "saw nothing" and "knows nothing" is a practical issue that can preclude surprise manufactured testimony and allow a prosecutor to impeach such an attempt. This differs from a witness coming forward who was not identified at the time of a crime or subsequently during the investigation.

Some people are willing, even anxious, to share what they know with law enforcement. They need only be asked. Some people do not realize they have relevant information, so they must be identified and then guided through a conversation or interview to learn of important facts. And some people hesitate to be involved or to speak with police, or they may be downright hostile to law enforcement. Each of these may require a different approach on the part of the officer seeking information. There are myriad variations of communication principles, rules, and techniques to elicit information from others by reducing barriers and asking the "right" question in the "right" way.

Setting the Tone

When the location of an interview can be controlled, it should be. Unfortunately, the time and place of interacting with victims and witnesses is often not something that can be predicted in advance. A brief discussion can take place and possibly move to a police station or other location to conduct a more in-depth interview. When this can be done it allows the interviewer to select an area with few distractions, perhaps comfortable seating and good lighting, a table and chairs to allow notetaking or recording. The use of notetaking or recording has considerations as well. Notetaking can be distracting to both participants, whereas recording allows a more free-flowing conversation and enables the detective to provide more eye contact and simply observe and listen more fully to the interviewee. Some investigators choose not to record interviews because they may become part of the *discovery* process where defense attorneys have access to the recording. Other investigators opt to have the recording so they are able to review the specific statements later and repeatedly if needed. Physical and psychological aspects of an interview setting can facilitate a more effective conversation that can focus on the content of the information-sharing. Interviews conducted immediately after a crime has occurred may also provide "fresh" information as well as comments from victims and witnesses that are potentially more unguarded and therefore more frank. Time to reflect may cause a person to hesitate in sharing thoughts or facts they know about for a variety of reasons, including those mentioned earlier in regards to reporting crime.

Guiding the Interview

The person being interviewed may, instead of being hesitant to speak, offer too much commentary and show a tendency to digress or wander aimlessly through a narrative about anything and everything. It is an important responsibility, and key skill, for the investigator to control and direct the interview. Overly blunt or terse questioning that shows the interviewer's frustration will likely impede the interview process and result in little or no information and future challenges to cooperation. Instead, the interviewer must work through a variety of techniques to move the interviewee past irrelevant or barely relevant monologues and

...t them to the main points and areas where the detective is seeking new or corroborating ...mation.

...formation obtained from any source should to the extent possible be corroborated by ...gators. If a victim or witness recalls the description of a suspect or a vehicle, the detec-...l want to note where the person was in relation to the suspect, vehicle, or event. Also, ...ytime or nighttime, how was the lighting, does the witness routinely wear glasses or contacts, and were they wearing them at the time? The investigator must also be alert for the person who offers false statements for a variety of motivations. Detectives will sometimes begin to suspect a deceptive witness when various other aspects of a case are significantly out of sync with the statements given by the witness.

In addition to the traits noted earlier for effective investigators, two others are tremendously important: self-awareness and other-awareness, combined with an adaptability that allows the detective to shift his approach based on what his awareness of self and others indicates. This awareness includes potential bias that virtually everyone possesses, though many fail to recognize this fact without training to include methods such as critical reflective thinking (Poos, van den Bosch, & Janssen, 2017). If an investigator knows that he has a generally blunt or frank questioning style and he is speaking with someone who appears to have information but who seems to be hesitant in sharing, the officer may need to alter his approach. A challenge here is that officers and detectives do not often have prior knowledge of someone before they interact. This presents a common challenge of not having a baseline of another person's behavior. If speaking to a witness that investigators believe to be lying, or when talking to the suspect, the process, intent, and method of the conversation transitions to one of interrogation.

Interrogations

If a person is identified as a suspect in an investigation, the discussion officers may have with him that centers on his actions and involvement is generally termed an interrogation. Basic demographic or work information can be voluntarily sought without invoking a person's right to be represented by counsel since the information is not considered to be potentially incriminating. Most people, again through television and movies, are familiar with a detective reading someone their rights or saying them from memory quickly to their fellow actor. This is the result and requirement by the Supreme Court's *Miranda* (1966) decision. The decision put in place protections for a suspect's Fifth Amendment right against self-incrimination. The admonition pertaining to the warning of constitutional rights must be given if the questioning of an individual is considered a custodial interrogation. A lone officer speaking to someone in a public place, and with no impression given that the subject is not free to leave, would not typically be considered custodial, whereas placing a person in the back seat of the squad car while a few officers stand around the door and ask questions would in all likelihood be seen as a custodial interrogation. Certainly most agencies as a matter of policy require the warning to be given subsequent to an arrest and prior to any questioning.

The appropriate way to admonish someone of his or her rights is to read them verbatim from a department-issued rights card or interrogation form. The form is then signed by the suspect, if he is willing, and witnessed by the officer and, ideally, a second individual who can be a disinterested citizen, agency employee, or another officer. There are also exceptions to the admissibility of statements made by a suspect who has not been provided the so-called

Miranda warning. Statements made by the suspect, such as those that may try to establish an alibi, can be used, as well as spontaneously made comments before the warning can be issued. The process can also be recorded based on agency policy or procedures, though this is not a current legal requirement in most jurisdictions.

Planning the Interrogation

Once again, as with interviews, preplanning is important to ensure the most effective interrogation. The investigator will want to speak to the victim if available, witnesses, and officers involved in the preliminary investigation before speaking to a suspect in the case. This preparation allows the detective to have a good idea of the sequence of events and to have some advance knowledge of what the suspect may claim, as well as possible motivations in the crime. All of this information can help devise strategies for questioning including specific questions to ask during the conversation. The detective may be able to verify certain pieces of information prior to meeting with the suspect and this will give additional insight once the interrogation begins if the suspect makes clearly false statements. Statements already made by the suspect in prior interactions, perhaps between an accuser and suspect, can play into how a detective prepares for the next stage of questioning (Mason, 2016).

Determining the setting in advance of the interrogation will also allow the investigator to arrange the physical setup of a room and minimize potential distractions. Many agencies have dedicated rooms or offices specifically for interviewing and interrogation of victims, witnesses, and suspects. In general, a room selected to conduct interviews or interrogations should be relatively uncluttered. Experienced interviewers recognize that just as voice and body language can influence the tone and conduct of a conversation or interview, so too can the amount of furniture in the room, and how many distractions can be eliminated will provide the right setting.

Interrogation Methods

There are many styles used or adopted by investigators as they conduct interrogations. Some of these styles and methods develop from an investigator's personal experience in communicating with victims, witnesses, informants, and suspects. Detectives will also decide which interrogations should be handled by one officer, by alternating officers coming in and out of the interrogation room, or by two detectives simultaneously. There are advantages to each of these approaches or the use of a combination. It is not uncommon for an interrogator to appear sympathetic to a suspect and offer rationalizations to the suspect for why they may have committed a particular crime. Alternately, a detective may appeal to logic and list the evidence that points to the suspect. Most approaches will involve encouraging the subject to talk at length so that as much information as possible can be gained while also listening for inconsistencies of story and observing for inconsistencies between statements and physical behavior.

Some methods are developed by researchers or consultants and presented in training seminars. Among the most popular methods used in the United States for many decades is the Reid interview method that combines statements to a suspect with questioning the suspect and all the while closely observing verbal and physical actions and reactions. In a series of steps the detective informs the suspect of his status as the primary focus of an investigation and then proceeds to talk about reasons or excuses why perhaps the suspect committed a crime. The investigator will be anticipating denials by the suspect as well as excuses the

...ect may offer. Through effective communications techniques the investigator will show ...re interest in all that the suspect says and again be analyzing the person's behavior to ...mine if a shift in the suspect's mind-set is bringing the conversation closer to an **admis-** ...**r confession**. These two are not the same thing. An admission is something short of ...ssion to the crime: "I saw the victim that day," "Sure, I've been at the victim's house ...," etc. A confession is generally stating full responsibility: "I waited until the clerk turned her back, and then I grabbed all of the money I could and I ran with it!"

If the interviewer sees behavior that indicates a growing likelihood of admitting guilt by a suspect, the detective will be prepared to offer choices to a suspect to maintain a cooperative communication. If an interrogation reaches this point an investigator will move to asking details of the crime and perhaps assuring the suspect that the information and their cooperation will be made known to the prosecutor in the case. If the suspect does confess it is best to secure the confession in writing even after the person has verbally told of their acts. This includes whether or not an interrogation was recorded by audio and video.

Interrogations like interviews are all about communicating. The process of communications, in a simplified form, involves a sender of information, a receiver of information, the information itself, and the method by which the information is sent. If I send you an email or a letter or text you by smartphone, you have the words presented to you but nothing else. You may or may not be able to determine whether the message is intended to be urgent, angry, merely informative, or any other characterization of an emotional or secondary component besides simple transmitting of a statement or question. When we speak by telephone the added dimension of voice further informs the communication. If instead we meet in person to exchange information we add further physical nonverbal components to our interaction.

It is at the level of nonverbal communication that many people believe they can determine whether or not someone is speaking truthfully. Though repeated studies have shown that even experienced interviewers are only successful "unaided" in detecting deception about 50% of the time, the effort is elemental in the work of a detective. It is not a simple yes/no proposition. Interviewing (or interrogation) is about gaining information. If thorough and substantial information and statements are obtained they can be of use as an investigation progresses. There are a variety of physical behaviors that can signal deception or at least stress that impacts the normal communication of most individuals. The way that we communicate through speech, physical movements, and facial expressions can be observed to help an investigator determine what the norm is for us in conversation. This usually takes place during the rapport-building phase of pre-interrogation activities by an officer.

The observation and analysis of body language is called kinesics. This includes facial, postural, and eye movements, hand gestures, and the pace, pitch, and modulation of a person's voice, referred to as paralinguistics. From the standpoint of the interrogator, knowledge of proxemics—the way people react to the space between themselves and others—can be used effectively as she varies the distance or relative position to a suspect. There are various other aspects of how individuals relate that influence communications. Haptics, for example, examines touch and perception as humans interact with objects and others. One of the difficulties in determining some comments as deception rather than merely reflecting stress is that an interviewer may not know the "normal" or "baseline" presentation of self and speech of the interviewee. This leads to an important concept in the technological aid of detecting deception.

How It's Done

GOOD COP–BAD COP

While not seen by experienced detectives as being particularly effective with longtime or serious criminals, the stereotypic technique of good cop-bad cop can be an effective strategy with juvenile offenders or inexperienced adult criminals. In this familiar interrogation technique, two investigators adopt differing personas—one appearing sympathetic or supporting of the suspect and the other coming across as edgy, hostile, or intimidating. The officers may alternately question or speak to the suspect while the other leaves the room, or they may remain together with the suspect, each speaking or acting in a way with the hope of gaining cooperation from the person being questioned.

Technology

The seeming omnipresence of smartphones illustrates one of the new dimensions of criminal investigations with the potential for tracking location, usage, and content of an electronic companion designed to help facilitate modern life. Where there were once budgetary struggles to ensure enough 35mm cameras were available each shift so that someone could take a crime scene picture when needed, now recording devices are relatively inexpensive and, when a procedure is followed, personnel can also record images on their personal devices to supplement observations in an investigation. Technology is very important in many criminal investigations, and there are opportunities to use it at different stages of a case and in different ways. In this chapter we note the type of technology that is most generally used by an investigator in contrast to lab-based and larger pieces of equipment that perform an analysis. Here the detective uses the technology as an extension of his examination of a person.

Detection of Deception Devices

Even Torquemada said that a tortured confession should be verified. Your mother convinced you that she knew when you were lying. Police, lawyers, and therapists have a generalized belief that they "can just tell" when someone is dissembling. Whether a parent's intuition based on a child not looking their mom or dad in the eye or the anecdotal success of a professional who has caught many people in a lie after listening to their words or watching their physical reactions and responses, a good many people feel confident in their ability to discern when they are not hearing the truth. Can a person really tell when another is speaking the truth?

Some aspects of detecting deceit or other thought processes can be learned through a combination of training and experience. Investigators are always evaluating the person in the process of the interview. How that person speaks and acts, including physical mannerisms the subject may not be aware of or able to control, can provide indications that they are under stress and may not be truthful. The study of body language is extensive and ongoing, and is used by interviewers the world over to aid in detecting deception as well as to assist in

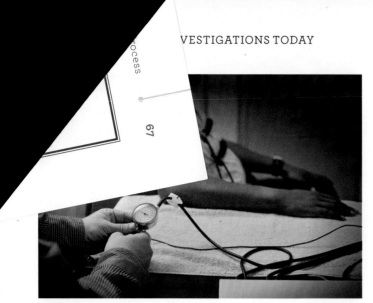

What are the benefits and limitations of detection of deception devices, such as this polygraph?

©iStockphoto.com/Sproetniek

Polygraph Device that measures physiological responses during questioning to detect stress levels.

Detection of deception device One of several pieces of technology that measure physiological response as a way to detect stress, which may, in turn, be caused in part by speaking or acting deceptively.

Computer voice stress analyser (CVSA) Pseudoscientific device that purports to measure stress from a person's voice.

sensing conflicting emotion and put interviewees at ease with the hope of gaining information. As we consider how our experience and perceived intuition inform our assessment of other people's statements, we recognize that such evaluations can be enhanced either mechanically through the use of technology, or at times by the triangulation of assessments by other observers or interviewers.

The **polygraph** has been used to aid investigations for quite some time. As with any investigation, the use of **detection of deception devices**, such as the polygraph, can also assist in eliminating individuals as suspects in the crime under investigation. The polygraph examiner acts in the role of an interviewer or interrogator; however, the manner in which they ask questions is markedly different from the often freestyle "conversation with a purpose" that typifies a detective questioning a suspect. Polygraph instruments were once, as with most technology, larger than the devices in use today. Contemporary units are typically computer-based software with specific sensing attachments integrated. The digitized files can be examined in comparison with previous interviews more rapidly and efficiently than scrolling through a lengthy role of printed paper that typified older equipment. Before the actual interview occurs the polygrapher will talk with the suspect and talk to him about the specific questions he will be asked. After the suspect and examiner have discussed the method and questions, the examiner will affix the sensing devices to the subject. These include hollow tubes connected to the new monograph, which monitors the breathing pattern of the subject and metal electrodes attached to the fingers of the subject's left hand to register the *galvanic skin response* of a person's perspiration through the use of electrical conductive any. Additionally, a *cardiograph* readout comes from a blood pressure cuff that also registers a subject's pulse rate. Once the examination begins, the polygrapher will ask yes or no questions of the subject in sets of questions spanning only a few minutes at a time. A popular approach is called the *control question technique* (CQT) that asks the subject to reply to *relevant* questions about crime involvement, and *control* questions that are not about the specific crime but that also elicit responses about deviant behavior.

Another device used with some frequency in the field of criminal justice is the voice stress analyzer, or VSA. The training for examiners is generally not as long as that required for polygraphers, and the devices are less expensive partly due to the fact that no attachments are required other than a microphone connected to a computer where the software resides. The VSA examination goes beyond yes or no questions and analyzes microtremors in the speech of the person being interviewed or interrogated. Originally developed in the 1970s, and now called a **computer voice stress analyzer** or **CVSA**, the device is not supported with sufficient peer-reviewed research to substantiate an acceptable level of accuracy that would allow results to be introduced in court.

EXPLORE THIS

Search the Internet for detective training. Internet search for a number of websites to learn about requirements for both criminal investigators and private detectives. Each state mandates what training is needed for certification. Most of the sites you will visit point out that criminal investigators usually are selected from experienced law enforcement officers who have a significant amount of experience. As with patrol officers, the trend toward possessing college education is present and important for the aspiring investigator.

SUMMARY

The process of criminal investigation involves a number of people in various roles: the victim who reports a crime or who is unable to do so, leading to some other complainant making the report; the suspect who committed the offense, perhaps witnesses, informants, or others who have information about the incident; and then there are the various official actors who can range from call-takers and dispatchers accepting an initial report, to the patrol officers, crime scene technicians, detectives, victim advocates, lab criminalists, supervisors, prosecutors, and more.

It is clear that while patrol officers are trained to seek information and facts, investigators typically acquire even greater depth of technique since their primary role is to *investigate,* where the patrol officers handle myriad tasks. The types of crimes investigated in a follow-up range from property damage and burglary to rape, robbery, homicide, as well as cybercrimes, arson, and many other offenses that require the detective to possess further expertise and training to effectively conduct an investigation. At the federal government level, United States Code (USC) gives numerous agencies the authority to investigate and enforce specific federal criminal statutes. There are not enough investigators to fully pursue all cases identified as having sufficient leads; the need for quality training of those detectives available is critical.

As the detective makes her way through the initial stages of a crime investigation, she is interviewing various individuals. Some may be eyewitnesses to a crime; many others will be considered "witnesses" but may not have actually seen a crime committed, though they may have important bits and pieces of information that go into solving a crime. Officers and detectives will interview complainants, victims, and witnesses to gain as much information as they can about the circumstances of an incident before interrogating a suspect if one has been identified. While both the interview and the interrogation are exercises in gaining information, the interview is considered nonaccusatory, but the interrogation focuses on someone believed to have participated in the crime in some way.

KEY TERMS

Actus reus 56
Admission 66
Biological theory 54
Blood feud 52
Computer voice stress analyzer (CVSA) 68
Confession 66
Confidential informant (CI) 62
Crime Victims' Rights Act (CVRA) 53
Detection of deception device 68
Differential association theory 54

Field identification 57
Hard surveillance 58
Interrogation 62
Interview 61
Intimate partner violence (IPV) 55
Mens rea 56
Miranda warning 57
Personal retribution 52
Polygraph 68
Probable cause 62
Psychological theory 54
Reliable confidential informant (RCI) 62
Routine activities theory (RAT) 54
Social control perspective 54
Social control theories 54
Social disorganization theory 54
Social learning theory 54
Soft surveillance 59
Strain theory 54
Subcultural perspective 54
Victim advocate 55
Victimology 54
Witness 60

DISCUSSION QUESTIONS

1. List and discuss the different actors in a criminal investigation.
2. How do victims and society respond to criminal acts, and why?
3. How has the role of victim changed in the United States in the last several decades?
4. What types of training do detectives receive?
5. Discuss the differences and similarities between interviews and interrogations.
6. What technological devices are sometimes used during interrogations? What are the challenges in using them?

SAGE edge™

- Get the tools you need to sharpen your study skills. SAGE edge offers a robust online environment featuring an impressive array of free tools and resources.
- Access practice quizzes, eFlashcards, video, and multimedia at **edge.sagepub.com/houghci**

PRACTICE AND APPLY WHAT YOU'VE LEARNED

▶ edge.sagepub.com/houghci

WANT A BETTER GRADE ON YOUR NEXT TEST?

Head to the study site where you'll find:

- **eFlashcards** to strengthen your understanding of key terms.
- **Practice quizzes** to test your comprehension of key concepts.
- **Videos and multimedia content** to enhance your exploration of key topics.

4 Managing the Criminal Investigation

RUNNING CASE: THE DETECTIVES INTERVIEW FRANK DENNEY, SUSPECT IN HOMICIDE AT THE FLORIDAN

Introduction

Investigation as a Process
    Crime Analysis

Preliminary Investigation

Case Assignment and Solvability Factors

How It's Done: Solvability Factors

Follow-Up Investigation
    Sources of Information
    Coordination With the Prosecutor

Supervising the Investigative Unit

Case Management Software

Pitfalls, Fallacies, and Sloppiness in Investigations
    Cognitive or Confirmation Bias
    Insufficient Time
    Framing and Groupthink
    Probability Errors
    Strategies to Avoid Criminal Investigative Failures

Developing an Ongoing Relationship With the Media

Improving Productivity
    Civilianization
    Increased Investigative Duties Performed by Patrol Officers
        Training
        Adapting to New Technology and the Changing Nature of Crime
        Benefits of Greater Productivity

Explore This

Summary

Key Terms

Discussion Questions

LEARNING OBJECTIVES

After reading this chapter, students will be able to

4.1 Describe the investigative process.

4.2 Contrast the preliminary and follow-up investigation.

4.3 Analyze solvability factors.

4.4 Describe the follow-up process in an investigation.

4.5 Outline supervisory duties in an investigative unit.

4.6 Explain the role of case management software.

4.7 Critique investigative decision-making with the potential for errors that may slow or misdirect an investigation.

4.8 Explain the symbiotic nature of detectives and journalists.

4.9 Evaluate the interplay of resources and investigative caseload.

Running Case: The Detectives Interview Frank Denney, Suspect in Homicide at the Floridan

Officer Carl Jayden, the patrol trainee, knocked on the interview room door and asked Detectives Bradley Macon and Richard Ashley to step out.

In the hallway, Officer Carl Jayden said, "Officer Thompson and I located a suspect sitting inside a car. He was parked several blocks from the crime scene. Thompson knocked on the window, and the man let it down. We didn't have a chance to say anything before he led with, 'I guess you guys want to talk to me about the shooting.'

"I read him his Miranda warning. The guy just sighed and said, 'Take me to jail.'

"We asked his name, and he replied 'Frank Denney.'"

"Where is he now?" Bradley Macon asked.

"He's in a secure room upstairs."

Bradley Macon nodded. Suspects were always kept in a different part of the building from where witnesses were interviewed.

Detective Macon and Detective Ashley looked at each other.

Richard Ashley asked, "Are you ready?"

Bradley Macon nodded his head and said, "Let's talk about it on our way."

As the two detectives walked down the hall, Bradley said, "I'll warm him up when we get there. I'll issue the Miranda warning form and set up the digital recorder."

"Okay," Richard said. "I will lean against the wall and look bored like I usually do."

While climbing the stairs, Bradley realized that this casual banter belied the nervousness he felt when approaching a two-person interrogation with an unknown suspect. After being assigned to the criminal investigations division, he had received advanced training in interviews and interrogations. Bradley's mind was running through what needed to be accomplished to establish that the suspect was the person responsible for Bill's death.

At the top of the stairs, the detectives stopped at the vending machine and bought a canned pineapple drink that Bradley would offer to Frank Denney when they sat down with him.

The two detectives entered the interview room.

"Hello, Frank," Detective Bradley Macon said and handed Frank Denney the cold, canned juice.

Frank looked surprised. He took the can and set it on the table in front of him.

Detective Richard Ashley put both of his hands in his pockets and leaned against the wall.

Detective Bradley Macon read Frank the rights form.

Frank signed the piece of paper and leaned back in his chair and crossed his arms over his chest.

Both officers knew that the physical act of crossing his arms didn't mean that he *wouldn't* talk with them, but they knew that individuals would close themselves off when they didn't want to talk and were trying to hold back information.

Frank stared at the can of juice but made no move to open it.

"Can you tell us what you did yesterday?" Bradley Macon asked him.

"I went to work. Then soccer practice. Then I headed over to the bar to go over some stats with Bill."

"What type of work do you do?" Bradley asked.

"I'm a high school gym teacher."

"And you play in the local soccer league?"

"Yeah. I've played in the league for 5 years."

"How often did you frequent the Floridan?" Richard Ashley asked.

"Not often. Mallet's downtown is more my scene."

"You know anyone who worked there? The bartender?" Bradley asked.

"No. I go there maybe once or twice a year."

"You say you went to meet Bill Johnson?" Richard asked.

"Yes. Bill and I needed to settle the score on which player scored the most goals."

"Why was this so contentious?" Richard asked.

"Because everyone knows Joe Wilson is the star of the team. Bill couldn't accept that it wasn't him."

This made no sense to Bradley. If Bill couldn't accept these scores, why was he the one who'd walked away from the fight? Why was he the one shot dead in Milton Way?

Richard Ashley pulled his hands from his pockets. "The gun you used? Was it yours or someone else's?"

Frank broke down. Between sobs, he said, "It was mine."

Do suspects frequently agree to talk with officers? What do you think might make a person, suspected of a serious crime, willing to answer a detective's question?

Introduction

A reality that many people are unaware of is that not all reported crimes are investigated by law enforcement. Given the lack of evidence or insufficient investigative resources or a combination of the two, agencies must decide which cases will receive follow-up investigation. Most members of the general public are not directly impacted by crime and so they are understandably uninformed about the number of crimes of various types and the process by which they are (or are not) examined. Each citizen may rightfully feel that their victimization merits an official response. We may mistakenly believe that the crime is solvable if only the police will devote the resources and cleverness shown in fictional television programs. The disconnect between various public beliefs about crime and criminal justice system responses leaves room for frustration on the part of everyone impacted by crime, including the police and the courts.

Local crimes covered by the evening news tend to be violent or shocking such as robberies, murders, sexual assaults, or kidnappings. The news report will comment on what the reporter knows at the time of the newscast and what information about suspects or investigative activities the department is willing to comment on early in the investigation. The combination of what citizens believe can be accomplished in a given investigation and the prominent placement of shocking crimes at the beginning of a newscast leads to a skewed impression of crime in a community. Surprising to many citizens is that the overall crime rate in the United States has continued to decline for the past two decades, though some communities have experienced an increase in violent crime in the last few years. Because the length of time has been short and there is an absence of a clear change in any major factor or social driver in the country, this means that researchers and others are not yet able to assert or support any cause(s) for a shift in any short-term crime patterns.

For most people, crime is local in nature and so is governmental authority for crime control in the United States. The modern information and technological era has resulted in the need to draw upon the central infrastructure and access to the resources of federal agencies. This has brought about less strict reliance on (smaller) local agencies when investigating many crimes that are facilitated through data, communications, and mobility. More traditional crimes remain the province of local agencies. This modern reality was hardly dreamed of when the Framers wrote the 10th Amendment to the U.S. Constitution that granted policing power to states and their subdivisions. And so much of the crime that is investigated is done within relatively small agencies. These agencies often perform a cursory examination of an incident looking for a named or obvious suspect and evidence to move a case forward. If information is not readily available, the case may go no further. The majority of reported cases are not cleared or solved, though persons' crimes generally have a higher clearance rate.

According to the 2013 Law Enforcement Management and Administrative Statistics (LEMAS) conducted by the Bureau of Justice Statistics (Reaves, 2015), almost half of law enforcement agencies employed fewer than 10 officers, and only about 5% out of approximately 18,000 employed 100 or more officers. This means that smaller agencies must rely on very limited resources as well as work cooperatively with neighboring jurisdictions or state police organizations to investigate major crimes.

Investigation as a Process

Recall that a case is closed or cleared if a suspect is identified and arrested or is unavailable to face the legal system (e.g., dead or fled the country to a place without extradition, or the cost to do so is prohibitive). This measure of investigative impact may not be an assessment of the overall effectiveness of an agency, but it is the clearest indicator of the singular function of detectives. Patrol officers are tasked with myriad responsibilities and are called upon to respond to a dizzying array of calls for service, much of which are not directly crime-related. The detective's efforts, on the other hand, are typically directed only at crime. An investigation may be proactive, such as in vice crimes of narcotics, prostitution, or gambling. More often the investigation is reactive, as investigators work to assess and pursue leads in crimes reported after the fact.

The modern conceptualization of a police detective is properly one of knowledge worker (Gottschalk, Holgersson, & Karlsen, 2009). While detectives have long been appraised based on their ability to gather information, the ability to parse voluminous data present in many contemporary investigations relies on coordinating and sharing information with other knowledge workers. Investigators search databases, comb through reports and witness statements, and work to assemble the story of a past event. Reconstructing a crime that has occurred requires investigators to sift the available information to come away with the relevant bits that describe and hopefully explain what happened in a given incident. This is no small task, and examining even relatively straightforward crime cases can illustrate how many factors need to be examined in determining accountability for a criminal act let alone compiling the evidence necessary to achieve a finding of guilt in court.

A criminal investigation is a process with different stages and many activities assigned to various people within those stages. Initially, after a crime has been discovered and reported, law enforcement will be notified and, at the local level, a patrol officer will likely be assigned to respond and meet with a complainant. This begins the preliminary investigation stage. It is at this point that an initial complaint may be found to be inaccurate, unfounded, or incorrectly categorized (e.g., calling a home burglary a robbery). The matter may be one for the civil court system, and the uniformed officer can provide basic information regarding how to contact the local court system or recommend that the complainant contact a private attorney. Other criminal justice personnel may be involved in interaction with victims, witnesses, and suspects. And it is at this stage that many cases are resolved by patrol officers (Horvath, Meesig, & Lee, 2001; Willman & Snortum, 1984). Some cases will move on to a follow-up stage and, possibly, the court stage, where a prosecuting attorney and defense attorney will work to present opposing views of what occurred, who is responsible, and to what extent. The primary detective and other personnel will remain important in preparing the case and testifying in hearings or trial.

The process of criminal investigation is one that must be managed at every stage. Resources and the efforts of detectives must be managed, and the investigators themselves

Law Enforcement Assistance Administration (LEAA) Was a federal agency that administered funding to state and local government entities for research, training, and programs.

manage many aspects of the activities that make up the overall process. In 1977 the **Law Enforcement Assistance Administration (LEAA)** awarded grant monies to the University Research Corporation to compile a manual to address the management of criminal investigations. In looking at key components of the investigative process, the authors noted benefits of "adopting modern management methods" and the following:

- An increased participation by uniformed personnel in a comprehensive initial investigation at the time a crime is reported.
- The establishment of a case-screening system that will remove non-solvable cases from the investigative process at an early point.
- The development of a police/prosecutor relationship that will result in better case investigation and preparation and greater likelihood of successful prosecution.
- The establishment of a management information system which provides agency administrators with appropriate information for managing the criminal investigative process and alerts them to emerging problems.
- A searching re-examination of agency structure to maximize the use of all personnel.
- The development of investigative management techniques for the improved use of detective personnel. (Cawley et al., 1977, p. viii)

Crime Analysis

Managing the criminal investigations process is also about managing the available resources to address all crime matters an agency is tasked with handling. Two parts of the equation are to know what resources you have and what the crime circumstances are. This second part connects with any patterns or trends that may trigger preventive or investigative measures geared to take advantage of the identified pattern.

What role does crime analysis play in the criminal investigation process?

Monty Rakusen/Cultura/Getty Images

Possession of information or data about an area or about ongoing crimes may be helpful if that information has been put into context. This typically is accomplished by an experienced analyst or individual with or without the help of computer software or an algorithm that may provide a starting point for how one incident or set of circumstances may relate to others. Information generated as the result of such analysis may be assigned to an investigator for follow-up. In addition to the potential help on a specific case, crime analysis is used for developing tactical plans to address ongoing crime such as a wave of robberies, or strategic plans such as supporting the need for additional resources by showing crime increases or changes in certain geographic areas.

The Bureau of Justice Assistance, part of the U.S. Department of Justice, conducts ongoing technical assistance for agencies to build their capacity in crime analysis and technology solutions. State agencies often provide local and regional information to law enforcement as well.

Preliminary Investigation

As discussed earlier in the book, the preliminary investigation generally encompasses the activities conducted by the first responding officer to a reported crime. The uniformed officer must identify and question witnesses and the victim, protect and possibly collect evidence for processing, and make notifications to others while creating thorough documentation of what happened and what actions the officer took. Managing the case from the beginning means managing the gathering of all available information. Whether the patrol officer effectively interviews the victim and provides timely information to other personnel are often determining factors in achieving a positive outcome (McDevitt, 2005). Research also suggests that in cold case investigations it is new information from witnesses that most often brings a case to closure (Davis, Jensen, Burgette, & Burnett, 2014). There are several reasons why the issue of witness interaction is so important. Initially, people may be upset or in shock and either unclear on some details or very clear on who committed the crime and why they think the crime was committed. Either circumstance may give the first officer a window of opportunity to hear a straightforward recitation of what the victim and witnesses know or believe and why and how they know or believe what they do. An important reason to focus on the preliminary investigation phase is that during this time rapport with the victim and witnesses will be established—or not. We have all heard that there is no second chance to make a first impression. While the homily can be debated, we do know that unprofessional, discourteous, or disinterested treatment of a victim or witness by a police officer may spell the end of cooperation or information-sharing, even at the cost of catching the person who victimized the individual. People want to be treated with respect.

Officers speak to multiple witnesses just after a crime took place. Why is the issue of witness interaction so important in an investigation?

Scott Varley/Digital First Media/Torrance Daily Breeze via Getty Images

It is at this beginning stage of investigating a reported crime that physical evidence may be most readily available. Once again, if the patrol officer conducts an effective securing and search of a scene, important items may be located and gathered, ultimately protecting their admissibility and use in a case. Quickly roping off a crime scene and controlling entry of individuals to the scene while beginning the process of documentation are all critical components of scene management that become part of the overall management of the investigation process. When additional personnel arrive at the crime scene they may be assigned to perimeter security, canvass of the area for witnesses or evidence, or numerous other tasks that assist the effort. Crime scene technicians and responding investigators must be briefed by the initial officer(s) to alert them to information and evidence as well as to help prioritize the gathering of information that may be perishable or sensitive in nature. The handoff of investigative responsibility occurs at this point, but the patrol personnel remain important in the incident. Some agencies have expanded the role of patrol officers to include limited investigative functions, but unfortunately the majority has not. Giving patrol officers enough time at crime scenes and initial report locations can improve case outcomes as well as enhance the role of the patrol officer (Womack, 2007). In smaller agencies the patrol officer may in fact be the only investigator of many reported crimes, including any evidence-gathering. Resources are a factor as we again consider that there are some 18,000 separate law enforcement agencies in the United States serving a wide array of communities with varying crime challenges.

Case Assignment and Solvability Factors

Because resources are not unlimited, not all cases are assigned to be followed up (Eck, 1983; Liederbach et al., 2011). Cases of a certain type, such as homicide, sexual battery, armed robbery, and other universally condemned or serious offenses, will almost always receive some amount of investigative effort. Less serious matters or ones with little or no evidence to follow-up may receive little or no follow-up save filing in a que to await any information that may arise through sources other than detective activity. The RAND study of the early 1970s that examined the criminal investigation process noted at that time that less than half of reported serious crimes received much investigative effort. This is (and was) not necessarily the result of investigator disinterest or a pattern of agencies across the United States systematically ignoring reported crime. Rather, many cases lack sufficient information to allow an investigation to proceed. Determining which cases will receive follow-up efforts can be aided by empirical examination of the likelihood of success.

Many if not most investigations lack significant physical evidence initially. Evidence may not have been left behind at the scene; victims and witnesses may not know the identity of the perpetrator; and time may have passed since the crime was committed, leading to further loss of physical or eyewitness evidence. A practical approach to determining the amount of resources devoted to any particular reported crime that takes into account the known clues in a given case is one that assesses a case's **solvability factors**. How likely is it that, given the known information, a suspect can be identified and a case can be resolved to the point of a clearance? Has a suspect been named? Did the victim or a witness get the tag number of a getaway vehicle? Were fingerprints or other physical evidence items left at the scene? The concept of solvability factors is more than the heuristic experienced investigators already used to gauge whether there was enough "there, there" to pursue a further investigation. Solvability-factor scoresheets were developed, partly from the research conducted in the mid-1970s by Peter Greenwood of the RAND Corporation (Greenwood, Chaiken, & Petersilia, 1977), Greenberg's work at the Stanford Research Institute (SRI; Greenberg et al., 1975), and the LEAA study completed in 1977. Both research studies were seminal as early efforts to examine detective work and each looked at the types of functions and amount of time devoted to various investigations by detectives, as well as time elapsed to close a case. More than 40 years later, most studies continue to emphasize the importance of patrol efforts in case solving and closure. This includes patrol officers gathering information on solvability factors (Horvath, Meesig, & Lee, 2001).

Solvability factors Those pieces of information or evidence that provide adequate information to move an investigation from the preliminary to the follow-up stage. Examples include pictures of a suspect or the tag number from a car involved in a crime, a viable way to identify stolen property.

How It's Done

SOLVABILITY FACTORS

The concept of solvability factors is one that most people easily grasp; if you have a number of solid clues or leads, the case merits further investigation to try and find the perpetrator. The Managing Criminal Investigations Manual (Cawley et al., 1977) noted that each jurisdiction should consider specific factors for determining continuation of an investigation but offered the following factors as indicative of a higher probability to solve a crime:

1. Immediate availability of witnesses.
2. Naming of a suspect.
3. Information about suspect's location.
4. Information about suspect's description.
5. Information about suspect's identification.
6. Information about suspect's vehicular movement.
7. Information about traceable property.
8. Information about significant M.O.
9. Information about significant physical evidence.
10. Presence of evidence technician who indicates an *a priori* judgment that good physical evidence is present.
11. A judgment by the patrol officer that there is enough information available that, with a reasonable investment of investigative effort, the probability of case solution is high.
12. A judgment by the patrol officer that there is sufficient information available to conclude that anyone other than the suspect could not have committed the crime. (p. 7)

Follow-Up Investigation

If a case is "serious" and/or merits attention by community acclaim, or the crime has adequate known facts to cross the threshold of assignment to a detective, a supervisor will assign the matter to an individual for a follow-up investigation. Some agencies must prioritize felonies over misdemeanors due to resource considerations. There will always be political consideration as well if someone convinces an elected or appointed official that his case is deserving of investigative effort, regardless of what the available evidence or information says. In addition to individual case factors that may indicate whether investigative effort will be well spent on solving a crime, cases are divided in many agencies by the nature of the crime. Mentioned earlier in the book are the persons' crimes unit that investigates serious individual crimes such as rape, robbery, or homicide, and the property crimes unit that investigates burglary, theft, white collar crime, and other crimes that focus on property. Still other specialized units may focus on crimes against children, arson, drugs and vice, or cybercrime. Investigations that involve multiple agencies or a task force must always be mindful of coordinating their efforts to ensure an effective case management strategy.

The use of specialized investigators may entail separate units such as a major case squad, homicide unit, or combination such as a robbery homicide unit, etc. The logic of specialized investigators is apparent when considering the need for extensive training and experience to properly investigate serious and complex crimes. The potential weakness of this approach may be in the limited opportunity for the specialist detective to assist in other types of investigations with which he is less familiar. Some investigations take more time to complete than others. Given this fact, caseload management for investigators is important in trying to maximize productivity and case solving. In any job where a caseload is involved (e.g., probation officer, social worker, school guidance counselor), an important factor in success will always be determining what a manageable caseload is and avoiding any increase past that point as well as having resources available to assist personnel when one or more cases are more than what the assigned individual can effectively handle. While the case screening may be handled by a variety of personnel, the responsibility for the results and assigning cases to detectives lies with the unit supervisor.

Managing the follow-up investigation continues the process of managing information as well as resources. It is at this stage of an investigation that a determination has been made that additional resources will be devoted to a specific crime. This may include submitting items of evidence to a laboratory for testing, assignment of additional investigators to seek out and interview persons with information about the case, and other follow-up activities that claim the time of the assigned detective(s).

During the follow-up investigation detectives perform a range of activities that appears to have changed relatively little over the decades. In the RAND report of the 1970s, Chaiken, Greenwood, and Petersilia (1977) summarized issues of both detective workload and productivity. The researchers noted that a large percentage of cases receive little or no investigative effort past the initial report, that much of a detective's time is spent on administrative duties, and that many of the activities revolve around trying to speak with victims, creating and reviewing reports, and putting all of the information together to pursue prosecution. It has been found that very little time is spent by detectives in performing "nontraditional" but promising activities such as building community rapport and devoting additional time to cases that might be solved with extra effort (Liederbach, Fritsch, & Womack, 2011). This highlights the reality that workload is central to productivity (Horvath, Meesig, & Lee, 2001) and that strategies to modify tasks to improve the *specific* task demands of detectives are important.

Some detectives work as partners handling major cases together but generally carrying independent caseloads as well. The **team policing** initiatives of the 1970s showed some success with a coordinated group of uniformed officers and detectives working in a coordinated fashion (Bloch & Bell, 1976), but contemporary group work is more often done by task forces.

Team policing Patrol officers and detectives permanently assigned to neighborhoods to improve police community relations.

Sources of Information

As individual detectives are managing specific cases they are drawing upon information from not only victims and witnesses, but from other human and data sources. While sometimes exaggerated in movies, the use of **informants** can propel a case forward. Some informants provide information only once, and that information is verified through additional sources. As mentioned earlier, this person is referred to as a *confidential informant* (CI). A person who has supplied verified information to law enforcement more than once may be classified as a *reliable confidential informant,* or RCI. If information supplied by an informant is to be used to support a search warrant, for example, the court may demand additional information from the detective or agency. Other individuals who are employees of

Informants Persons who give information to authorities.

various agencies may also have information relevant to an investigation. Database searches and official and private records are all sources of information that must be considered in many cases.

Coordination With the Prosecutor

While more will be said on this topic in the final chapter of the book, overall management of the criminal investigation must include coordination with the prosecutor's office and the potential consideration of the eventual handoff or further work and collaboration to ready a case for court. In some jurisdictions on-call prosecutors may be available to respond to the scene of major crimes such as homicide to assist investigators. Often prosecutors can be of help in guiding detectives in securing search and arrest warrants, ensuring proper lineup procedures, and conferring on important information to seek in a suspect interrogation. While overall consideration of criminal investigation in court is addressed in a later chapter, it is important to recognize that many aspects of the prosecutorial function are important in managing a criminal investigation.

Supervising the Investigative Unit

A small agency may have few or no personnel assigned solely to follow-up investigation duties. The chief of police may decide which reported crimes will receive follow-up. If there are one or more investigators (but still just a few), they will likely be general assignment detectives, meaning they will handle any and all follow-up investigations with no specialization of duties based on specific crime types. In smaller agencies one investigator may be considered the supervisor of the investigative function, while in other departments there may be a sergeant or lieutenant who supervises all law enforcement personnel and is therefore considered the investigative supervisor. Larger agencies will likely have more personnel assigned to the role of investigator and will typically have unit heads, and supervisors of various ranks up to that of deputy chief or deputy superintendent responsible for the overall management of the agencies' investigative functions, which may include support components such as the crime scene unit, property and evidence section, and laboratory, among others.

An important managerial function within criminal investigations is the selection of detectives and ensuring their training once they are in the role (Horvath, Meesig, & Lee, 2001). Different agencies approach this in different ways, but many mid-sized and larger modern departments utilize written tests, interview boards, and a thorough review of the officer's work in patrol. The trait approach to selecting personnel can be subjective and therefore unreliable. But examining the arrest and conviction performance of an officer along with the quality of her written reports and pertinent information from her personnel jacket can combine with the results of testing and interviews to arrive at good selections for investigators. Sometimes attracting candidates for the investigative role can be more challenging than one might imagine. Agencies do not always provide clothing stipends to investigators to help them dress the part as compared to patrol where uniforms are typically provided by the agency. In addition, the irregular nature of investigative work often precludes the detective from planning a secondary or part-time job or to attend college.

Some agencies rotate supervisors between units, which can result in someone being assigned to oversee the investigative function that has not performed in the role of detective previously. In some cases, the supervisor of the investigative unit or function will also carry a caseload in addition to supervising or monitoring the progress of other investigators. In addition to the challenges of workload, supervising investigators can sometimes be difficult based on the investigators' significant amount of experience for years on the job and the self-perception of being quite important. Detectives should be responsible enough to work largely without close supervision and professional enough to understand the reasons and need for accountability of their work efforts and interaction with fellow detectives and supervisors. The span of control, or number of investigators supervised by one individual, may vary based on the types of investigations being conducted and available personnel.

It is important to comment on the function of *management* itself and skill sets that should accompany exercise of managerial oversight. People often think of management as a generic or homogenous product or function. This is typically not the case, as the type of functions or resources subject to the managerial efforts of an individual or an organization are unique and may require insights about the function to properly, or at least efficiently and effectively, manage. There is often a debate on this point, and whether the CEO of one corporation can move on to a different company and manage effectively. This often obscures differences among leadership, management, and supervision, and what the expectations are for someone leading a large organization as opposed to managing or supervising organizational components within a department. The importance of the competency of detectives as well as those who supervise them cannot be overstated. Supervisors, as key people in the process, shape the direction of investigations and the practices of investigators (Stelfox, 2011). With the volume of information, complexity of issues, and various legal and procedural concerns, management of criminal investigations can be aided through technology when reviewing and tracking case information and activities.

Case Management Software

Paper record-keeping has been around since the time of the pharaohs. Organizing files and information has long been seen as important to the success of many endeavors, whether in the government or private sector. As time passed, systems of files developed to manage most organizations, not the least of which were agencies involved in the criminal justice system. Files are maintained by individual investigators and various other personnel involved in an investigation. It is important that the team members have access to at least view all information and actions regarding their assigned cases. Computer software has improved and facilitated this access and coordination of criminal investigations. An example of such software to assist agencies in the performance of complex modern investigations, Pennsylvania's Justice Network (JNET) acts as "information broker." Municipal, county, state, and federal agencies provide access to their respective data sources and facilitate tracking of information and offenders as they move through the criminal justice system (Commonwealth of Pennsylvania, 2018). Some of the available information on JNET include data-sharing of police agency records, birth certificate information, various data sources for people wanted or owing money due to court order in family situations, facial recognition and photo searches, and user training through an online **learning management system (LMS)**.

In many jurisdictions, complainant files in the Computer-Assisted Dispatch (CAD) systems have become linked to the **Records Management Systems (RMS)** that compile

Learning management system (LMS) Software to assist in the delivery and documentation of training and education.

Records Management System (RMS) Generally a relational database system for record storage.

incident reports, traffic citations, field interview cards, and more in agencies. The integration of these databases in what are often called criminal case management systems (CCMS) or **case management software (CMS)** allows investigators to more efficiently and effectively search for information about suspects, witnesses, or victims as well as geographic or place information that may bear on a crime investigation. Case management software further expands and integrates the management of data and information and facilitates case and status review by detectives, supervisors, prosecutors, and others. These software programs have made strides in organizing and making searchable the mass of information that a case can generate. Investigators, prosecutors, and supervisors can link and examine incident and supplemental reports, audio and video files, tips given to law enforcement, information on evidence, and other pertinent documents. Given the improvements in software flexibility in the last two decades, custom reports can be generated to assist those in roles that require review of cases.

An international example of a system used country-wide is HOLMES (Home Office Large Major Enquiry System), the too-clever name of the computerized system used by the police in the United Kingdom for major case management. Cook and Tattersall (2016) list the five major aims of the HOLMES system:

- Provide an accurate record of the state of the enquiry,
- Determine the current status of the enquiry (e.g., number of actions generated, messages received),
- Provide an accurate record of all data collected,
- Assist in the cross-referencing of information and intelligence,
- Assist legal decision-making and disclosure issues. (p. 137)

Cook and Tattersall (2016) correctly point out that human error must always be considered in any system that relies on data input by people. This is axiomatic for information management but also highlights a critical area in the investigation of crimes where the process can stall and a case go unsolved, even though buried within the mass of information lie key items that could move a case forward. Detectives, like patrol officers, are gatekeepers of the criminal justice system. If a case is not initiated or is allowed to cease, no further action is taken by any other component. This systems effect is well known and can be kept in check through supervisory case monitoring and external reviews of investigative efforts.

Information must also be shared or passed to external agencies. While most criminal justice agencies have the ability to search as well as upload information to regional, state, and national databases, they do not have seamless capability of searching or sharing incident reports and the like. With the reminder of some 18,000 separate law enforcement agencies in the United States comes the realization that nearly as many methods or systems exist for managing an individual agency's files. While large agencies may handle all aspects of an investigation in-house, many more departments must rely on cooperative agreements with other local and state organizations to effectively handle major cases (Hough, McCorkle, & Harper, 2019). In addition, software must be coupled with a purposeful review of the system of assigning cases and the manner in which tasks are allocated to achieve the greatest improvement in workflow, use of investigative time, and outcomes.

> **Case management software (CMS)** Allows the coordination and management of digitally collected and stored information.

Pitfalls, Fallacies, and Sloppiness in Investigations

As with any human undertaking, criminal investigation may suffer from a number of weaknesses generated by or allowed by the people in the process. Rossmo (2009) devoted an entire book to cataloging some of the difficulties faced by agencies and individuals in the business of investigating crime. Aptly named *Criminal Investigative Failures*, the book generally pursues three themes the author believes are responsible for most investigative failures: **cognitive biases**, such as tunnel vision; **organizational groupthink**; and **probability errors**. Cognitive biases are manifestations of the subjectivity with which we all interpret our reality. Many fields of research and endeavor realize how important it is to train and educate those in their respective fields about the causes and consequences of developing a mind-set that impairs judgment or impedes the work a professional is conducting. While investigators are admonished to keep an open mind so as not to ignore or dismiss information not in keeping with an investigator's or agency's assumptions about a case or suspect, routine discussion of the dangers inherent in misperceptions is an important supervisory and managerial function.

Cognitive or Confirmation Bias

Working to inoculate officers and investigators against cognitive or **confirmation bias** (Rassin, Eerland, & Kuijpers, 2010) is important because criminal investigation is so completely about the business of gathering and making sense of information. Each person over their life and career develops certain *intuitive* shortcuts referred to as *heuristics* that facilitate assessments and decision-making about the world. From an evolutionary standpoint we can understand how this can often keep people safe as they rapidly identify and avoid danger. More benignly (yet important), from infancy through adulthood we learn to read the facial expressions and body language of others and form quick assessments of any number of emotional intents from those we observe. *Rational* reasoning, on the other hand, is the type of thinking we do with reflection and while gathering information to reason our way to a conclusion about something. Kahneman (2011) uses previously established labels of System 1 and System 2 thinking. Kahneman makes the distinction between these two systems through examples that illustrate how System 2 requires that we *pay attention* to mental operations under this rubric. Rather than the involuntary or virtually automatic reaction or response of System 1, an investigator or any one of us must pay attention to the thought process we are undergoing if we have a hope of avoiding the tunnel vision or cognitive bias that leaves one singularly focused on perhaps the wrong suspect or interpretation of the available information (Groenendaal & Helsloot, 2014).

Rossmo (2009) cites many examples from existing psychological research on some of the various ways cognitive bias influences investigative matters. One of these is referred to as anchoring, or being influenced as to where we begin our thinking about a matter. Picture yourself as a detective arriving on the scene of a crime and having a victim, witness, or patrol officer urgently or confidently tell you who the suspect is and why they believe it is so. They may be right, of course, but what if they are not? Can you imagine your brain snapping to attention, focusing on the individual identified and beginning to efficiently put together information in support of the original assumption? We can certainly trot out another familiar saying here—never assume. Rossmo also gives us the example of tunnel vision, where an officer considers few options and ignores potentially important information or alternative

---

Cognitive bias Each individual's perception of social reality. Includes the information processing shortcuts that allow faster response to previously encountered circumstances.

Organizational groupthink Group members maintaining the status quo at the possible cost of effective decision-making or positive outcomes.

Probability errors The probability of making a wrong decision, which can result from rejecting a true hypothesis or failing to reject a false hypothesis.

Confirmation bias Interpreting evidence as confirming an established belief.

theories of the crime. Assembling information in this selective manner to support the assumption we have made is called *confirmation* or *verification bias*.

Insufficient Time

Another concept Rossmo mentions that is quite familiar to public administration and management folks is the idea of "**satisficing**." The term was coined by Nobel laureate Herbert Simon (1956) to describe decision-making under conditions that make it unlikely that an optimal solution can be achieved, so the decision-maker chooses the first option that will "satisfy" and "suffice." Applied across a variety of domains, Simon's concept underscores that reason and rationality are not always the predominant drivers of our decisions, regardless of what the architects of various deterrence and sentencing schemes would like to suppose. *Availability* has to do with what first comes to our minds as opposed to making decisions based on all of our knowledge and experience if we were able to access them at the point of decision-making. Most people can think of an example where they have made just such a decision and upon reflection lamented that they could not go back and change that decision. Such is decision-making for humans.

Satisficing In decision-making, accepting the first adequate alternative.

Framing and Groupthink

Another common concept included in Rossmo's work is that of **framing**. News organizations present information within a frame that can influence how we interpret the "facts" presented. Each of us may present information within a context more likely to influence people with whom we are communicating. And we often estimate what may have happened (or what may happen) based on previous similar circumstances; this is referred to as **representativeness**. It is not difficult to see where this type of cognitive bias may lead a detective to follow the wrong path in a case.

Framing The filters individuals apply when viewing situations or that organizations or individuals may use to present an issue in a particular way.

Representativeness Decision-making based on how alike one event is to similar events. This can lead to decision-making errors.

A unit or agency may hold a particular view about a case, and all members are either captive to the rightness of the notion or no one wishes to rock the boat with an alternative theory. This is an example of groupthink, and Rossmo enumerates the symptoms of this phenomenon as power over estimation, closed-mindedness, and uniformity pressures. Power over estimation is believing that the group cannot be wrong and the corollary that any decisions made by the group cannot be immoral or unethical. Closed-mindedness in an organizational sense speaks to rationalizing decisions and viewpoints as well as criticizing opponents of the group's view. And uniformity pressures lead to ostracizing members who fail to support the group consensus or who are seen as agreeing with the group through their individual silence. Friction within agency subdivisions and between agencies with different mandates or jurisdiction is an all too familiar set of organizational dynamics found in many if not most organizations of any type.

Probability Errors

Probability errors occur in a variety of ways, including what is known as the law of small numbers that addresses how we may erroneously draw conclusions from evidence or examples such as past incidents that are too few to rely on as predictive of the current incident or case. Another error of probability can happen when an investigator links cases or offenders based on similar facts but insufficient analysis to take into account differences in cases. Similar to these would be not understanding commonalities and differences in, for example, who may have committed a crime if we don't have a grasp of the *base rates* of what type of person committed similar crimes previously. Every stage of an investigation and every activity within each of those stages holds the potential for a misstep. Errors and omissions can damage or derail an investigation or sideline valuable evidence based on procedures that render the information inadmissible in court.

Strategies to Avoid Criminal Investigative Failures

Supervisors or agencies may have insufficient resources to adequately investigate a crime or they may simply fail to assign sufficient resources that are available. Failing to monitor and review investigative efforts at set times may result in cases simply falling off an investigator's radar screen as it is pushed out by newer and perhaps more promising cases. Occasionally detectives may suffer from burnout or distraction that leads to ignoring information or cases, thereby doing victims, the agency, and the whole community a disservice through laxity or negligence. Each of these actions or situations in myriad others may result in less than optimal productivity for investigators and agencies.

Rossmo (2009) offers the following list of strategies to help agencies and investigators avoid criminal investigative failures:

- Use case studies in training.
- Encourage an atmosphere of open inquiry, and ensure investigative managers remain impartial and neutral.
- Defer reaching conclusions as much as possible until sufficient data have been collected.
- Avoid tunnel vision. Consider different perspectives and encourage cross-fertilization of ideas.
- Follow the evidence in the data.
- Organize brainstorming sessions and seek creativity rather than consensus.
- Ensure that investigative managers are willing to accept objections, doubts, and criticisms from team members.
- Encourage investigators to express alternative, even unpopular, points of view.
- Consider using subgroups for different tasks, and facilitate parallel but independent decision-making.
- Recognize and delineate assumptions, inference chains, and points of uncertainty.
- When appropriate, obtain expert opinions and external reviews and give them proper consideration.
- Conduct routine systematic debriefings after major crime investigations.
- Encourage and facilitate research into criminal investigation failures. (pp. 351–352)

External and summative reviews and the use of expert advisers are carried out as a matter of course in various countries around the world including the United Kingdom and the

Netherlands (Cook & Tattersall, 2016; Salet & Terpstra, 2014). As in many professions and industries, this type of review can help spur law enforcement practitioners to do competent work adhering to proven practices to avoid most pitfalls. These types of reviews or even the use of internal devil's advocates or "contrarians" (Salet & Terpstra, 2014) should be selected carefully and likely be senior and respected members of the investigative unit.

Developing an Ongoing Relationship With the Media

Law enforcement and media organizations have different roles when it comes to homicide. While police departments gather information they also must work to protect its sensitive nature and avoid compromising an investigation or the privacy of victims or witnesses. Departments also know there is a need to have an ongoing relationship with the media (Hough & Tatum, 2014). Experienced investigators recognize that their interactions with the media are necessary and may provide an opportunity to gain information to further an investigation as well as to help inform the public. The media remain the major way that the general public gets information about crime and criminal justice efforts.

Media outlets of all sorts have goals of providing crime stories as well as increase viewership or readership. Unfortunately, trying to quickly get stories out ahead of competitors can lead to conflict with agencies protecting the information that may impact a case. Experienced detectives and agency heads know of journalists whose actions have put cases at risk. Similarly, law enforcement can be seen as secretive or intentionally thwarting legitimate efforts of news media reporting.

Major case investigations typically involve interactions with media representatives. The lead law enforcement agency typically holds the responsibility for providing basic information about a case to news outlets. Law enforcement agencies are not the only agencies routinely involved in cases such as homicide. A medical examiner (ME) or coroner's office, state law enforcement agency, or others may assist a local department during a high-profile or large-scale investigation. Law enforcement agencies therefore must coordinate with those other organizations throughout the course of such investigations. Department personnel are discouraged from speculating about issues in a case to anyone, let alone the press. It is also not advisable to project time frames for concluding an investigation. In much the same way that police agencies decide which resources will be devoted to cases, news agencies make similar decisions about how and to what extent various crimes will be covered. Media outlets may be applying their own version of case solvability, or at least public interest factors, to this decision about devoting resources.

Media organizations can help the public through education about the risk to the public and whether citizens can provide helpful information. News reporters may find themselves in possession of information they develop that may assist in solving a case. All in all, law enforcement agencies and media have separate goals, but also some common ground to work with.

Improving Productivity

With the research done over the years regarding the investigative function, what opportunities have arisen to boost productivity? Many disciplines have worked to include policies and practices arising from the ongoing *evidence-based research* in their respective fields

Civilian crime scene technician photographs evidence. What are the benefits of having civilian employees assist with an investigation?

Michael Scott/Alamy

"grounded in dependable scientifically derived evidence which establishes improved investigative process outcomes" (McClellan, 2008, p. 44). As part of criminal investigation research, agency practices do present opportunities to shift some functions away from investigators. Two approaches (or a combination) are civilianization and increased investigative duties performed by patrol officers.

Civilianization

Civilians in law enforcement have been valuable members of departments and contribute greatly, including in the investigative function. While some of the earliest civilian positions were those in the communications center as call-takers and dispatchers, crime scene technicians filled by highly trained but nonsworn employees have become commonplace in the field. As the result of both experience and research many agencies have experimented with or adopted practices that utilize the talents of civilian employees to perform a wide variety of tasks that aid the investigative objectives of an agency by allowing detectives to focus on investigative functions and thereby boost productivity (Liederbach, Fritsch, & Womack, 2011).

Increased Investigative Duties Performed by Patrol Officers

As for patrol officers, it has long been recognized that the majority of arrests are accomplished by patrol officers. A study by the Police Executive Research Forum (PERF) concluded that patrol officers contributed as much to burglary and robbery investigations as detectives did (Eck, 1983), though many of these cases were not pursued past two days, which underscores the low clearance rate overall of many serious crimes. While most of these arrests come in cases that are not the most complex or involved, the actions of patrol officers during the preliminary investigative stage are often critical to the eventual success of an investigation, let alone the rapport with victims, witnesses, and the overall relationship between an agency and its community. In small agencies, patrol officers may be the only personnel available to carry out most investigations. Patrol officers may work with investigators in ad hoc task force configurations or even in some agency-directed joint investigations reminiscent of the team policing of the 1970s. The generally higher level of expertise and training of detectives can aid in the effectiveness of such overall efforts but also lend to the development of the other personnel involved. A New Perspectives in Policing paper from the Executive Session on Policing and Public Safety, a joint effort of Harvard's Kennedy School of Government and the National Institute of Justice (NIJ: Braga, Flynn, Kelling, & Cole, 2011), noted the expertise of investigators as compared to patrol officers and others includes:

> interviewing skills; developing and managing informants; conducting covert surveillance, including the use of advanced surveillance technologies; identifying and locating potential witnesses and sources of intelligence; preserving and developing evidence; preparing cases for prosecution and liaising with prosecutors in the

lead-up to, and conduct of, a trial; protecting, managing and preparing witnesses for trial; sequencing of investigative steps in an inquiry, so as to optimize chances of success; maintaining knowledge of, and in some cases relationships with, criminals and criminal groups. (p. 3)

TRAINING

Training generally within law enforcement has improved tremendously over the decades and training opportunities for in-service officers in even small agencies are more plentiful with various face-to-face and online options. Regional training academies and public and private institutes offer advanced training in topics ranging from laws to interviews and interrogations, to each of the many specific crime types investigated by agencies. The Bureau of Justice Assistance, a component of the U.S. Department of Justice, sponsors the National Resource and Technical Assistance Center for Improving Law Enforcement Investigations (NRTAC). This entity has provided funding to a number of additional partners to develop and provide training and technical assistance to law enforcement and other agencies at the local, state, and tribal level. Some of the training and technical services they are able to provide include:

- Comprehensive assessments and recommendations related to agencies' existing investigative policies, practices, and resources
- On-site consultation services
- Current research on best practices for investigating specific types of crimes
- Investigative trainings
- Development of investigative training curriculum
- Peer to peer networking
- Resources to develop model policies and procedures on investigations
- Tools to help agencies better track, supervise and organize investigations
- Local partner agency convenings to improve coordination and information sharing on active investigations (NRTAC, policefoundation.org)

Kingshott, Walsh, and Meesig (2015) conducted a small survey of relatively large agencies (29 responded) examining the type and amount of training that investigators receive. While not representative of the nearly 18,000 departments at the local, county, and state levels, the results did find that the agencies responding reported using similar delivery methods and topics, as well as having similar agency needs for detectives. Of the few who responded, most (69%) said they provided training for investigators, but fewer than 40% provided investigative training for uniformed personnel. The researchers did believe that their results indicated little change in overall type and delivery mode for detective training and that more use could be made of online training. The FBI conducted a training needs assessment of police agencies during the 1980s and listed 20 topics that appeared frequently as being important to agencies (Phillips, 1988). These topics were:

- Handle personal stress,
- Maintain appropriate level of physical fitness,

- Conduct interviews/interrogations,
- Collect, maintain, and preserve evidence,
- Drive vehicle in emergency/pursuit situations,
- Promote positive public image,
- Develop sources of information,
- Fire weapons for practice/qualifications,
- Testify in criminal, civil, and administrative cases,
- Search persons, dwellings, and transportation conveyances for illegal drugs,
- Write crime/incident reports,
- Investigate conspiracy to illegally import, manufacture, distribute controlled substances,
- Protect crime scene,
- Detect, gather, record, and maintain intelligence information,
- Investigate possession with intent to distribute or sell illegal controlled substances,
- Search and document crime scene,
- Develop and maintain control of informants in drug investigations,
- Use effective supervisory philosophies and leadership styles, use undercover techniques in drug investigations,
- Handle domestic disturbances. (p. 12)

ADAPTING TO NEW TECHNOLOGY AND THE CHANGING NATURE OF CRIME

Thirty years on, we find that a number of investigative needs identified by agencies of various sizes resonate as ongoing priorities. The nature of some crimes has changed and technology has facilitated crimes based in or augmented by that technology. And even in the 1980s the study supported the need to "explore alternative training technologies" (p. 16).

Technology continues to aid investigative efforts as well as the management of investigations. In this chapter our focus on technology is largely that which comes from the use of computers and software to help organize information, conduct **case review**, and prepare and present a final case for court. There are many technologies that may be available to even small agencies to assist in the investigation of crime. Many capabilities are accessible through larger local agencies or from state and federal organizations. The power of computing and relational databases has enhanced the search capabilities and information-sharing of agencies and investigators every year since their creation. Case management software allows investigators, administrative personnel, supervisors, and prosecutors to stay up-to-date on issues and progress in cases.

Case review Set time periods to evaluate all information gathered and actions taken during an investigation.

BENEFITS OF GREATER PRODUCTIVITY

In the 1990 book *Beyond 911: A New Era for Policing*, Sparrow, Moore, and Kennedy mused about what uniformed patrol officers might do if they had time beyond that spent running from one call for service to the next. The law enforcement agencies of the United States were beginning to engage in their various versions of community policing and the authors provided a thoughtful discussion of suggested ways to improve the ways in which policing was being done. What of investigators? If productivity gains were achieved it would likely not be the result of burden-shifting to civilian employees or deferring to artificial intelligence algorithms. But if tasks were examined critically and reimagined to share functions among patrol officers, civilian employees (even volunteers), and technology, could investigators engage in potentially beneficial or innovative activities? The answer must surely be yes. Liederbach, Fritsch, and Womack (2011) spoke of detectives performing community-policing-like things including "problem solving, crime analysis and meetings with community residents" (p. 61). And consider the term *beneficial*. This need not be solely measured by case closure, as we have repeatedly asserted. If investigator and investigative unit activities improved relations with community members, this in itself would be one measure of success. And such improved rapport may result in the more concrete contributions of cooperation and information from the community to aid in solving and preventing crimes.

EXPLORE THIS

Visit the website of the Federal Law Enforcement Training Centers and navigate to the "Training at FLETC" page at https://www.fletc.gov/training-catalog?page=2. Search through the topics and look at several. How do agencies at the local and state levels access training opportunities through this federal organization?

SUMMARY

As a process, the criminal investigation must be managed. The beginning of a case comes with the discovery and reporting of a crime, and soon a patrol officer is on the scene conducting a preliminary investigation to determine whether or not a crime has actually occurred; whether there is evidence to protect and collect; and whether there are victims, witnesses, or others with whom to speak.

If the preliminary investigation shows that a crime has been committed and through the work of the initial reporting officer and others there are sufficient leads to follow, the case may be assigned to an investigator for follow-up. During the follow-up phase of the investigation the detective will gather all available information and coordinate with others, including the prosecutor, to move toward case resolution. This assigning of a case and monitoring of its progress are two of the functions of the investigative supervisor, if there is one. For an investigative unit that consists of multiple investigators or is divided into specialized investigative areas, an important role for the supervisor is to acquire resources for the detectives to do their jobs.

The work of both detectives and supervisors is very much involved in information acquisition and knowledge management. This can be aided by an organized approach to the work, procedures in place to review case progress, and in the case of many agencies the use of case management software to track information and activity in cases. Some departments continue to use a predominantly paperwork approach to compiling case files and sharing information, but this may not be efficient nor as effective with the reality that cases can generate an enormous volume of information to be managed.

Officers and investigators as humans hold the potential to make errors. Human employee errors do not automatically render a case unworkable, or the conviction of an offender unattainable, but such errors can complicate case management. The training of investigators and supervisors can go a long way to minimize errors as well as to improve case quality and positive outcomes. Periodic case review as well as external reviews of the investigative process of an agency can also prevent cases stalling out, or following the wrong direction due to tunnel vision, groupthink, or other errors not seen by those closest to the investigation. An open approach and an emphasis on working collaboratively with others in the investigative process can boost case and agency productivity. Ensuring that patrol officers are invested in the criminal investigation process and that both officers and detectives receive ongoing training in various aspects of investigations are productive areas of focus for agency administrators.

KEY TERMS

Case management software (CMS) 85
Case review 92
Cognitive bias 86
Confirmation bias 86
Framing 87
Informants 82
Law Enforcement Assistance Administration (LEAA) 78
Learning management system (LMS) 84
Organizational groupthink 86
Probability errors 86
Records Management Systems (RMS) 84
Representativeness 87
Satisficing 87
Solvability factors 80
Team policing 82

DISCUSSION QUESTIONS

1. What are solvability factors and how do they help make decisions in managing detective caseloads?
2. What steps are taken in the follow-up investigation and how does this build on the preliminary investigation?
3. How does case management software benefit the work of investigators and supervisors?
4. Agencies often do not have adequate resources to effectively address all priorities. How may resource limitations affect assignment of cases and caseloads?
5. What factors impact investigative decision-making?
6. Why is the role of uniformed officers at the initial report and scene of a crime important to a successful investigation?

SAGE edge

- Get the tools you need to sharpen your study skills. SAGE edge offers a robust online environment featuring an impressive array of free tools and resources.
- Access practice quizzes, eFlashcards, video, and multimedia at **edge.sagepub.com/houghci**

PRACTICE AND APPLY WHAT YOU'VE LEARNED

▶ edge.sagepub.com/houghci

SAGE edge™

WANT A BETTER GRADE ON YOUR NEXT TEST?

Head to the study site where you'll find:

- **eFlashcards** to strengthen your understanding of key terms.
- **Practice quizzes** to test your comprehension of key concepts.
- **Videos and multimedia content** to enhance your exploration of key topics.

Jim West/Alamy Stock Photo

5 Searches, Seizures, and Statements

RUNNING CASE: THE DETECTIVES BEGIN DOCUMENTING THE FLORIDAN CASE AND APPLYING FOR SEARCH WARRANTS

Introduction

Fourth Amendment

Search Warrants
- Search Warrant Exceptions

How It's Done: Most Common Search Warrant Exceptions
- Voluntary Consent
- Exigent Circumstances and Emergency Searches
- Stop and Frisk and Plain View
- Open Fields
- Abandoned Property
- Protective Sweeps
- The Carroll Doctrine
- The Exclusionary Rule

Making the Arrest

Fifth Amendment

Suspect Statements

Lineups and Showups

How It's Done: Live Lineup

Sixth Amendment

Explore This

Summary

Key Terms

Discussion Questions

LEARNING OBJECTIVES

5.1 Discuss key aspects of the Fourth Amendment that are relevant to criminal investigation.

5.2 Explain the search warrant requirements.

5.3 Describe the conditions that must be met before an officer can make an arrest.

5.4 Discuss key aspects of the Fifth Amendment that are relevant to criminal investigation.

5.5 Explain the issues related to suspect statements relevant to criminal investigation.

5.6 Critique methods of conducting lineups.

5.7 Discuss key aspects of the Sixth Amendment that are relevant to criminal investigation.

Running Case: The Detectives Begin Documenting the Floridan Case and Applying for Search Warrants

Following the investigation, Detectives Macon and Ashley walked back to their unit. In the hallway, Lieutenant Kimberly Kellan, the Criminal Investigations Division (CID) Commander, and Corbin Kelsea, Assistant State Attorney (ASA), a prosecutor for the state attorney's office in their jurisdiction, came up behind them.

"I'll be around your office later, if you need to talk about any aspects of the case," Corbin Kelsea said.

Bradley nodded. Kelsea had been prosecuting cases a long time. Bradley would check in with him later to make sure he had the necessary warrants to search the suspect's car.

Detectives Macon and Ashley walked into the empty interview room to go over the facts of the case and plan their next moves. While they spoke, Richard Ashley entered the data into the computerized case management system (CMS). Every officer and technician entered their initial or supplemental reports into this database. In addition to prompting the lead investigator to complete all tasks, the system also provided a timer function to route the case to the appropriate supervisor or manager to review case progress at set intervals.

"Frank probably didn't have time to return to his apartment after the incident," Richard said, typing away.

"From what Officers Jayden and Thompson said, he just sat in his car a few blocks from the scene," Bradley said.

"A gun was found in the nearby dumpster, but we need the ballistics test to determine it was the weapon used in the murder. That will take a few days to a few weeks, depending on where the nearest lab is and how backlogged they are," Ashley said. The computer keyboard clicked under his fingers.

"There could be other evidence inside the car to tie the suspect to the scene or to the victim," Bradley added.

"Go ahead and put together the search warrant affidavit to take to Corbin Kelsea's office to review," Richard told Bradley.

Bradley knew he would need to make sure that enough information was present in the search warrant application to show *probable cause* for the judge to issue the search warrant. Even though Frank's car had been taken to the police impound lot, the warrant application would include where the car was initially found, the fact that Frank had been seated in it, and what specific items the detectives believed might be found in the car. Bradley knew that the vehicle could be searched even without a warrant as part of doing an *inventory* before lawfully towing it from the public street, but he also knew that it would be helpful to have the prosecutor, and then the judge, review and grant the search warrant because it would be less likely to be challenged later in court than simply performing a search by inventory.

"I'll review everything that has been done in the investigation up to this point," Richard said. "I'll ensure the suspect's statement was fully recorded and was properly taken with consideration to his Fifth Amendment rights."

"And Frank confessed," Bradley Macon said, "but his statements will need to be corroborated, if possible. I can conduct a lineup with the hopes that our witnesses,

Bubba Paul and Miley Denis from the Floridan, will be able to identify Frank as the man they saw at the bar last night."

"Good thinking," Richard said. "I'll contact Officers Thompson and Jayden to make sure they completed their reports and that those reports completely described all they learned and every action they took, including documenting the chain of custody for the gun they collected from the dumpster."

Bradley nodded. "Lieutenant Kimberly Kellan will appreciate having all that information at her fingertips when the media come calling."

Richard agreed. "And they always come calling."

After you have read the case study, consider how the requirements for a search warrant can strengthen a criminal case. Also, why is it important that an officer read a suspect his rights and have him sign a waiver if he wants to speak?

Introduction

Criminal investigation is founded in the law. Law enforcement officers are challenged to uncover and gather information and other evidence about the commission of a crime and who is responsible while obeying the law and legal guidelines about how that information and evidence must be gathered to remain admissible in court. Working with prosecutors through the criminal investigation process is often helpful and sometimes crucial to the successful resolution of a case. The United States and most countries are concerned with both crime control and the rights of the accused, which can sometimes be in conflict. How evidence and information are to be gathered and used in determining the guilt or innocence of an accused person follows laws that evolve to account for changes in society to include new technology, new illicit drugs, and new crimes.

Law enforcement officers must constantly be aware of the broad guidelines that have arisen out of the U.S. Constitution and the Bill of Rights as well as specific constraints on their methods of gathering information that are formed as the result of case law that interprets what is and is not permissible. These rights include ones that bear directly on protections afforded to those accused of a crime. The Fourth, Fifth, and Sixth Amendments to the Constitution are those most frequently considered in a discussion of due process and the rights of the accused during criminal investigations. The 14th Amendment extended the applicability of many rights afforded to individuals charged by federal authorities to those charged by state and local law enforcement and prosecutors. This broader application of rights to those accused resulted in a more consistent and systematic understanding of rights across the country.

Fourth Amendment

The **Fourth Amendment** to the U.S. Constitution provides "[t]he right of the people to be secure in their persons, houses, papers, and effects, against unreasonable searches and

Fourth Amendment
To the U.S. Constitution: prohibits unreasonable searches and seizures.

Probable cause The level of information needed for a court to issue a warrant, for an officer to conduct certain searches or arrests, and the standard for a grand jury indictment.

seizures, shall not be violated, and no Warrants shall issue, but upon **probable cause**, supported by Oath or affirmation, and particularly describing the place to be searched, and the persons or things to be seized."

Investigators seeking evidence and information must be aware the main intention of this amendment is about privacy and therefore protecting people from unreasonable searches and seizures. It is notable that the Fourth Amendment does not protect citizens from *all* seizures and searches, but just those which are deemed by the court to be *unreasonable*. It has generally been the case that searching the home or private property of a citizen is on more solid ground through the use of a **search warrant** approved by the court as opposed to *consent* or other mechanisms that provide exceptions to the search warrant requirement. An officer may search locations, vehicles, and people when the search is conducted in a reasonable manner and is based upon *probable cause* or one of the established search warrant exceptions.

Search warrant Court order authorizing law enforcement to search a particular person, place, or vehicle for evidence of a specific crime.

The concept of *probable cause* is the foundation for official authority to search places, seize evidence or things that may be unlawful, and physically arrest a person, which is also considered a seizure within the meaning of the Fourth Amendment. To satisfy the requirement of probable cause an officer must possess sufficient information that would warrant a *reasonable person* to believe that a crime has been or is being committed by someone. The Fourth Amendment also gives rise to the idea of **reasonable suspicion**. This refers to information or conditions leading an officer to suspect that a crime is being committed or is about to be committed, or that has already been committed. The contrast between probable cause and reasonable suspicion turns on the ideas of *believe* or *suspect*. Both of these concepts are based on an approach known as the *totality of the circumstances*. Officers may rely on all the information and evidence they have legally obtained to establish their suspicion or belief about the commission of a crime.

Reasonable suspicion Lower standard of proof than probable cause based on specific and articulable facts.

Search Warrants

To seek a search warrant, an investigator submits a statement of probable cause, known as the *application,* or affidavit, to a judge where he articulates all of the facts known to him at the time that form his belief about a crime and why the place or person to be searched may have evidence of the crime. If the judge determines that sufficient grounds exist to establish probable cause, and the officer swears under oath that the contents of his statement are accurate, the court issues a *signed search warrant*. The warrant must, in keeping with the Fourth Amendment, *specifically* describe the location, vehicle, or person to be searched, and the search must be narrow in its scope. In other words, search warrants are not provided by the court that allow officers to search large areas or places to find and then use anything that may indicate someone has committed some unspecified criminal act. The majority of search warrants require officers to *knock and announce* their presence and intention to conduct a search (*Wilson v. Arkansas*, 1995). If investigators present compelling information to the court of a significant physical danger to the officers or the belief that a suspect may destroy evidence, the judge may specify the warrant as "no knock," thus allowing officers to conduct a surprise entry of the location. Once the search warrant has been served the investigator will inventory and list all items seized on what is called a *search warrant return* and file this document with the court. A *seizure* is when the officer or official removes an item pursuant to the search that is believed to be relevant to the investigation or unlawful.

93 (Rev. 01/09) Search and Seizure Warrant

UNITED STATES DISTRICT COURT
for the

Eastern District of Pennsylvania

In the Matter of the Search of)
(Briefly describe the property to be searched)
or identify the person by name and address)) Case No.
THE CONTENTS OF THE SERVER ASSIGNED IP ADDRESS)
207.106.6.25 MAINTAINED BY JTAN.COM)
)
)

SEARCH AND SEIZURE WARRANT

To: Any authorized law enforcement officer

An application by a federal law enforcement officer or an attorney for the government requests the search of the following person or property located in the _____ Eastern _____ District of _____ Pennsylvania _____
(identify the person or describe the property to be searched and give its location):
THE CONTENTS OF THE SERVER ASSIGNED IP ADDRESS 207.106.6.25 MAINTAINED BY JTAN.COM

The person or property to be searched, described above, is believed to conceal *(identify the person or describe the property to be seized)*:

SEE ATTACHED RIDER.

I find that the affidavit(s), or any recorded testimony, establish probable cause to search and seize the person or property.

YOU ARE COMMANDED to execute this warrant on or before _____
                                                                                                                      (not to exceed 10 days)

☑ in the daytime 6:00 a.m. to 10 p.m.   ☐ at any time in the day or night as I find reasonable cause has been established.

Unless delayed notice is authorized below, you must give a copy of the warrant and a receipt for the property taken to the person from whom, or from whose premises, the property was taken, or leave the copy and receipt at the place where the property was taken.

The officer executing this warrant, or an officer present during the execution of the warrant, must prepare an inventory as required by law and promptly return this warrant and inventory to United States Magistrate Judge ███████
███████.

☐ I find that immediate notification may have an adverse result listed in 18 U.S.C. § 2705 (except for delay of trial), and authorize the officer executing this warrant to delay notice to the person who, or whose property, will be searched or seized *(check the appropriate box)* ☐ for ___ days *(not to exceed 30)*.
                                              ☐ until, the facts justifying, the later specific date of _____.

Date and time issued: _____   _____
                                                                                                                           Judge's signature

City and state: _____ Philadelphia, PA _____   ███████████████, U.S. Magistrate Judge
                                                                                                                         Printed name and title

Search warrant.

United States District Court, Eastern District of Pennsylvania

The use of a search warrant by investigators has three separate components: the application where the officer articulates in writing for the court the probable cause that the place or person to be searched contains evidence of crime; the signed search warrant, issued by the court following review of the application; and the search warrant return, an inventory of items seized and the warrant supporting documentation filed with the relevant clerk of court.

Search Warrant Exceptions

While it is always advisable to obtain a search warrant, it is not always practical to do so, and the law recognizes a number of exceptions to the search warrant requirement for searches and seizures. Officers must always remember that when one of the exceptions to the search warrant are used that the burden is on the officer to articulate that the conditions for that particular exception existed. After-the-fact review by the court of evidence obtained through a warrant exception might result in that evidence not being admitted if proper procedures were not followed or the court determines that the exception does not apply.

How It's Done

MOST COMMON SEARCH WARRANT EXCEPTIONS

Consent. Topping the list of most common exceptions to the requirement of a search warrant is the consent someone grants for a law enforcement officer to search his or her person or his or her property. Officers do not need to have probable cause or even reasonable suspicion to simply ask someone if he or she would consent to the search. The officer does not need to inform the person that he or she does not have to grant permission, but the officer must be able to articulate that the consent was clear to withstand a later court challenge. It is important to note that citizens can subsequently withdraw their consent for a search.

Stop and Frisk. Even absent an arrest, an officer encounters a person and has a reasonable suspicion that the person is engaging in criminal activity. A straightforward example is someone who runs from police after they are approached in a known narcotics-dealing area.

Exigent Circumstances. Immediate police action may reasonably stop a suspect from destroying evidence or posing a continuing harm to someone.

Hot Pursuit. As with exigent circumstances generally, a suspect's flight on foot or by vehicle may have the intention of hiding or destroying evidence, or gaining an advantage to assault the officer or someone else to where the suspect is fleeing.

Incident to Arrest. This exception to a warrant allows arresting officers to search for weapons as well as evidence the person may destroy. If the person is arrested in or near a car, for example, police may search the vehicle. Likewise, if the person is arrested in their residence, the immediate area of their arrest may be searched.

Plain View. Whether serving a warrant or otherwise present conducting a lawful duty, officers sometimes observe or take note of the evidence or proceeds of crime unconcealed. Any such item in *plain view* is subject to seizure or confiscation.

Vehicles. The courts have established that people generally have less of an expectation of privacy in their vehicles than they do in their homes. Vehicles encountered by officers may present dangers that require a search to mitigate, and they are mobile and may be driven or taken away from the investigating officer's jurisdiction before a search warrant might be obtained, which also necessitates searching without the benefit of a warrant.

One of the ways that a search may be conducted without a warrant is following the lawful arrest of a person. The fact that probable cause existed to make an arrest allows the arresting officer to search the arrestee to ensure that he does not have any weapons he might use against the officer or evidence on his person that he may try to discard or destroy. These two considerations, safety and evidence, were recognized by the U.S. Supreme Court in *Chimel v. California* (1969), which established the scope of a search incident to arrest. The Supreme Court determined that the area within the immediate control or arms reach of an arrestee was the area that a person might hide evidence or reach for a weapon. Given that the human officer may fail to discover a weapon or evidence after arresting an individual, it is common practice for any other officer taking custody of the arrestee to perform another search.

VOLUNTARY CONSENT

An individual may also grant consent to an officer to perform a search. This **voluntary consent** may be to search the individual, her vehicle, a home, or other property. Much like a consensual encounter where someone agrees to speak with an officer, neither reasonable suspicion nor probable cause has to be present for a search based on consent. Once again, the burden will be on the officer to show that the consent was knowing and voluntary if the individual later asserts that she did not consent to a search. This may be shown through a signed consent form witnessed by a second officer or other responsible adult, or through the testimony of others who were present when the consent was given by the subject, or even through an audio- or video-recorded interview with the subject when she grants consent for the search. Obviously, establishing the knowing and voluntary nature of the consent will be critical to whether evidence obtained will be admissible in court.

Voluntary consent Permission by the person with proper authority to control property to allow the search of that property.

Challenges have been made to consent when given by a juvenile, by someone under the influence of drugs or alcohol, by someone found mentally or developmentally impaired, or by someone who may not have the authority to grant consent for the area searched (such as a third party who is an occupant of a residence). And while deception in some forms and in limited ways may be utilized by law enforcement officers in the course of an investigation, obtaining consent for a search through deception or certainly coercion may render the search and any evidence found inadmissible. Officers also may not claim to have lawful authority to search without consent or to threaten or intimidate someone into granting consent. And, as previously noted, a person may withdraw consent at any time. An investigator may not be able to get the evidence she seeks, or she may be able to continue through a search warrant.

EXIGENT CIRCUMSTANCES AND EMERGENCY SEARCHES

At times a warrantless search may occur because officers perceive that an emergency exists, known as **exigent circumstances**. The need for officers to act immediately may be caused by a threat to others or the risk of the loss of evidence. Officers will have to show that the emergency circumstances were sufficient to prevent obtaining a search warrant before evidence might be destroyed or the situation might deteriorate. Hot pursuit of a fleeing suspect is an example where exigency may be shown as well as preventing someone's escape. Threats to the public or individuals may create a situation where an officer perceives danger if he does not immediately act to search for a potential weapon. Different from the exigency of searching for evidence of specific criminal activity is an emergency search by public safety employees at a location to deal with an emergency (perhaps a fire or medical crisis) that placed an officer in a position to observe evidence of criminal activity, thus rendering the situation the

Exigent circumstances May justify law enforcement entering a building without permission if people are in danger, evidence is at risk of imminent destruction, or if a suspect may escape.

Stop and frisk Temporarily detain a person and pat-down their outer clothing for the presence of weapons.

Plain feel exception If, during a pat-down, an officer feels something that he can immediately identify such as a weapon, the officer may confiscate or seize the object.

Plain view If an officer is in a location he may legally be in, and sees something he recognizes as evidence, he may seize it.

same as the *plain view exception* to search warrant requirements. Once evidence is observed, it may be advisable to secure the location after the emergency is dealt with and obtain a search warrant to proceed with an investigation on what was discovered during the emergency presence of police.

STOP AND FRISK AND PLAIN VIEW

Another search warrant exception that focuses on the potential for weapons being present is the **stop and frisk**. The "stop" and the "frisk" are separate actions and must be justified separately. A lawful stop does not automatically imply the authority to frisk someone. A frequent circumstance that leads to a stop is an officer investigating a call of suspicious circumstances or the officer's own initiative discovering a situation that rouses suspicion. This gives rise to what is known as an *investigative detention*, which is a brief period of time to allow an officer to investigate further what he has encountered. A vehicle or person that appears to possibly match a description broadcast as wanted may be the impetus for a stop as well. A pat-down or cursory search of someone to discover weapons was recognized by the Supreme Court in *Terry v. Ohio* (1968) as important to help ensure officer safety and is therefore considered an exception to the Fourth Amendment prohibition against an unreasonable search. This does not mean that anyone an officer encounters is subject to such a frisk or pat-down search, but rather when an officer considers the totality of the circumstances whether he has reasonable suspicion that a subject may be armed after the officer initiates the stop. The subject's verbal and nonverbal behavior, the location time and circumstances in which the person was found, and perhaps previous knowledge of the subject may all factor into the officer's determination to frisk the individual. The officer does not have to know that a person is armed, but he must be able to articulate why he feared that the person might be carrying some type of weapon. The frisk is limited to a person's outer clothing not including their pockets. If the officer feels what he believes based on his training and experience may be a weapon or perhaps drugs, then he may search the pocket or area of clothing. This concept of the stop and frisk is referred to as the **plain feel exception** (*Minnesota v. Dickerson*, 1993) and the *plain smell* when, for example, an officer smells marijuana emanating from a stopped car.

What circumstances would need to be present to make this officer's stop and frisk justified?

Hill Street Studios/ DigitalVision/Getty Images

Similar to the stop and frisk, an officer who is lawfully in a place and sees an illegal item in **plain view** they may seize it and use it as evidence. These types of circumstances can arise with some frequency from traffic stops where an officer observes weapons or drugs on the seat or console of a car, or at a home where the officer has been called to interact with the people there and sees illegal drugs or other items of contraband plainly visible as they speak to the subject in the residence. The item must be immediately recognizable and not require further investigation to determine what it is, as this would expand the scope of an officer's actions to an actual search. Seizing items in plain view can also occur while officers are serving a search warrant and trying to locate different items of evidence. Again, the key is whether the officers are legally in a place where they may observe the illegal items or evidence. The use of technology to enhance observation or inspection of an area is considered a search and will typically require probable cause and a search warrant, as with other electronic surveillance methods.

OPEN FIELDS

Another exception that allows for a warrantless search is *open fields*. The basis for this exception is that there is no expectation of privacy in an open field, including one where crops are grown. This is distinguished from an area, including buildings, attached to where a person lives such as structures in someone's yard including sheds, barns, detached garages, and the like. The Supreme Court case that addressed this concept was *Oliver v. United States,* 466 U.S. 170 (1984), which involved law enforcement searching a field located more than a mile from where Oliver lived. The field was planted with marijuana, and the Court declared that Oliver's reasonable expectation of privacy did not extend to an open field. Buildings, other structures, and the area around a home or yard, referred to as *curtilage*, are considered private and therefore officers need to obtain a search warrant for such places unless one of the exceptions apply. Areas considered curtilage by the courts are determined in part by how close they are to the home of the owner, what the use of the area is, whether it is shielded from public view, or other factors that would indicate exclusive use of the area by the owner and the owner's intention to keep it private. While precisely defining an open field can be challenging at times, the protection of one's home afforded by the Fourth Amendment does not extend to the open field.

ABANDONED PROPERTY

A search may also take place on *abandoned property*, which may be the home or vehicle, a temporary living space such as a motel, or personal property. The courts evaluate whether property is abandoned based on the totality of the circumstances. Another circumstance that does not require a search warrant is the inventory of a car that is being impounded. This is an administrative procedure intended to document the contents of a vehicle so that later claims of damage or theft are not successfully lodged against an agency or officers. The inventory provides accountability ensuring protection of someone's property while also protecting the agency and possibly the public from anything dangerous located within the vehicle. The law enforcement agency should have in place a policy that establishes the use of inventories and how they are conducted. If these are not adhered to, the court may find that the inventory was pretense to a search.

PROTECTIVE SWEEPS

Some wonder at the legality of roadblocks or checkpoints established to check drivers for sobriety or some other reason. Sobriety checkpoints have been in use for decades in the United States and have been challenged as tantamount to a search without probable cause. The U.S. Supreme Court in *Michigan Department of State Police v. Sitz* (1990) held that such stops are seizures under the Fourth Amendment but that the brief nature of the stops made them acceptable as limited intrusions on citizens given the greater benefit of protecting the public from the damages caused by impaired drivers. Border roadblock and checkpoint searches were determined by the Court to be a limited intrusion in the national interest and therefore constitutional (*United States v. Martinez-Fuerte*, 1976). Other limited search actions have been determined to fall outside the Fourth Amendment's protection, such as *protective sweeps* performed by police when securing a person or place to ensure no other immediate threats are present (*Maryland v. Buie*, 1990), searches in schools, searches of parolees or probationers, and searches inside jails and prisons.

THE CARROLL DOCTRINE

The Court also noted a distinction between structures and vehicles for determining the requirement to obtain a warrant before searching. Known as the *Carroll Doctrine* from the case *Carroll v. United States* (1925), the warrantless search of a vehicle rested upon the logic that the vehicle could be driven away, likely out of the jurisdiction, while officers were trying to obtain a search warrant. The Supreme Court specified that probable cause must still exist and that evidence is likely to be found in the vehicle, that exigent circumstances exist, and that there would be insufficient time to obtain a warrant before the vehicle left. Another issue that courts have addressed for Fourth Amendment implications involves searches by drug dogs. Courts have generally held that the use of drug-sniffing dogs does not violate a defendant's constitutional rights so long as the officer has a legal basis for the stop (*Illinois v. Caballes*, 2005).

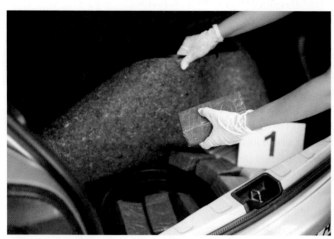

What are some common exceptions that would eliminate the need for a search warrant to legally obtain this evidence?

Mihajlo Maricic/EyeEm/Getty Images

Exclusionary rule Prohibits the use at trial of evidence that was not obtained legally.

Fifth Amendment To the U.S. Constitution: enumerates certain rights such as the right against self-incrimination, the right to due process, and the protection against being tried twice for the same crime.

Good faith exception When faced with the potential blocking of evidence due to the Exclusionary Rule, if officers relied on a search warrant they believed to be valid, evidence seized illegally may be admitted by the court.

The Exclusionary Rule

If during the course of a search and seizure, a stop and frisk, an arrest, or an interrogation of a suspect, evidence is obtained unreasonably and in violation of the Fourth Amendment, then the courts will exclude it from consideration. This **exclusionary rule** was established at the federal level more than 100 years ago in the case of *Weeks v. United States* (1914) and was later applied to the states in the case of *Mapp v. Ohio* (1961). These and subsequent Court decisions reinforce the protections of people and their homes provided by the Fourth Amendment as well as the **Fifth Amendment** protections against self-incrimination. An extension of the exclusionary rule is what is known as the *fruit of the poisonous tree*. This concept means that in addition to the initial illegally obtained evidence or information being excluded, any additional evidence flowing from the original evidence will also be excluded. Both the exclusionary rule and the fruit of the poisonous tree doctrines are considered to be limits on police power. Knowledge of the consequences of an unlawful search is also important to the crime scene technician who may need to verify with the detective whether the scene has been held so long as to now need access with a search warrant. This is another example of the importance of communicating with the office of the prosecutor.

The exclusionary rule also has exceptions that may make some evidence admissible even if the search and seizure was unlawful for one reason or another. An exception may be granted if the court believes that the error made by police was an unintentional or honest mistake. This so-called **good faith exception** is the one that most often allows evidence to be accepted following an error or when police did not realize they violated a Fourth Amendment protection (*United States v. Leon*, 1984). Another exception granted in some cases is called *inevitable discovery* and means that the court rules that it is likely that law enforcement would have legally discovered the evidence that they uncovered in a technically illegal search (*Nix v. Williams*, 1984). A somewhat related concept is the basis for the independent source exception. If evidence was obtained in a manner ruled inadmissible, the same information arrived at by means not involving the illegal act may allow the information to be admitted. Another exception to the exclusionary rule balances the illegal actions of law enforcement officers against voluntary behavior of a suspect after the originally inappropriate act by police, which is then considered to have acted to achieve a purged taint of the police actions.

The search warrant itself must follow requirements and guidelines that have been established over time. The document must be authorized by a competent court with jurisdiction. The warrant must particularly describe the items sought and authorize only the search for those objects. The court ensures that the warrant is issued based on sufficient probable cause presented by the investigating officer. The warrant must list the name of the officer who has made the request along with the reasons for the request. It will include the place to be searched and the items to be seized. And, of course, the issuing judge must sign the warrant.

Making the Arrest

If a person has been made aware that he or she is not free to leave the presence of an officer conducting a stop, search, or inquiry, he or she may be considered detained or *arrested*, a seizure under the Fourth Amendment. And, as all actions under the Fourth Amendment, the officer must have probable cause to make the arrest. The officer may indicate the arrest by telling the person he or she is under arrest or by handcuffing or restraining someone. A person may attempt to walk or run away, defeating the arrest, but being guilty of *resisting* arrest. When a person is arrested, *Chimel* provides the scope of the area around or within arm's reach of the person that may be *incident to arrest*. In *New York v. Belton* (1981) this expanded to include searching a car if the person arrested had been in it. This ability to search a person's vehicle incident to their arrest included the passenger compartment as well (*Thornton v. United States*, 2004).

To make an arrest the officer must again possess sufficient information that would warrant a *reasonable person* to believe that a crime has been or is being committed by someone. The determination of probable cause may be when an officer is confronted with a situation she believes is a violation of law based on her own observations or information she has gathered; the determination may be in advance, as when a court issues an arrest warrant for an individual based on facts presented by law enforcement (and usually required to be served during daylight hours), and certainly the courts may make an after-the-fact determination. Retrospectively, an officer may have believed that he had probable cause to make an arrest and the court determines that he did not. In this event it is likely that any evidence obtained as a result of the faulty arrest will be inadmissible.

Completing an actual physical arrest itself is directed by agency policy and procedures and is also subject to scrutiny by the courts in subsequent actions where a plaintiff may question the arrest and the manner it was carried out. Courts will examine whether officers acted in an objectively reasonable manner in initially making an arrest, and then, in what became a two-pronged test, whether the use of force employed by officers was objectively reasonable (*Saucier v. Katz*, 2001). The first prong of the test established by *Saucier* is whether "the facts alleged show the officer's conduct violated a constitutional right," and if so the second prong is to determine whether the constitutional right was "clearly established" (*Saucier v. Katz*, 2001). Determining such arrest issues in a civil proceeding subsequent to the criminal case from which it arose often seeks the court's determination of whether the officers are entitled to *qualified immunity*—that is, they acted reasonably based on what they knew at the time.

Fifth Amendment

The *Fifth Amendment* to the Constitution reads:

> No person shall be held to answer for a capital, or otherwise infamous crime, unless on a presentment or indictment of a grand jury, except in cases arising in the land or naval forces, or in the militia, when in actual service in time of war or public danger; nor shall any person be subject for the same offense to be twice put in jeopardy of life or limb; nor shall be compelled in any criminal case to be a witness against himself, nor be deprived of life, liberty, or property, without due process of law; nor shall private property be taken for public use, without just compensation.

A number of the rights of citizens implicated in cases brought by the government are derived from the Fifth Amendment. The Legal Information Institute (LII) of the Cornell Law School lists the individual rights this way:

> 1) right to indictment by the grand jury before any criminal charges for felonious crimes, 2) a prohibition on double jeopardy, 3) a right against forced self-incrimination, 4) a guarantee that all criminal defendants have a fair trial, and 5) a guarantee that government cannot seize private property without making a due compensation at the market value of the property. (Wex free legal dictionary and encyclopedia)

The main Fifth Amendment interest during a criminal investigation is the right against self-incrimination in a custodial interrogation. We turn first to the consideration of statements made by the suspect.

Suspect Statements

Most people are familiar with the Fifth Amendment right against self-incrimination and how officers are generally required to read these rights to a suspect in custody before questioning him. Earlier in the book we examined issues of technique during the interrogation process. In this chapter we take a look at the case law guiding the conduct of officers talking with suspects. Once someone is considered by police to be a suspect of a specific crime, the ways that officers communicate with the individual change. In an initial part of the investigative process officers may seek information from everyone involved without having a complete sense of how each person is involved. This information can range from name, telephone number, and address to work information and additional background that a person may voluntarily provide. When a person is considered a suspect of a specific crime an officer may conduct an interrogation that is a conversation with the intention of eliciting admissions or a confession from the suspected individual. If the setting is considered *custodial*—that is, the person believes he is not free to leave—then the interrogating officer must make the suspect aware of his rights before he is questioned. Casually talking with someone in a public place would not necessarily be considered custodial, and so the conversation would not require a warning against self-incrimination. If several officers approached a person and stood around him in a manner that made the individual feel he could not simply walk away from the encounter, a custodial setting has likely been created.

An officer may speak with and even question someone who may be a suspect without advising the individual of their Fifth Amendment rights, as long as the person is not in custody and understands that they are free to leave at any time. Typically, however, when a person becomes a suspect, officers will read the rights or have the person sign a form acknowledging their rights so that statements will not be later ruled inadmissible. The courts look carefully at the issue of *voluntariness* of a statement or confession.

The 1966 landmark Supreme Court decision in *Miranda v. Arizona* established the Fifth Amendment's protection against self-incrimination for anyone accused of a crime. The case additionally addressed the provision of the **Sixth Amendment** that "the accused shall . . . have the assistance of counsel." Most agencies quickly adjusted their procedures and training to reflect that officers who were conducting custodial interrogations must inform the person of their right against self-incrimination. Department forms were developed that allowed officers to have suspects also read and sign an acknowledgment of their right. This admonition should be read to the individual and, if a form is used, it should be signed by a witness, ideally not just the officer conducting the interview or interrogation. Once again, numerous exceptions have developed to account for certain circumstances regarding this area of criminal law. If an officer arrives on the scene of an incident, for example, and asks the onlookers, "What happened here?" and one person blurts out, "I did it, I shot him," that **excited utterance** may be used against the person. The officer was not intentionally or specifically focusing on someone as a suspect; rather, he was trying to understand the circumstances of what he was trying to deal with. A suspect may also make statements in trying to establish an alibi that can be countered if they are inconsistent with statements made by the person but not previously admissible because law enforcement officers failed to advise her of her rights. Essentially, you cannot have your earlier statements suppressed and then make up a new story as if you had not said something previously.

The Supreme Court in its holdings in *Miranda* established the following safeguards for the Fifth Amendment: "the person in custody must, prior to interrogation, be clearly informed that he has the right to remain silent, and that anything he says will be used against him in court; he must be clearly informed that he has the right to consult with a lawyer and to have the lawyer with him during interrogation, and that, if he is indigent, a lawyer will be appointed to represent him." (pp. 467–473). If after a suspect is advised of his rights he simply does not speak, or waits an inordinate amount of time to begin speaking, investigators can still ask questions of him, effectively conducting an interrogation before he waives his rights (Morris, 2012).

Even after the admonishment of rights a conversation may begin between investigators and the suspect, and it may yield important information about the circumstances of the case. An individual may provide officers with information that establishes an alibi or otherwise guides them to a more productive avenue of investigation or more viable suspect. The suspect may also engage in the conversation to try and find out what the police know, to appear cooperative, or simply to mislead investigators by lying in a variety of specific ways

Sixth Amendment
To the U.S. Constitution: guarantees the right of a speedy trial, an attorney to represent the accused, the ability to face one's accusers, and the selection of a jury of one's peers, among other protections.

Excited utterance
A spontaneous comment or statement by someone as a reaction to a startling or shocking circumstance. Usually an exception to the hearsay rule.

(Strömwall & Willén, 2011). The discussion between the suspect and investigator is guided by legal constraints and various techniques developed through experience and training. One technique that is misunderstood by many people is the use of deception by officers during the interrogation (Slobogin, 2017). A variant of this method is the false-evidence ploy in which officers tell the suspect that police have evidence or a witness who saw the suspect do something incriminating. In fact, the police do not have the evidence or witness, but they hope to convince the suspect and get him to admit to certain facts or confess outright. In a series of cases, the U.S. Supreme Court has ruled that the use of deception does not taint an otherwise voluntary confession (*Frazier v. Cupp*, 1969). While jurors will judge the value of a confession, there has been some research to suggest higher rates of conviction even after an individual claims his confession was false and only offered due to law enforcement deception (Forrest & Woody, 2010). Past law enforcement behavior in some cases also led to the Supreme Court ruling in *Brown v. Mississippi* (1936) that precludes the use of a coerced or involuntary confession. Jurors are not expert in the dynamics or methods of police interview and interrogation, and in trial experts may be called by either side in a case to help the jury understand the science and research around police–suspect communications.

Lineups and Showups

Lineup A presentation of the suspect and filler individuals all at once or one after the other for a witness or victim to identify the offender.

Showup Viewing a suspect by a victim or witness within minutes of a crime. Also known as a field identification.

Although eyewitness identification has been relied upon around the world and for most of recorded history, in an era of scientific verification and comprehensive gathering of evidence and information, we are aware that such testimony by victims and witnesses can be faulty or completely mistaken. One area of a criminal investigation that is susceptible to procedural errors or inadvertent or sometimes intentional influence by investigators or misrepresentation by prosecutors is that of **lineups** and **showups**. As covered earlier in the book, human observation and perception may suffer from a number of deficiencies. A person may not have been in a position to completely view a suspect, or the lighting conditions may have compromised the witness's ability to clearly see a perpetrator. A victim or witness may also be influenced by another witness with a stronger though mistaken opinion of a suspect's description, and he or she may be influenced by subtle or not-so-subtle comments or behaviors of an investigator when shown a physical lineup containing a suspect or when looking at a photo array or lineup containing a potential suspect.

While the lineup is familiar to most people, a showup is when the victim or witness is taken immediately following a crime to where a suspect may have been stopped. While having a witness view someone encountered or detained close to the scene of a crime and within a short period after it was committed may fit within our understanding of an exigent circumstance, the courts will examine closely whether the totality of the circumstances may make an identification under the circumstances too suggestible to accept as reliable. For example, seeing a person either handcuffed or in the back of a squad car may create a strong impression in the mind of the witness that law enforcement

Suspect lineups can be live (like this one) or use photographs. What are the pros and cons of each?

Jason V./E+/Getty Images

officers would not have this person if he were not guilty. Even if time made it impractical to arrange a live lineup containing a suspect, the court may be skeptical about accepting an identification in the field of only one individual. Officers can minimize the suggestibility of the showup encounter by ensuring that only one witness views the individual at a time and, if possible, the suspect is not handcuffed or in the back of a patrol car. As with lineups, officers conducting a showup should keep their comments neutral and take great care not to say or indicate anything suggestive about the suspect that may influence a victim or witness to make an identification of a suspect. Consciously or unconsciously, the witness may want to "please" the officer; he or she also might assume he or she would not be asked to view somebody if he or she were not the guilty party.

Lineup accuracy might be considered as a positive identification of a suspect by a witness (*hit*), which is subsequently verified by other evidence to provide a level of confidence in the identification. The hit can be contrasted with a false identification in which one of the nonsuspect individuals (*fillers*) is identified erroneously by a witness. And of course there are many lineups where the witness is unable to positively identify anyone. Various efforts to reduce suggestibility in conducting lineups have been more widely accepted with academic researchers (Carlson, Carlson, Weatherford, Tucker, & Bednarz, 2016; Mu, Chung, & Reed, 2017; Wixted et al., 2016) and professional groups such as the Police Executive Research Forum (PERF) and the International Association of Chiefs of Police (IACP); with PERF and the IACP calling for consistent use of guidelines. Scientific advances such as DNA analysis have helped the criminal justice system and society recognize the not insignificant number of wrongful convictions, sometimes based on flawed and inaccurate identification of a person as having committed a crime (Innocence Project, 2016). The methods law enforcement use to conduct lineups should matter as much as conducting an unreasonable search (Oliver, 2016). Protocols that include unbiased and consistently delivered pre-lineup instructions and the use of double-blind lineup administration can reduce the number of false hits (Wells, Steblay, & Dysart, 2012, 2015).

Recent Florida legislation is an example of a state enacting procedures through law to reduce incorrect or improper identifications through potentially faulty lineup methods. The "Eyewitness Identification Reform Act" requires officers in the Sunshine State to conduct blind lineups in which the officer conducting either a live or photo lineup does not know which person in the lineup is the suspect. A portion of the law's language addresses using alternative lineup methods if an investigator is not available who does not know the suspect's identity, noting the intent is that the alternative "achieves neutral administration and prevents the lineup administrator from knowing which photograph is being presented to the eyewitness." Possible alternate methods include photos placed in randomly numbered folders that the person conducting the lineup does not see, or using a computer to present photos to witnesses without the lineup administrator intervening. The mandating of such procedures by state legislatures may be seen more across the country. An article in the *Orlando Sentinel* points out that according to the executive director of the Innocence Project of Florida, 9 of the 14 DNA-based exonerations of people in the state were cases that had misidentification of a suspect by an eyewitness (Doornbos, 2017). Florida's certifying body for officers, the Criminal Justice Standards and Training Commission, has been instructed to assemble training materials for agencies on how to conduct lineups based on the new law that went into effect October 1, 2017.

A main theme of this chapter is to identify critical points in the process of a criminal investigation where applicable rights and representation are mandated or offered to a person who is a suspect or has been charged with a crime. The live lineup is an investigative tool that may be critical in providing evidence in the form of eyewitness testimony against a person. For this reason the Supreme Court ruled that a person has the right to have an attorney present at a lineup, though they may not actively participate or intervene and may only observe (*United States v. Wade*, 1967). If officers use a photo lineup, suspects do not have the right to have an attorney present (*United States v. Ash*, 1973).

How It's Done
LIVE LINEUP

Eyewitness identification of a suspect remains an important piece of evidence in many criminal investigations. If a victim or witness was in a location or position to clearly see the person who committed a crime, he or she may be asked to later pick that suspect out of a live lineup of individuals or a photographic lineup or array, in the UK referred to as an identity parade.

Most agencies use either a simultaneous presentation of the individuals or photographs to witnesses, or a sequential showing. Most lineups are photo lineups because of the ability to put together similar-looking individuals (called fillers) without consideration of a suspect or filler individual's height (not reflected in the photo). Additionally, the logistics are far less than acquiring five other people to serve as fillers for a six-person live lineup. Video lineups can also be utilized, and these can be presented simultaneously or sequentially as well.

Sixth Amendment

The wording of the Sixth Amendment is as follows:

> In all criminal prosecutions, the accused shall enjoy the right to a speedy and public trial, by an impartial jury of the State and district wherein the crime shall have been committed, which district shall have been previously ascertained by law, and to be informed of the nature and cause of the accusation; to be confronted with the witnesses against him; to have compulsory process for obtaining witnesses in his favor, and to have the Assistance of Counsel for his defence. (Bill of Rights, United States Constitution, 1789)

Of the various rights addressed within the Sixth Amendment, officers and investigators must be concerned most often with the right to counsel at critical stages of criminal proceedings. The amendment states in part "in all criminal prosecutions the accused shall enjoy the right to . . . have the assistance of counsel for his defense." The circumstances of whether a suspect either did or did not have access to or conversation with an attorney will feature significantly in whether the court later rules information gained from a suspect admissible. You can see the important connection between the Fifth Amendment's prohibition against being compelled to make statements against self-interest without due process, and the Sixth Amendment's specifying that benefit of counsel, and the right to a public trial, are key parts of protecting due process. *Miranda v. Arizona* (1966) established that prior to custodial interrogation suspects must be informed they have the right to consult with an attorney before questioning begins. And while, as we mentioned earlier, it is best to read the rights warning verbatim from a department-issued card or form to ensure thoroughness, the specific wording of the warning is not mandated (*Florida v. Powell*, 2010) as long as they are not limited in any way (*California v. Prysock*, 1981).

In addition to the right of counsel, the Sixth Amendment calls for a speedy and public trial, an impartial jury, clear notice of the charges leveled against a citizen, and the right to confront one's accusers. A speedy trial may be waived by a defendant, which is often done by agreement between the prosecution and defense. A public trial is not always granted because the courts recognize that the defendant's ability to have a fair trial may be impeded through the overpublicity or logistics involved in some trials. During a process called ***voir dire***, prospective jurors are questioned by attorneys to attempt to ensure an impartial jury, though the jury selection process is also used by attorneys to try to filter out potential jurors who may appear unhelpful to one side or another. Clear articulation of what facts constituted a specific statute must be contained in an indictment to achieve the requirement of notice of charges. And the right to confront accusers generally bars *hearsay evidence*, as opposed to the actual person with direct knowledge of a crime, from being used in court proceedings.

Voir dire Usually refers to the initial procedure of selecting jurors for a trial.

EXPLORE THIS

Search the Internet for information on police lineup procedures. Read through two or three academic or government documents on how to conduct lineup identifications. What are some of the pitfalls that can lead to misidentification described in the various sources?

SUMMARY

There are many legal protections and requirements within the U.S. criminal justice system. The origins of the United States saw early colonists leaving England and arriving in North America in the hope of a more just society independent from many of the constraints of England. The amendments discussed in this chapter act to protect the rights of the accused against unreasonable action by the police and the state, while still allowing officials to investigate and prosecute crime to protect the community.

The Fourth Amendment to the U.S. Constitution is currently applied in the justice system to protect the privacy rights of citizens against unlawful intrusion. This is most often seen when the courts examine and determine if officers had probable cause to search a person or place, or to seize something in relation to a criminal inquiry. The exclusionary rule may bar admission of evidence illegally seized by police. The Fifth Amendment works to protect citizens from inadvertently incriminating themselves in a criminal proceeding. Citizens are most familiar perhaps in seeing the application of the Fifth Amendment when officers read suspects their warning of constitutional rights before being questioned as a suspect in a crime. This is often referred to as the Miranda warning, which also involves the Sixth Amendment. This amendment identifies critical points in criminal proceedings where citizens have the right to be represented by an attorney.

Contemporary times have myriad examples of the impacts of technology on the lives of citizens. This is certainly true in the process of criminal investigation as offenders use technology to avoid detection and officers use increasingly sophisticated and sensitive equipment to detect crimes and evidence. The protections of the amendments mentioned above extend to the use of technology by law enforcement.

KEY TERMS

Excited utterance 109
Exclusionary rule 106
Exigent circumstances 103
Fifth Amendment 106
Fourth Amendment 99
Good faith exception 106

Lineup 110
Plain feel exception 104
Plain view 104
Probable cause 100
Reasonable suspicion 100
Search warrant 100

Showup 110
Sixth Amendment 109
Stop and frisk 104
Voir dire 113
Voluntary consent 103

DISCUSSION QUESTIONS

1. Describe the key aspects of the Fifth Amendment that impact the interrogation of a person suspected of a crime.
2. What role do prosecutors play over the course of an investigation? Describe how they may be of assistance to police.
3. Compare and contrast reasonable suspicion and probable cause. Provide examples.
4. In what ways do errors occur in the conduct of lineups?
5. How does an investigator obtain a search warrant?
6. List and discuss two or more exceptions to the search warrant requirement.

SAGE edge

- Get the tools you need to sharpen your study skills. SAGE edge offers a robust online environment featuring an impressive array of free tools and resources.
- Access practice quizzes, eFlashcards, video, and multimedia at **edge.sagepub.com/houghci**

PRACTICE AND APPLY WHAT YOU'VE LEARNED

▶ edge.sagepub.com/houghci

SAGE edge™

WANT A BETTER GRADE ON YOUR NEXT TEST?

Head to the study site where you'll find:

- **eFlashcards** to strengthen your understanding of key terms.
- **Practice quizzes** to test your comprehension of key concepts.
- **Videos and multimedia content** to enhance your exploration of key topics.

©iStockphoto.com/D-Keine

6. Homicide, Death, and Cold Case Investigations

RUNNING CASE: THE DETECTIVES ATTEND CRIME SCENE INVESTIGATION IN THE MORGUE ON FLORIDAN VICTIM AND BEGIN PREPARING FOR COURTROOM TRIAL

Introduction
  Scope of the Problem

Definitions and Legal Classifications of Cases

Medical Death Classifications

Preliminary Investigation and First Officers on the Scene
  First Officers Responsibilities
  Managing Suspects
  Securing the Crime Scene
  Managing Witnesses and Completing Neighborhood Canvass
  Identity

Arrival of the Detectives

Medical Examiners, Coroners, and the Autopsy
  Medical Examiner or Coroner (ME/C)
  Autopsy
  Time of Death
  Other Specialists

Criminal Homicide Methods
  Firearm, Edged Weapon, and Blunt Force Wounds
  Asphyxiation
  Poisoning

Suicide

Challenges in Death Investigations
  Equivocal Death
  Clearing Homicide Cases
  Working With the Media and the Public
  Death Notifications and Survivors

Cold Cases
  Cold Case Investigations

Serial and Mass Murder
  Serial Murder
  Mass Murder

Explore This

Summary

Key Terms

Discussion Questions

LEARNING OBJECTIVES

After reading this chapter, students will be able to

6.1 Contrast the legal elements of murder and manslaughter.

6.2 Distinguish the difference between legal and medical death classifications.

6.3 Distinguish between preliminary and follow-up activities in homicide investigations.

6.4 Describe the role of the detective in homicide investigations.

6.5 Explain the role of medical examiners or coroners in death investigations.

6.6 List and discuss criminal homicide methods.

6.7 Discuss the importance of properly classifying suicide.

6.8 Devise investigative strategies to address challenges in homicide investigation.

6.9 Explain the challenges unique to cold case homicides.

6.10 Differentiate between serial and mass murder.

Running Case: The Detectives Attend Crime Scene Investigation in the Morgue on Floridan Victim and Begin Preparing for Courtroom Trial

At 3 p.m., Lieutenant Kimberly Kellan, the Criminal Investigations Division (CID) Commander, asked Detective Richard Ashley, Detective Bradley Macon, Sergeant Mike Joseph, and the crime scene supervisor, Charla Lynne, to meet in the conference room of the Detective division to review actions taken in the case so far.

Lieutenant Kellan knew that both experience and research showed that timely and ongoing case reviews helped prevent damaging mistakes in major investigations such as a homicide.

"Any updates?" she asked.

Detective Richard Ashley nodded. "The medical examiner's office notified me that the autopsy will take place tomorrow morning. As lead detective, I'll be there, as will Brad. I imagine that someone from the crime unit will also be joining us."

Lieutenant Kellan reviewed all the steps taken in the case so far and ensured that all activities and reports had been entered into the computerized case management system.

"Let's call it a day, then," Lieutenant Kellan said to the detectives. "You'll need your full attention tomorrow for the autopsy and other laboratory work."

Bradley Macon sighed inwardly with relief. He had now been on the case for almost 12 hours and the initial adrenaline rush after getting the dispatcher's call had worn off hours ago. They'd already tracked down every lead and had made excellent progress. All he wanted to do now was go home and sleep.

As if she could read his mind, Lieutenant Kellan said, "Yeah, I know. I used to be a homicide detective, too. Getting rest and time away from the stress of the investigation is critical."

Detectives Ashley and Macon shook Kellan's hand, settled their affairs on their desks, then headed out to their unmarked cars to drive home.

When Bradley finally returned to his apartment building, the retired guy who watered the flowers out front greeted him. "Solve any crimes lately?"

Bradley just smiled and said, "Not today." So many of his friends and relatives thought they knew all about forensics and how technology affected criminal investigations. They followed the myth depicted in TV shows, that a DNA test could come back in

48 hours, and a crime could be solved during a commercial break. Bradley laughed to himself. Solving crimes wasn't that easy.

*

At 8 a.m. Detectives Ashley and Macon met outside the morgue suite at Central Hospital downtown. Finishing their coffees, they dropped their cups into the recycling bin and went inside. They donned gowns, gloves, and facemasks, then entered the examination room. Dr. Dianne Arnold, the associate medical examiner assigned to the case, greeted both officers and began to question them about statements the suspect had made regarding how far he stood from the victim when he fired the fatal shot. Because the identity of the deceased victim was not in question, the major areas of interest to Dr. Arnold related to the suspect's and victim's movements and positioning during the confrontation that ended in Bill's death.

As Dr. Arnold worked, Crime Scene Investigator (CSI) Chip Traci moved methodically around the examination table. He took pictures of items of clothing and areas of the victim's body that Dr. Arnold pointed out. This information would be shared with the medical examiner's office as well as included in the police department's case files. Dr. Dianne Arnold's assistant, Sam Ryan, assisted in the physical examination of the victim and occasionally recorded notes. CSI Tracey documented and bagged several pieces of evidence, including the victim's clothing. Laboratory testing would later reveal the presence of gunpowder residue on the victim's shirt to help determine the approximate distance between Frank, the suspect, and Bill.

Dr. Arnold's report would specify the *cause* of Bill's death—gunshot trauma. Dr. Arnold would also specify the *manner* of the death, which was homicide—one person killing another.

Once the autopsy investigation was complete, Detectives Ashley and Macon said goodbye and walked out. Dr. Macon knew that the team of professionals from across the different agencies would be working hard to fit the pieces together to determine if the case would move forward and potentially result in a courtroom trial.

Television shows often incorporate the work of medical examiners in murder cases. What are the ways in which these forensically trained doctors interact with detectives and crime scene technicians?

Introduction

It is likely that you begin this chapter as most would, with a keen interest in how law enforcement and others investigate the most important of crimes—murder. And while homicide rates are lower by far now than they were 20 years ago, that does not lessen the importance and seriousness with which law enforcement and society take such violent acts. Citizens are rightfully concerned when it appears as though the number of incidents of one person killing another increase in their community. The task of investigating death cases falls to investigators of the local or state law enforcement agency with jurisdiction at the location where it is determined someone was killed. Investigation of such cases on federal property is often conducted by federal law enforcement or military investigators if on a military installation or property.

Regardless of the location where a death investigation takes place, investigators generally follow accepted investigative processes and approach cases in a similar manner. This certainly includes the initial response by uniformed patrol officers, the employment of crime scene specialists to search for and collect evidence at the scene, and the involvement of specialist investigators to thoroughly conduct the follow-up inquiry in cases of criminal homicide or questioned death cases. The need for in-depth training and experience for investigators is quite important given that many jurisdictions experience no criminal homicide cases in any given year, or very few. The investigative process used for most crimes will apply to homicide events as well, but with special considerations.

The *manner* of death (homicide, suicide, accidental, natural, or undetermined) is an important determination as law enforcement officers encounter death investigations. This determination, typically made by a **medical examiner** or **coroner** (ME/C), helps focus the efforts of investigators but also relies on shared information between the medical examiner and law enforcement. After identifying a victim and determining both the *cause* and *manner* of the person's death, identifying a motive in a case of criminal homicide is helpful in understanding how the incident came about and who may be responsible. If the remains of a murder victim are not discovered for some period of time, the investigation may face a significant challenge due to the decomposition of the deceased and the impact on determining time of death and recovering other evidence from the body.

The efforts of all those involved, from the first responding patrol officer to the crime scene technician, medical examiner personnel, and law enforcement investigators, go together in preparation of a criminal case against a suspect coordinated by the prosecutor. While law enforcement officers have direct experience in assessing potential criminal incidents, the prosecutor will ensure that adequate evidence of each element of a specific crime is present as formal charges move forward. Each of the professionals performing tasks in the homicide investigative process faces challenges. Specialized training courses are available from various governmental, private, and public education organizations to familiarize personnel in basic and advanced techniques of death and homicide investigation.

Scope of the Problem

To understand the problem of homicide, let alone attempt to devise ways to prevent it, we must first know about how and when and where it occurs, as well as the victims and the perpetrators. Research into homicide is accomplished by specialists in many fields and relies on data gathered from a variety of sources. Practitioners in law enforcement, health services, social work, and others need access to these data and a way to extract the information they need to implement effective practices. Research is not simply a counting of homicide victims or people arrested for murder. It is important to understand serious violence including precursors to violence, factors affecting an increase in violent behavior by individuals, and the issues that accompany violence, dangerous places, people, and circumstances.

Medical examiner Trained pathologist and medical doctor appointed at the county, regional, or state level in many U.S. states.

Coroner Elected or appointed officials who inquire into the cause and manner of death.

Although an act as serious as one person causing the death of another would seem to be an event that is quickly noted and properly documented, this is not always the case. Crimes reported immediately to law enforcement officials will likely end up documented and tabulated in that department's reports to state-level agencies and subsequently passed on to the FBI for inclusion in the uniform crime report (UCR) program data. One of the immediately identified challenges in the system lies in the unique parochial nature of American law enforcement. There are some 18,000 separate law enforcement agencies in the United States, and each has the ability to follow its own reporting methods while varying widely in the resources available to enhance or integrate those reporting mechanisms.

The Uniform Crime Report revealed the following numbers about violent crime in 2017:

- In 2017, an estimated 1,247,321 violent crimes occurred nationwide, a decrease of 0.2% from the 2016 estimate.

- When considering 5- and 10-year trends, the 2017 estimated violent crime total was 6.8% above the 2013 level but 10.6% below the 2008 level.

- There were an estimated 382.9 violent crimes per 100,000 inhabitants in 2017, a rate that fell 0.9% when compared with the 2016 estimated violent crime rate and dropped 16.5% from the 2008 estimate.

- Aggravated assaults accounted for 65% of violent crimes reported to law enforcement in 2017. Robbery offenses accounted for 25.6% of violent crime offenses; rape (legacy definition) accounted for 8%; and murder accounted for 1.4%. (https://ucr.fbi.gov/crime-in-the-u.s/2017/crime-in-the-u.s.-2017/topic-pages/violent-crime)

This murder scene with evidence will be important for technicians and detectives.

©iStockphoto.com/D-Keine

The FBI's 2017 Crime in the United States report showed a decrease in murder and nonnegligent manslaughter numbers nationwide from 17,413 in 2016 to 17,284 in 2017 (FBI, 2018). The 2017 figure results in a rate of 5.3 such killings per 100,000 of population.

In addition to the UCR data, other sources of the FBI include the Supplemental Homicide Report (SHR) of the FBI, and the National Incident-Based Reporting System (NIBRS), used in only some states. The National Violent Death Reporting System (NVDRS), a program funded by the CDC, gathers data from states to help local stakeholders make decisions about prevention strategies. The state-based system gathers information from medical examiners, coroners, law enforcement, and vital statistics offices to give breadth and depth to the picture of homicides and suicides. In 2016 federal funding allowed the NVDRS to expand to collect data from 40 states, the District of Columbia, and Puerto Rico. The National Center for Health Statistics (NCHS) of the Centers for Disease Control and Prevention (CDC) maintains information open to researchers, policy makers, the media, and the public. The SHR is used to request additional information about homicide events from reporting agencies. The initial UCR data are brief, and the expanded information (when agencies supply it) can be helpful in analyzing the factors of criminal homicide events. But without a mandate to submit such information as contained in the SHR, smaller agencies or those without sufficient clerical or analytical personnel simply cannot or will not provide the supplemental data. The FBI has long been working on the National Incident-Based Reporting System, or NIBRS, which is designed to gather far greater information on most reportable crimes, but the same resource challenges persist.

Definitions and Legal Classifications of Cases

Although the term *murder* is often used interchangeably with *homicide*, they are not one and the same thing. Any person who kills another person has committed *homicide*. Causing the death of another person may be through the act or omission of an individual. Yet the act may be by combatants in war, a state-sanctioned execution, or an act of self-defense. None of these are criminal offenses and would not be categorized as a *murder*. *Criminal* homicide cases are those of greatest interest to law enforcement investigators, though often police are called to the scene of suicides, suspicious circumstance deaths, or other incidents that must first be assessed before determining further involvement of authorities.

Law enforcement officers may initially become involved in a variety of deaths, some of which are determined not to require police action. Those that are most frequently addressed in-depth by police are generally categorized as *murder* or *manslaughter*. Murder cases are those in which evidence indicates there was "malice aforethought." That is, the person committing the homicide *specifically* intended to kill the deceased. Proving this intent can be challenging, and the type and amount of evidence will affect what *degree* of murder a person may eventually be charged with by the prosecutor's office. Many people have heard the language of *premeditation* as being one component of a deliberate and intentional killing that will likely result in a charge of *first-degree* murder. This charge may also be the one selected by prosecutors if the victim and crime scene show indications of a particularly brutal killing including excessive violence done beyond what was needed to kill the victim. A criminal homicide that appears intentional but lacks evidence of premeditation may result in a charge of *second-degree* murder. A charge of *third-degree* murder may be one where neither premeditation nor intent is apparent or can be proven. Although states differ in some aspects of these criminal charges, first-degree murder will carry a heavier sanction or sentence, potentially life in prison or the death penalty if allowed in a particular state. Second-degree murder will obviously carry a less severe penalty. In each of these examples of a murder charge, the *element* illustrated is to knowingly cause the death of someone else.

Manslaughter is often charged in cases that do not provide clear evidence of malice aforethought, but may come from reckless action causing the death of another person. An example of a case of *voluntary manslaughter* lacks premeditation or possibly even the intention to kill someone but nonetheless does so either in the heat of passion or spontaneous emotion such as a parent killing the abuser of their child. In many states, however, this might be charged as a second-degree murder. *Involuntary manslaughter* does not involve planning; nor is it shown that the killer had the intent to kill the deceased. This charge is often leveled against someone who commits an act that was so completely careless or negligent that it resulted in a death that goes beyond the bounds of being a simple accident. Some incidents involving the operation of a vehicle, mishandling of firearms, unsafe practices in a workplace, and other human mistakes may result in charges of involuntary manslaughter, or *negligent homicide*.

Law enforcement may also be called to investigate deaths resulting from accidental circumstances, perhaps a person running out in front of a moving vehicle or moving underneath a heavy object set in motion by another person. These may result in a determination of *excusable homicide* because of the *victim's* negligent actions. *Justifiable homicide* may include killing another person in *self-defense,* or while defending someone else against imminent serious injury or death. As mentioned earlier, killing an enemy in wartime would be another example of justifiable homicide, as well as actions of law enforcement officers in the course of duty or corrections officials carrying out a state-sanctioned execution.

Motive for murder, while individual to the offender, is generally one of several well-known reasons. Financial benefit, such as in a robbery, murder for hire to collect the proceeds from an insurance policy, or killing a loved one to receive property or an inheritance, has long been known and indeed forms the basis for the plots of stories since before Shakespeare. Jealousy and the powerful emotions involved in the relationships between intimates has also long been the reason for violence including murder. The category generally underlying the greatest number of criminal homicides is that based on confrontation or argument. Of homicides where the cause is known, *confrontational* or argument-based homicide is accountable for the largest number for as long as such figures have been tracked. And while this last category illustrates the difficulty of law enforcement playing any meaningful role in prevention, the confrontation or argument often provides information as to the identity of the perpetrator. Of course, some murders such as gang-related ones may also be argument based, but the victim and perpetrator may have only known one another on site based on gang affiliation, and witnesses may be nonexistent or too frightened to come forward with any information they may know about the killing.

Medical Death Classifications

It is critically important to the homicide investigator to understand as soon as possible both the *cause* and *manner* of someone's death. Information gathered at the crime scene from witnesses and evidence can provide the initial indications necessary to guide the investigation, but it will be the medical examiner who classifies a death based on the mechanism that caused life to cease (cause) and the character of how that came about (manner). The manner of death is generally classified as one of the following:

Natural. The death resulted from a naturally occurring disease or condition such as a heart attack or the body's failing in old age. Very often the deceased will have been under the care of a medical doctor who may certify in a death certificate that the person's death was expected, as in the case of a terminal patient, or that a cardiovascular condition and subsequent examination of the deceased resulted in a cardiac death.

Accidental. Traffic crashes, drowning when the undercurrent pulls a swimmer underwater, a construction worker falling from a building, and other such examples are clearly unintended, but investigators must always work to ensure that an apparent accidental death was not staged to cover a criminal homicide.

Suicide. A person may take his own life through several mechanisms and methods. In many states suicide is criminalized under statutes such as "assisting self-murder," and the individual attempting suicide may be charged as well as anyone helping the person with the suicide attempt. Some cases of criminal homicide may be staged as suicide and some cases of auto-erotic asphyxiation may be mistakenly classified as suicide, so investigators must carefully investigate any and all death cases with an open mind.

Homicide. One person killing another person, criminally or otherwise. As noted, not all homicide cases are criminal. News accounts will often say something such as "the medical examiner ruled the case a homicide," which can lead many people to jump to the mistaken conclusion that the incident involved murder.

Undetermined. If there is insufficient evidence to classify one of the above, the death may be categorized as undetermined. While frustrating to medical and legal authorities, this is a reality of some cases. Regardless of the initial assessment of a death scene, officers must proceed cautiously and let the facts lead them to any conclusions they may reach. When a call comes into an agency reporting a dead body, the dispatch center will generally respond by sending more than one officer if they are available. The responding officers arrive as quickly as possible, taking great care to approach the scene safely, see to any injured people, and detain the perpetrator if he is at the scene.

Preliminary Investigation and First Officers on the Scene

Major criminal investigations such as murder or manslaughter begin with the first call placed to police. A recurring theme noted throughout the book is the importance of the initial responders to the potential reports of crimes and the first actions taken once a case begins. This is certainly the case in death and homicide investigations. Patrol officers from the smallest to the largest of agencies know that, in addition to their own safety and helping injured people at the scene of a crime, securing the crime scene against interference with evidence and quickly working to identify witnesses and suspects are paramount tasks. If the information received is that the incident has just occurred, or whether there is a risk of the loss of evidence, it is important that patrol officers arrive quickly to stabilize the situation and protect the crime scene.

First Officers Responsibilities

Crime scene duties and the identification of witnesses and potential suspects are important tasks for the assigned patrol officers. It is a standard law enforcement response to send multiple officers to a reported homicide and the primary officer assigned or a supervisor will determine many of the activities to be performed and who is to perform them. It is also common to assign any officers in training who are working on a particular shift to respond to the scene and gain experience in what might be a major case investigation. Arriving safely and checking the scene for an active suspect or dangerous condition is the responsibility of the first responding officers. At the same time, they are determining whether anyone is in need of medical assistance and communicating with the dispatch center to send EMS and to provide as much information about injuries as they can while using the training they have had to try and help the injured. And all of these things are occurring as the officer enters a situation with many unknowns. If officers do not locate a suspect at the scene of an assault or death they must always be aware that the individual may return while the officers are conducting their investigation. In fact, officers have been assaulted and killed as a result of these type circumstances.

If EMS or fire department personnel arrive on scene to render aid to victims or to verify that someone is deceased, officers are responsible for guiding these personnel into the scene and cautioning them against inadvertently altering the scene or damaging evidence. This includes documentation by the officers present of who entered the scene, where they walked, and any other changes that may have occurred to the scene of the crime. If medical personnel transport a suspect or a victim to the hospital they should be reminded to document anything said to them, as it may assist in the investigation. A victim or a suspect may make what is referred to as a dying declaration or a statement about what happened while the victim or person has a well-founded belief that he or she is not going to survive his or her injuries. Once again, any such statements or comments should be written down and included in the case report, taking care to quote as directly as possible what was said.

Managing Suspects

If the suspect is actually at the scene when the initial officers arrive a number of things may happen. Certainly the individual may try to flee or otherwise resist the attempts by officers to detain him while the investigation occurs. Alternately, an individual may immediately start speaking and in fact admitting or confessing to whatever has occurred. After the officer provides the warning of constitutional rights or Miranda warning to a suspect, the uniformed officer will not necessarily move into an in-depth interrogation of the individual. Rather, this should wait on the arrival of the detective who will be responsible for the case. Any spontaneous statements made by the suspect should, however, be thoroughly documented by any of the officers who heard the remarks. It is certainly appropriate to move a suspect away from the immediate vicinity of the crime scene so that he does not attempt to alter it in any way. Likewise, there may be the immediate need to gather physical evidence from the person of the suspect. If a suspect is not on the scene but there are witnesses who can provide information describing the person or vehicle used to leave the scene, officers will quickly summarize that information to be dispatched to other officers via a "be on the lookout" (BOLO) broadcast.

Securing the Crime Scene

After handling the twin tasks of ensuring a safe scene and getting medical assistance for anyone who is injured the officer moves deliberately into establishing and securing the crime scene. At some crimes scenes there are many things going on that come at the officer through multiple senses and that can therefore cause distraction from immediately noting some of the items of evidence. Establishing a perimeter warns most unauthorized persons to stay back and focuses the attention of investigators and technicians on the area most likely to contain evidence. It is advisable to set the crime scene perimeter broader rather than narrower in case there is evidence spread farther than what is initially expected. The dimensions of the scene can always be contracted as warranted by adequate searching.

Law enforcement may need to seek a search warrant before completing the crime scene search, so the perimeter is critical to protecting the integrity of the scene until the warrant is acquired. If evidence is subsequently located outside of the established crime scene perimeter, officers will need to expand the scene or isolate the newfound evidence and try to determine the path that brought the item to that spot. Similarly, if items of evidence are located later, technicians or officers return to the scene, collect the evidence using appropriate methods, and document the circumstances of locating the additional evidence. In addition to medical personnel already mentioned, law enforcement officers themselves can also inadvertently compromise a crime scene, so it should be standard procedure to allow as few personnel as possible into a scene. Too many people inside the perimeter who think they have a reason to be there may increase the likelihood that evidence can be disturbed. However many official actors are present must be organized and directed in their efforts. An officer assigned as scribe to log people into and out of the scene is important, and requiring supplemental reports from everyone who performed any function is good practice. Even if personnel performed no specific duties at all, their presence at the scene should be noted in the case report and the log to demonstrate scene integrity if the case goes to trial.

Managing Witnesses and Completing Neighborhood Canvass

Outside of the crime scene there are also duties to perform. Officers should speak with witnesses present and, where appropriate, conduct a **neighborhood canvass**. An often productive tool, the canvass of the local area can produce information from nearby residents or business owners about the activities of a location or the people who lived, had visited, or were seen there. Patrol officers are often assigned to such canvassing, and they should be reminded what an important function this can be. Sometimes if no one is home at a particular house in the area it will be written off as irrelevant. This can be a mistake, as each such neighbor may be a source of information in a homicide investigation. Officers should note in their individual reports which addresses they went to, which they located someone to speak with, and which had no one there and should be noted for a follow-up visit to try and locate someone home. This work, like all activities in the overall investigation process, must be coordinated, and responsibility for organizing the effort should rest with one individual.

Neighborhood canvass Door-to-door contact with businesses or residents to gather information.

Identity

To begin an investigation in earnest it is critical that officers know the identity of the victim. Knowing who the victim was allows investigators to determine who may have intended the person harm, what the victim's most recent activities were, whether the location where the deceased was found and its condition are significant, and many other avenues of investigation that provide context once the person's identity is known.

If the victim is found by relatives at the home, this may provide an immediate answer to the question of identity. Not all death investigations begin with the certainty of the identity of the deceased. Experienced death investigators also know that identity should be verified in more than one way if it all possible. For instance, decomposed remains with a wallet and identification inside cannot be accepted as certain if foul play or a faked death is possible. Decomposed bodies or skeletal remains even when found in a location with limited access to only a few known people can raise questions as to the identity of the deceased. Decomposed bodies may yield sufficient biological evidence to determine blood type and perhaps a DNA profile that can hopefully be matched to known samples from possible victims. The deceased can also be fingerprinted if decomposition or desiccation is not pronounced. Skeletal remains may yield information such as the age, gender, and possibly the ethnicity of the deceased. Dental work can aid in establishing the identity of the victim if previous dental X-rays are available for comparison.

Arrival of the Detectives

The response of patrol officers to the report of a serious crime is typically the first stage of any major investigation. This means that handling the crime scene and the involvement of investigators becomes a second stage in the ongoing investigation. In a death case requiring a law enforcement response it is typical that an investigator will respond soon after the responses of patrol officers. In smaller jurisdictions without a homicide detective or perhaps any detectives at all, there may be an agreement with a state investigative agency or a larger local agency to provide investigative support. If the agency lacks investigative expertise or resources it will likely also request crime scene assistance from a larger agency with greater experience working crime scenes (Hough, McCorkle, & Harper, 2019).

Even some homicide investigations do not require the large number of fictional television detectives who seem to arrive in droves along with so many crime scene technicians one would expect they would trip over one another. Still, there are studies that point to greater case clearance if certain numbers of investigators can respond initially (Wellford & Cronin, 1999). Of course, each person must have specific tasks assigned and there has to be coordination of information-gathering, not merely large numbers of personnel idly standing about. With that said, beginning a case with multiple investigators assigned has clear benefits. As personnel involved in the death case carry out crime scene examination, neighborhood canvass, and interviewing, one or more detectives or a supervisor can coordinate the various tasks to be carried out and information coming in. Coupled with the use of checklists and case management software, this approach provides greater likelihood of a competent and successful case outcome.

Solving homicides is a task that all communities and their police agencies take seriously. While clearance rates in the United States have dropped over time, renewed research interest points to actions by investigators as perhaps being more important than previously believed (Brookman, Maguire, & Maguire, 2019; Lum & Nagin, 2017). Twenty years ago, Wellford and Cronin (1999) expressed the view that adequate training of detectives, and proper resources, could allow detectives greater success in homicide investigations. Adding resources to any investigations carries costs, but the importance placed on death investigation has generally provided the support needed to fund such cases. Having sufficient numbers of detectives, training enhancements, and the use of a problem-oriented approach may be some of the ways case clearances may be increased (Braga & Dusseault, 2016). Clearly supported is the need for agencies to have specific policy and procedures for homicide investigations.

The primary case detective will likely coordinate with crime scene personnel in searching, documenting, and collecting evidence. With the potential in many cases for multiple crime scenes, these responsibilities may be spread among several investigators. Additionally, a detective may need to go to the hospital where a still-living victim may be able to provide important information. And then, of course, if a suspect is known or has been detained, arrangements for and the conduct of an interrogation will be of primary importance. These and many other duties require close coordination and teamwork.

Medical Examiners, Coroners, and the Autopsy

It is important knowing not only the mechanism by which a person died, but perhaps having information or insight into how the assault unfolded. For example, does the victim's body exhibit the defensive wound marks that would indicate facing her attacker as opposed to a single blow to the back of the head? Determining whether more than one weapon was used to wound and ultimately cause the death of someone may be critical in determining whether there was more than one offender and how that may be of assistance in questioning suspects.

Medical Examiner or Coroner (ME/C)

In another common misconception of terms, many people believe that coroners and medical examiners are the same. While both individuals are typically responsible for examining the bodies of deceased under certain circumstances and within their jurisdiction, there are differences between these officials. A medical examiner is a medical doctor who also has advanced training in forensics. The coroner is often a locally elected official and in many states need not even be a medical doctor.

Autopsy

Not every case brought to the attention of the medical examiner or coroner is such that it requires a full autopsy including an internal examination. Some death cases require full anatomical examination, both internally and externally, as well as submission of samples for toxicological results that may indicate abnormalities. The office of the medical examiner may employ death investigators who can begin their inquiry at the scene where a person's body has been found. Information gathered by the investigator including that which is supplied by law enforcement will make up part of the basis upon which the medical examiner will determine cause and manner of death.

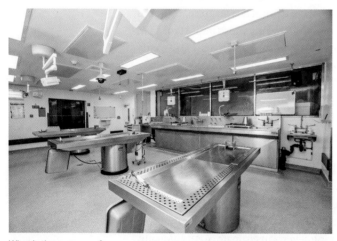

What is the purpose of an autopsy?

LEWIS HOUGHTON/Science Source

The autopsy continues the investigation that began at the crime scene as one or more medical personnel work to locate physical indications and evidence of what happened to the deceased and who may have been involved in whatever did occur. Personnel examine any clothing or remnants of clothing on or near the body and search the surface of the body as well as body cavities to locate potential evidence. As an autopsy takes place, usually at the office of the medical examiner, personnel will document pertinent observations about the internal and external examination of the body. Photographs will typically be taken during the autopsy depicting the overall condition of the corpse as well as close-ups of wounds or other specific items of relevance. X-rays are also often done to determine whether there are internal skeletal injuries or foreign bodies revealed by the X-rays.

Time of Death

In some homicide cases the issue of the time of death of the victim holds little relevance. When there are witnesses to the act, a suspect apprehended at the scene or a victim who dies subsequent to an initial assault and while being treated by paramedics or emergency room personnel, the time of death will be known. Where the question of time of death may have implications for suspect and victim movements as well as suspect alibis and witness testimony credibility, there are a variety of methods available to investigators and medical personnel to assist in determining a time frame in which someone died.

When a person dies a number of biological functions cease, but some continue as the body, a structure of biological matter, begins to decompose. Because researchers have studied the rate of decomposition, an assessment of such processes can help in determining a range of time for when a person died. Changes in the body affecting the determination of time of death are also affected by a number of factors outside of the body itself. The temperature and humidity of the environment affect how fast or how slow a body will change. If the environment is dry or cold the rate of decomposition may slow to a halt or take far longer than if the body is in a warmer or wet environment. It is typical for detectives and crime scene technicians to contact a local meteorological office and note these measurements at the time the deceased is discovered. Additionally, the number and type of insects in the area will affect how quickly the body is broken down. Bacteria present in our bodies while alive continue to act after respiration and heart functions cease in a person. The observable and measurable changes brought on by decomposition help investigators and scientific personnel to gauge an approximate time of death.

The most important factors in the decomposition process following human death are temperature and moisture. Medico-legal death investigators may insert a thermometer into the rectum of the deceased to gauge temperature. The ME or coroner may also insert a thermometer into the liver, if they receive the body in a timely fashion. In addition to the **lividity** affecting skin coloring through blood pooling, the eyes of a deceased person will immediately lose their sensitivity to light, and then take on a milky or cloudy color over a period of 8 to 10 hours.

Lividity Pooling of body and the blood that can help determine body position after death.

Because blood stops circulating throughout the body as soon as the heart stops, gravity will begin to draw the liquid to the lowest point of the body. At such points that come into contact with the ground or any surface, *lividity* stains will appear on the skin with purple coloration. This *livor mortis,* or postmortem *lividity,* begins almost immediately but is obvious over about 4 hours following death and fixed about 12 hours after death, when the coloration will remain until decomposition breaks down the body itself. If the body is moved prior to the fixation of the lividity staining, investigators may see more than one pattern, which underscores the importance of documentation including photographs. The lips and fingernails will also lose color. Many people are also familiar with the stiffening of muscles that occurs within the body following death. This process, called **rigor mortis**, affects muscles throughout the body, with smaller muscles stiffening before the larger ones. Once again, the process takes time and may not be noticeable for many hours, with complete stiffening by around 24 hours for the entire body. Following this point the rigor goes away gradually over the next 12 to 36 hours, dependent once again in large part on temperature and other conditions in the surrounding environment.

Rigor mortis Temporary stiffening of muscles and joints following death.

The living body's normal temperature of 98.6 °F changes after death (**algor mortis**) to match the ambient temperature up to approximately 24 hours after death. The chemical breakdown occurring in the body may also increase temperature in areas of the corpse. Once again, measurement of the temperature of a body recently dead can be helpful in determining the approximate time of death. The complication of time-of-death determination for a body found in water is compounded because marine life may have eaten away portions of the deceased, making it difficult to discern **premortem** or **perimortem** injuries

Algor mortis Body temperature change after death.

Premortem Before death.

Perimortem At, or near, the time of death.

Postmortem After death; also refers to the examination of a dead body.

from **postmortem** scavenging. During the autopsy an examination of the stomach contents of an individual may also provide indications as to time of death. If it is known that the victim consumed a meal at dinner time and the contents of the victim's stomach are undigested, a time frame of a few hours can be established because the food had not yet been digested.

Other Specialists

Most people are at least somewhat aware of the existence and function of medical examiners and coroners. There are a number of other individuals with specialized knowledge who are called upon to assist in understanding various aspects of a death investigation. These include forensic anthropologists, forensic entomologists, forensic odontologists, and toxicologists, among others. The forensic anthropologist is sometime called upon to examine the skeletal remains of a person to help determine the identity of the deceased or to examine damage to understand what happened to the individual. The forensic entomologist may be helpful in determining a range of time when the deceased was killed based upon insect activity in remains. The forensic odontologist may assist in identifying the remains of a deceased through dental records as well as potentially providing insight regarding bite marks on a victim. The forensic toxicologist works in the laboratory environment to test bodily fluids and tissues for the presence of chemicals.

Criminal Homicide Methods

Homicide detectives and death investigators must be familiar with the variety of ways in which a person can die. This allows for proper interpretation of evidence at a crime scene or on a body, as well as offers clues to possible suspects or motives. When the medical examiner or coroner examines a victim, they are often able to determine sequence of wounds or injuries, as well as which were the fatal ones.

Firearm, Edged Weapon, and Blunt Force Wounds

The human body may be gravely and terminally affected by various forces including ones directed at it through the bad intentions of others as well as carelessness or negligence. Without the availability of immediate eyewitnesses it may be difficult for the initial investigator to distinguish among wounds caused by various methods including bullets, knives, or even blunt force. Because of the elasticity of human skin a penetrating wound may partially or completely close over as either a bullet passes through or a blade is withdrawn. Damage may be obvious but not the circumference or dimensions of the actual object that broke the skin.

Most homicides involve the use of a firearm. In 2016, 11,004 firearms were used in homicides, up from a total of 8,897 in 2012 (FBI, 2017). A firearm used at close range may additionally leave markings on the skin resulting from bits of burned and unburned gunpowder expelled from the barrel of the weapon and adhering to the area of the wound. This is referred to as *stippling* or *tattooing* and is part of the characteristics of the abrasion ring where the skin was torn and stretched at the entry point. If the barrel of the firearm was pressed against the victim's body, there may also be a tearing or eruption of the skin around the wound as the gases that escape the end of the barrel force themselves out from under the dermis. Further away from the skin when the gun is discharged will result in reduced marking until at specific distances there will be no mark but the entry wound from the bullet itself. Gunshot residue (GSR) consisting of the particulates from the spent cartridge carried by and partly composed of the gases expelled may be discovered on the victim, but also on the skin or clothing of the person who fired the gun. A collection using adhesive materials can gather this residue if it is done right away and before a suspect may change or dispose of clothing or wash away residue from their skin. The diameter and appearance of the wounds from a firearm are affected not only by the distance the weapon was from the victim but also by the caliber of the bullet as well as the bullet's design, including the velocity or speed at which the projectile traveled to its target. Sometimes such wounds are obvious, but with multiple wounds some may be harder to distinguish if they have passed through clothing, hair, or more than one portion of the body. And while some bullets enter *and* exit a body, usually leaving a larger hole upon exit, some projectiles may remain within the body and be discovered during the autopsy process.

Edged or penetrating weapons can take an almost unlimited form. While knives are generally thought of in this category, any sharpened object may result in a slicing or penetrating *incision* wound, and anything with a relatively sharpened tip can result in a penetration referred to as a *puncture* wound. Such wounds that result in death are frequently from multiple stabs or punctures and generally occur in the area of the head, neck, or torso. The examination of the homicide victim may also note defense wounds on the arms and hands of the victim indicating attempts to block or ward off an attack. Measuring and examining wounds may give an idea as to the type and dimensions of the blade or object, though skin elasticity once again may complicate definitive identification of the specific object. One or more weapons used in an assault may be left at the scene or taken away by the perpetrator. The presence of one apparent weapon does not rule out the use of more than one, and any conclusion should wait until after the autopsy results. The involvement of a specific weapon may be determined if a portion of the blade or object broke off and remained behind inside the wound.

Severe injury or death can also be caused by what is known as **blunt force trauma**. A victim may bleed to death internally or externally from damage sustained by being struck by another individual's personal weapons of fists or feet, or some other object. Medical examination in conjunction with the investigation may help determine whether wounds from blunt force trauma were the result of accident or inflicted intentionally. Everyone is familiar with the bruising that results from some type of impact force on the skin. This is referred to as a *contusion*. An *avulsion* is the ripping or tearing of skin that may occur from a sufficient impact. The physics of such trauma involved the weight, speed, and angle of the object that struck a body as well as the position and movement of the individual's body when struck. Damage may occur to underlying tissues, bones, and other structures of the body and partly be affected by whether or not the person was leaning against a hard surface. If the individual's skin and other parts of their body have no ability to flex with the area below, then some of the damage may come in the form of compression or absorbing more of the energy of the impact that otherwise may have dissipated.

Blunt force trauma Nonpenetrating injury to the body.

Asphyxiation

Asphyxiation, or insufficient oxygen for the brain, may be the result of suffocation, chemicals, or **strangulation**. These various mechanisms of asphyxiation may have occurred accidentally or through homicide, and investigators and forensic medical personnel may see a number of indicators that are clues to the specific cause of death. The previously mentioned toxicologist may discover a marked lack of oxygen in the red blood cells of the victim. But

Asphyxiation Suffocation resulting in loss of oxygen.

Strangulation Compression of the neck.

investigators and crime scene technicians as well as the medical examiner may have already observed specific discoloration of the skin with blue or purple around the nail beds of the victim's fingers and toes as well as the skin around the lips. An individual may suffocate from being smothered, such as an infant as the result of a co-sleeping parent who inadvertently blocks the nose and mouth of the child with a pillow or the parent's own body. An elderly person with little strength may also be intentionally smothered by a caregiver or other criminal holding a cloth, pillow, or cushion over their nose and mouth until they stop breathing.

If an individual died as the result of strangulation, there will typically be markings in the area of the neck such as redness or bruising from manual strangulation by a killer, or the chafing or indention of a cord or rope, referred to as a **ligature**. Choking is not the same as strangulation, and law enforcement officers should ensure that they use the correct language when describing both injuries they discover as well as descriptions given by victims or witnesses. If a husband or boyfriend puts his hands around the neck of a spouse or girlfriend and squeezes, that is properly referred to as strangulation. If something gets caught in a person's throat and prevents the person from breathing, that is appropriately referred to as choking. I investigated the case of a young man working at a cement plant who fell into a sand silo that had insufficient safety guards. The victim choked to death on the sand that he sank into before co-workers noticed he was no longer standing on the walkway at the top of the silo and only later recovered his body by emptying the silo of all the sand. This particular investigation received an immediate law enforcement response, and I was required to notify his pregnant wife and mother of their three children. Ultimately, we secured the scene until investigators from the regional office of the federal Occupational Safety and Health Administration (OSHA) arrived to investigate the physical site and company for workplace code violations rather than a charge of criminal homicide. It is important to note that the state could make a charge of involuntary manslaughter in such cases, or even voluntary manslaughter if the firm's management had been previously cited for safety violations.

Although drowning is not a common form of homicide, it does occur. The medical examination will likely discover water and various waterborne particulates in the lungs of a victim, indicating that the material was drawn into the lungs. If, rather than drowning, a person's body was placed into water after he was already dead, waterborne materials may be found in the mouth or sinuses but likely not in the victim's lungs. Similar to drowning, hanging is rarely the result of homicide. Some individuals commit suicide by this method, and some people die accidentally while engaging in autoerotic asphyxiation, which would then be classified as an accidental death. In this practice the individual experiences a heightened state of sexual arousal as a result of reduced oxygen to the brain, *hypoxia*. Typically, the ligature used is padded and a mechanism releases the individual's body weight at a certain point so as to avoid unintentionally passing out and subsequently resulting in death. Indicators that investigators look for include the type of device used, indications that the scene was set with sexual fantasy in mind, and the victim being naked or partially exposed.

Poisoning

The use of poison to do someone in seems somewhat exotic to many people, and more the stuff of novels. In reality, poisoning as a form of assassination and murder has been with us for a very long time indeed. A few notorious cases of attempted or successful mass poisonings made no effort at concealment and employed poisoning for its utility in affecting many people at once. Down through history, poisoning has also been a tool of assassins. Some

Ligature Something used to tie or bind.

incidents of poisoning are intended to mask criminal homicide as a natural or accidental occurrence, as when a victim is a patient at a medical facility, or create the appearance of an accidental overdose. Farrell (2017) notes, "Most poisonings avoid physical or mental confrontation because the victim is oblivious of their plight" (p. 7). Distinction between attempts at concealment and efficiency of poison as a mechanism of multiple murder, regardless of discovery, is an aspect of the investigation that will also combine an assessment of whether the poisoner chose a specific or random victim and perhaps the degree of spontaneity or advance planning involved in the poisoning (Farrell, 2017). Modern science has greatly enhanced the ability to discover the presence of toxic compounds, chemicals, or poisons, likely making murder by poisoning a less common crime. While limitations of data exist on homicides committed using poison, this mechanism of murder has a more equal gender distribution than most other categories, which are dominated by males.

Suicide

Most states criminalize the attempt or act of suicide, classifying it generally as self-murder as well as declaring it a crime to assist anyone in attempting suicide. According to the National Institute of Mental Health (NIMH), in 2017 it recorded a total of 47,173 suicide deaths, and the three most frequent methods of suicide overall were (in order of frequency) use of a firearm, suffocation, and poisoning (Table 6.1). For female victims of suicide, poisoning was the second most frequent method.

Investigators must always consider the possibility that what appears as a suicide may in fact be something else. Following a murder, for example, someone may try to stage the crime scene instead to make it appear as if the victim took his own life. Another death scene that may be misinterpreted by those who discover the deceased, including first responders or inexperienced investigators, could involve autoerotic asphyxiation. As mentioned earlier, in this instance the victim intended to enhance a sexual experience through partial compression of blood vessels to reduce oxygen to the brain and induce added euphoria. The person using this method may employ devices or rigging to allow for quick release to avoid unintentional asphyxiation. Sometimes these safeguards fail and a person dies accidentally. The scene may present as one where a person hung himself, and the ligature may simply appear elaborate or bizarre to the person who finds the victim.

TABLE 6.1 Suicide by Method, 2017

SUICIDE METHOD	NUMBER OF DEATHS
Total	47,173
Firearm	23,854
Suffocation	13,075
Poisoning	6,554
Other	3,690

Source: National Institute of Mental Health, 2019.

As with any death investigation, the conclusion of suicide should not be rushed. Investigators work carefully to differentiate suicide from murder. In regards to suicide notes, for example, while estimates vary widely, significantly fewer than half of people who commit suicide leave a note, and there may be differences in the individuals who do and do not leave a note (Paraschakis et al., 2012). Investigators will also examine the body for hesitation wounds made by the victim as they tested their resolve (and often the weapon) leading up to taking their life. I investigated a suicide where a young man had driven his car to a secluded area, written out a suicide note, and before shooting himself with a .22 caliber rifle, tested the weapon by firing it into the windshield of his car. A burglary detective, who had arrived on the scene before me, mistakenly thought the man had been murdered by someone who fired through his windshield. The detective had not seen the note on the car seat, nor any other aspect of the physical evidence. This is one example of the equivocal death challenge mentioned in the next section.

Challenges in Death Investigations

It is important that law enforcement agencies assign those best suited to conducting criminal investigations. The status of a detective has long been considered somewhat glamorous and desirable within law enforcement. Detective selection in homicide investigations is not only of great importance but is generally viewed as the pinnacle of criminal investigation. We therefore note that carefully selecting those who conduct homicide investigations is quite important.

Caseload was discussed in an earlier chapter, but it is important to reiterate that if an investigator is straining with the number and complexity of assigned cases she is handling, investigations may suffer as a result. Individual investigators, investigative units, and agencies must also have available adequate tools and personnel to properly identify and collect evidence as well as pursue leads in major cases. In addition to having enough time and resources to properly investigate death cases, there are many additional challenges.

Equivocal Death

As already mentioned, investigators must be thorough and work in concert with the medical examiner or coroner in determining suicide as the manner of death. If the death case can be interpreted in more than one way, these *equivocal* cases may present questions that obscure the manner of death and sometimes even the cause of death. Death case may initially present with information that makes it potentially a homicide, suicide, or an accidental death. This may be the result of insufficient physical or testimonial evidence to identify a motive for a suspect or a mind-set for a victim. The investigator must be cautious in examining an apparent suicide to rule out the possibility of homicide staged as suicide. In-depth examination of the victim's circumstances and lifestyle, termed *victim apology,* may be particularly important when trying to move a case from equivocal definitively into another category.

Equivocal death
Manner of death open to interpretation.

The category of **equivocal death** cases includes those of sudden infant death syndrome (SIDS), which are those of a child not yet 1-year-old whose death after an investigation and full autopsy remains unexplained. The SIDS label is applied when all other possible causes have been ruled out. In contrast, a sudden unexplained infant death, or SUID, is generally a term applied when an investigation is just beginning. In some death cases, especially children, a parent or caregiver will bring about or fake the sickness and even death of a child or someone in their care as a way of gaining attention for themselves as the person who was long suffering in trying to care for a loved one. This attention-seeking can be the manifestation of a psychological disorder called **Munchausen Syndrome by Proxy (MSP)**.

Munchausen Syndrome by Proxy (MSP) Also referred to as factitious disorder, MSP is a mental illness in which a caregiver claims or induces illness of someone whose well-being he or she is in charge of so that the caregiver may receive attention.

Clearing Homicide Cases

Although many crimes are never reported, and taken together most reported crimes are not solved, the fact of usually having a deceased person means a death investigation will be undertaken and the majority of these are resolved. If a death case is classified as a criminal homicide then the desired outcome is to identify the person or persons responsible for the death and bring them to account through the criminal justice system. This is the most commonly recognizable and hoped-for conclusion to those cases considered murder or manslaughter.

The FBI's UCR program also tracks several categories of crimes "cleared" by law enforcement officials each year. Case clearance is either by arrest (arrested, charged, turned over to the court system) or by "exception," meaning that while the offender has been identified, law enforcement is unable to arrest or formally charge them. Such circumstances of identifying and gathering sufficient evidence to charge a person without arresting them would include the death of the offender, victim lack of cooperation, or denial of extradition from a jurisdiction where the offender fled. In 2016, 59.4% of murders were cleared (FBI, 2017). In 2015, the clearance rate was 61.5% of murder cases; in 2010, the rate was 64.8%. This relatively low clearance rate for such a serious crime may be surprising to some. Research continues to examine the factors involved in clearing homicide cases and taking into account that different types of murder and manslaughter involve dynamics implicated in why the death happened as well as how to approach solving the case.

Working With the Media and the Public

One of the ways that the media can participate and contribute to homicide investigations is when they assist in publicizing efforts to review cold case homicides. Media outlets are also obviously greatly interested in homicides that have just occurred or are part of an ongoing investigation. Media reporting on homicide investigations may also hinder an investigation or cast investigators or agencies in a negative light. Media may choose to characterize an agency as obstructing their reporting when investigators are trying to protect the integrity of an investigation. Law enforcement officials can, however, adopt a less than cooperative approach to working with the media based on their prior experiences. It is important to recognize that the media have an agenda that does not match that of the homicide detective or police department. Cynically, this may be the desire for increased viewership or readership under the guise of servicing the public's

What are some ways that investigators can work with the media during an investigation? What are the pros and cons of this?

©iStockphoto.com/ carminesalvatore

right to know. Law enforcement, for their part, have legitimate concerns about finding and bringing to justice the individual who ended another person's life, and release of a variety of details or information about a case may tangibly damage those efforts. As noted by Hough and Tatum (2014):

> Law enforcement agencies must coordinate with other organizations during the course of a death or homicide investigation. Knowing what is sensitive in a case and understanding the roles and responsibilities of various agencies and personnel are key parts of an effective homicide detective's knowledge base. Officers should never speculate about issues in a case to the press nor project time frames for bringing resolution to an investigation.
>
> The media can help to educate the public on the realities of crime and the risk to the public of that crime. News reporters can assist law enforcement by sharing information they develop independently that may bear on the solving of a case. Calling attention to questionable investigative methods or commenting on the progress of a case can lead to agencies reexamining procedures and making changes in criminal justice policy. (p. 82)

Death Notifications and Survivors

An important and sensitive duty that often falls to law enforcement personnel is that of notifying loved ones of the violent death of the victim. This responsibility comes with the need in unresolved suspicious death cases to be observant of the responses of those notified to try and discern whether those notified may have somehow been involved in the death. In cases where the manner of death has been determined to be noncriminal, other personnel including victim advocates, clergy, or those from the office of the medical examiner or coroner may appropriately handle the death notification function. Survivors, sometimes referred to as co-victims, are frequently in need of emotional support and information in the wake of the violent death of a loved one. The development and availability of the victim advocate function over the past 30 years has greatly enhanced the ability of agencies to respond to these needs. In addition, organizations such as the Parents of Murdered Children (POMC) play a vital role in supporting family members and others impacted by violent death. The organization, founded in 1978, is dedicated to providing support and information to people who have lost a loved one to murder. The group holds an annual conference that brings together family, friends of those murdered, and professionals from various disciplines to share information and attend seminars to help understand different aspects of the crime of murder. The organization's mission is: "POMC makes the difference through on-going emotional support, education, prevention, advocacy, and awareness."

In addition to ongoing availability of resources and support, the annual conference has offered seminars such as The Courage to Grieve, DNA & Solving Cold Cases, Overcoming Survivor's Guilt, and Ask the Medical Examiner. Your author has co-presented at two of the POMC conferences on Homicide Investigations Best Practices and Understanding the Criminal Justice System.

Cold Cases

If a homicide case has run out of leads to follow with no suspect identified or indicted, it may become what is considered a **cold case**. Reopening such cases, sometimes referred to as warming them up, may occur because new leads have come to light, new technology is available to re-examine physical evidence retained in the case, public or media attention returns to a past murder, or sometimes because an agency receives funding to devote resources to past unsolved cases. Agency resources are not unlimited, but the public is often left somewhat unsettled when a murder case is not resolved. This leaves agencies prioritizing review of cold cases when they are able. Process for such reviews is varied based on agency size and available resources, including the potential to participate in a task force that examines cold cases.

Cold case Generally, a crime that is no longer being actively investigated, usually due to a lack of leads or evidence.

How long since the murder took place will partly shape the way an investigation is conducted but the people involved, the places thought to be relevant, and the actions previously taken by investigative personnel will be reviewed. Time passing can mean the loss of witnesses who have moved away, but can also mean that relationships among involved parties have changed and someone may now be willing to speak more freely than when the crime originally occurred.

Cold Case Investigations

Although the term *cold case* can apply to *any* older investigation that is not actively being pursued, the media, public, and law enforcement associate the label with homicides. An unsolved homicide nominated for new investigative attention may come about through new information, the anniversary of the crime, efforts by family members, or availability of funds to an agency through a grant. However it comes about, and regardless of whether examined by a single agency or a multi-agency task force, the following steps are generally included:

- The assigned investigator(s) will read all available case reports, supplemental reports, and evidence and forensic analysis reports. This will include all witness statements, notes made by previous investigators, and all photographs or video.
- The detective will, if possible, visit the crime scene(s) and any other relevant locations thought to be involved in the original incident.
- Re-conducting physical interviews with prior assigned personnel, family members of the victim, witnesses, or canvassing the area where the crime occurred for others with information will help create a more tangible reconstruction for the investigator. This step includes discussing any original forensic information with the medical examiner's office.

Science and technology have helped resolve many cold case investigations in recent years. With the application of DNA identification methods, old evidence in many cases has been submitted for searches against the CODIS database to locate a match to an offender identified in a different crime. More recently, and highly publicized, have been several cases solved through working with genealogy database companies. Close genetic matches to a person who submitted their DNA sample to help trace a family tree can lead to suspects related to the family member. Genealogical detective work combined with DNA testing is now responsible for identifying perpetrators, as well as previously unidentified victims, in dozens of cases.

Serial and Mass Murder

While garnering a disproportionate amount of media attention compared to its occurrence, serial, mass, and spree murder remains a source of fascination for the public. This is undoubtedly played to by novelists and Hollywood moviemakers. Still, these types of murder set them apart from the confrontational or **intimate partner homicides (IPH)** that are most frequent. The characteristics of the cases and offenders who kill in a series or in large numbers at one time bring specific challenges to an investigation.

Intimate partner homicide (IPH) Fatal assault often the end result of an ongoing current or previous abusive intimate relationship.

Serial Murder

When law enforcement officials determined that a series of murders are linked, investigators most often rely on an expanded approach, partnering with additional agencies including state-level organizations or the FBI's behavioral science unit (BSU). The coordination of efforts is multiplied and complicated when a task force or multiagency approach is utilized. Think about the logistics of sharing information, handling evidence and investigative leads, and generally coordinating among agencies and investigators who do not report through the same chain of command to one official.

The serial killer, unlike the mass murderer, typically is concerned about being caught. The serial killer wants to continue his pastime of murder, and so the investigation is typically far more challenging. What distinguishes the two is the period of time between killings carried out by the serial offender.

Mass Murder

When three, four, or more victims are killed in a short period of time or in one place at one time, it is considered a *mass murder*. Hickey (2015) refers to changing location during the killings as "bifurcated," beginning in a private place and continuing in a private place or vice-versa. And if the offender kills several people at multiple locations, this is considered a *spree* killing. Workplace killings and school shootings are generally of this type, but in contemporary U.S. society we have seen mass killings at a variety of locations. Places of worship have been targeted, such as the Charleston church shooting on June 17, 2015, when a white supremacist murdered nine African American church members hoping to create widespread violence between races. In June 2016, in what has been described as a terrorist attack and a hate crime, a 29-year-old in Orlando, Florida, killed 49 people and wounded another 58 at a nightclub before being killed by police. The gunman claimed to be doing the shootings on behalf of ISIL, though no evidence of a connection to the Salafi jihadist militant group was found. The nightclub was patronized by members of the LGBTQ community, and hatred of gays appears to have been a major motivation for the shootings as well. And then on October 1, 2017, a 64-year-old man in Las Vegas began firing on a crowd of outdoor concertgoers from the window of his 32nd-floor hotel room overlooking the venue. This murderous assault killed 58 people and injured another 851 before the man took his own life. In February 2018, a 19-year-old used an AR-15-style rifle to kill 17 students at a high school in

South Florida. The murders resulted in yet more outrage from citizens and legislative action in Florida addressing some aspects of background checks and weapon accessories.

In the case of the Charleston church shootings, the gunman fled the scene and was arrested the next day, readily confessing. The latter two examples ended with the Orlando nightclub shooter killed by police and the Las Vegas killer turning a gun on himself. Suicide or being killed by first-responders are the two most common outcomes of such tragedies, and in some cases the killer will surrender to law enforcement (Hickey, 2015).

EXPLORE THIS

Visit the website of the Florida Department of Law Enforcement at http://www.fdle.state.fl.us/. Navigate to the page "Help Solve Unsolved Cases in Florida." The cases listed have been submitted by law enforcement agencies from around the state and provide a means to submit tips and information to authorities in cold case or unsolved murder investigations.

SUMMARY

The criminal homicide of one person by another is of great concern in a civilized society. While such crimes have decreased as a rate in the United States, in the last several years there has been an increase in violent crime in general and murder and nonnegligent manslaughter in particular. The recent increase does not yet offer an explanation, but this illustrates the continued need to examine the methods and motives of killers and constantly devote resources to training criminal justice personnel and improving investigative practices.

The investigation of homicide and death cases is labor-intensive but has also increasingly involved the use of technology to discover, collect, process, and analyze evidence including trace evidence previously hard to collect. Investigating such incidents is truly a team effort, with many people playing roles in the process. Patrol officers must arrive safely and enter the scene cautiously. Safety and assisting potentially injured victims come first before the officers transition to a focus on the crime scene. Detectives and crime scene technicians generally arrive shortly after a confirmed suspicious or unknown-cause death case, and so begin many more actions in the process. Evidence collected at the scene may go on to a laboratory setting for analysis to properly understand and situate the results in the explanation of the crime. Personnel from the office of the medical examiner or coroner for the jurisdiction will be involved for the autopsy and other medical test procedures, and a death investigator may also respond to the initial scene to work alongside police personnel.

KEY TERMS

Algor mortis 129
Asphyxiation 131
Blunt force
    trauma 131
Cold case 137
Coroner 120
Equivocal death 134

Intimate partner homicides
    (IPH) 138
Ligature 132
Lividity 129
Medical examiner 120
Munchausen Syndrome by
    Proxy (MSP) 135

Neighborhood
    canvass 126
Perimortem 129
Postmortem 130
Premortem 129
Rigor mortis 129
Strangulation 131

DISCUSSION QUESTIONS

1. How should a patrol officer approach and enter a reported homicide scene, and why?
2. What are some of the functions of detectives once they become involved in a homicide case?
3. Discuss the work of the medical examiner in a suspected homicide case.
4. What are the characteristics of an equivocal death case?
5. How do agencies approach the challenge of cold case homicides?
6. What complicates the investigation of suspected serial murders?

SAGE edge™

- Get the tools you need to sharpen your study skills. SAGE edge offers a robust online environment featuring an impressive array of free tools and resources.
- Access practice quizzes, eFlashcards, video, and multimedia at **edge.sagepub.com/houghci**

PRACTICE AND APPLY WHAT YOU'VE LEARNED

▶ edge.sagepub.com/houghci

SAGE edge™

WANT A BETTER GRADE ON YOUR NEXT TEST?

Head to the study site where you'll find:

- **eFlashcards** to strengthen your understanding of key terms.

- **Practice quizzes** to test your comprehension of key concepts.

- **Videos and multimedia content** to enhance your exploration of key topics.

Flying Colours Ltd/DigitalVision/Getty Images

7 Assault, Battery, Robbery, and Sex Crimes

RUNNING CASE: ROBBERY AT THE 7–10 SPLIT BOWLING ALLEY

Introduction

Assault and Battery
- Aggravated Assault
- Documentation

How It's Done: Documenting the Crime Scene
- Confrontational Violence

Robbery
- Documentation
- Robbers' Methods of Operation
- Robbery Statistics
- Victimology

Sex Crimes
- Megan's Law
- Rape and Sexual Assault Typologies

How It's Done: National Protocol for Sexual Assault Medical Forensic Examinations
- Investigating Sexual Battery/Rape
- Sexual Abuse and College Rape
- Criminal Deviant Sex Acts

How It's Done: Sexual Assault Examinations

Explore This

Summary

Key Terms

Discussion Questions

LEARNING OBJECTIVES

After reading this chapter, students will be able to

7.1 Discuss factors that distinguish assault from battery.

7.2 Explain the role of solvability factors in the investigation of robbery.

7.3 Describe the sexual assault investigation, and discuss the considerations in the interview of sexual assault victims.

7.4 Discuss the considerations in the interview of sexual assault victims.

7.5 Explain and give examples of sexual paraphilias.

Running Case: Robbery at the 7–10 Split Bowling Alley

Jeremy let the 1997 sedan roll down the south side of the city.

"No name place over here," Jeremy said. "Not one of those fancy, rich neighborhoods with one of those names. No North Slope or Harbor Town over here."

"No," Will agreed from the passenger seat. "You could call this area Trash Side." He watched a plastic bag drift down the street. Will was 20 years old, and though he thought of himself as an adult, Jeremy was 12 years older and had been in prison before. He had the kind of experience Will didn't necessarily want to repeat, but definitely admired.

Jeremy didn't drive too fast or too slow. "Just keep a nice steady pace. That way no one notices you."

Will listened. Jeremy drove because Jeremy had to control everything.

"Now keep your eyes peeled, looking for cops, for cameras. Don't ever stop in a place with a crowd of witnesses. Look for some place with a dark and deserted parking lot," Jeremy said. He nodded at the 7–10 Split Bowling Alley across Drummond. "Now we'll drive up and down, pass this place a few times, then we'll jump into action."

Will nodded. He was there to learn from Jeremy, but mostly he was attracted to the excitement.

*

A few blocks away, Officer Carl Jayden idled in his assigned patrol car in a convenience store parking lot. Carl had just completed his field training officer (FTO) program 3 weeks before and he still felt new and inexperienced.

The police radio crackled, and Carl heard the familiar voice of Rayna, the dispatcher, alert him and two other units to respond to "7–10 Split," the bowling alley. Someone had reported that a man had robbed a couple as they left the bowling alley. When the male victim tried to protect his wife, the robber roughed him up.

Officer Carl Jayden noted that the 7–10 Split must have just closed. Patrons or employees leaving closing businesses were at risk to be victims of robbery attempts, especially if they were the last ones out. Carl drove into the parking lot and notified Rayna that he was on the scene. Rayna acknowledged this and told him that an ambulance with paramedics to check the victims was on its way. Carl saw a crowd of people near the end of the sidewalk waving him over.

The victims, Sherri and Sam, came forward. Sam had a purple bruise spreading across the left side of his face.

Officer Jayden listened to their account.

"We were walking toward our car," Sherri said. "A man stepped out from behind the corner of the building."

"He was holding a pistol," Sam said. "He said, 'Give over your purse, wallet, and cell phones.'"

"Sam gave over his wallet and phone. Then the guy tried to grab my purse," Sherri said, still shaking from the shock of the situation. "Sam said 'stop' and raised his hand toward the guy."

"He hit me," Sam said, pointing to his sore left cheek and jaw. "With his handgun."

"I let him snatch my purse."

"We saw him run to a car sitting over there with its light off. The vehicle sped off on Drummond," Sam concluded.

Officer Jayden calculated that the robbery had happened 6–7 minutes before. He relayed the vehicle description to Rayna, who put out a "be on the lookout" or BOLO for the car and its two occupants.

*

Jeremy and Will abandoned the sedan on a neighborhood side street where the residents might not notice a 20-year-old car right away. Then they jogged four blocks to the strip mall where they had parked Will's beat-up VW bug and slammed into it. With adrenaline affecting his emotions and body, Will tried to drive calmly to Jeremy's house. He bounced up and down in the driver's seat gripping the steering wheel too tight and repeating different bits of the robbery. Jeremy divided up the cash and stuffed some into Will's pocket.

At Jeremy's apartment, Will asked, "What's the next job going to be?"

"It's not cool to be so excited like this," Jeremy said. "You gotta cool down. I'll find you when I need you."

Will drove away, his euphoria ebbing.

How does robbery differ from theft? What evidence can law enforcement officers look for at the scene of a robbery? What information might the victim or a witness offer?

Introduction

Criminal investigation of crimes committed against the actual individual, as opposed to property, is categorized collectively as *persons crimes* investigation. Generally, these crimes include **assault**, **battery**, **robbery**, and sex crimes, including **rape** or **sexual battery**. Among these offenses, the victim and offender know one another or are acquainted in the majority of assaults and sexual batteries, but most robberies are committed by strangers. If a detective division is large enough, this is often one of the specialized units. The persons unit often includes investigation of death and homicide cases as well, though we devoted a separate chapter to such investigations. Likewise, agencies, if large enough, may designate a special unit to investigate homicide or "major cases." While injuries and dollars stolen are tabulated for the crimes that are reported each year to law enforcement, the untold psychological harm and terror caused to victims of personally violent crime should be considered by everyone

Assault An attempted physical attack.

Battery One person's unlawful touching of another person.

Robbery Taking the property of another by force or the threat of force.

Rape or sexual battery Sexual assault involving penetration.

in and out of the criminal justice system. Given this consideration, agencies and individual officers must continually examine how they train and prepare to support and interact with victims and witnesses.

In 2015 there were an estimated 1.2 million violent crimes committed in the United States, which increased for a second straight year in 2016 to an estimated 1,248,185 (FBI, 2016, 2017). The 2015 figures were higher than 2014 yet lower by 0.7% from 2011 and down significantly (16.5%) from 2006. Disappointingly, in 2016 violent crime was estimated to have risen 4.1% from 2015 (FBI, 2017), though 2 such years of increase are not readily explained and do not necessarily indicate a trend. In 2016, 45.6% of violent crimes were cleared. This included 53.3% of aggravated assaults, 40.9% of rape offenses, and 29.6% of robberies.

Assault and Battery

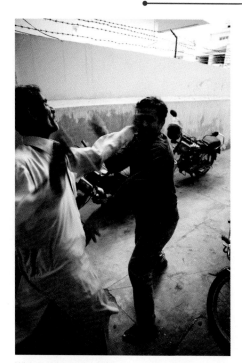

What would you need to know to determine if this fight is assault or battery?

©iStockphoto.com/ danishkhan

Aggravated assault or battery If the person committing an assault or a battery does so with the intention to inflict severe injury or by using any item as a weapon to cause injury.

Movies and television often pair the terms *assault* and *battery* when describing a fictional crime in a story. In many states these are two separate crimes. Assault in many states combines the elements of a credible and immediate threat coupled with a well-founded fear by the intended victim. If someone says to someone else across the room, for example, "Someday if you keep this up I might just punch you" and then they laugh, there is no crime of assault. If a heated argument has just ensued and one of the people in the argument screams, "That's it!" while balling up his fists and lunging at the other person 3 feet away—and the other person is convinced she is about to be struck—there are probably grounds for an assault charge.

In our previous example, if the person who commits the assault makes it as far as the intended victim and intentionally *strikes* the other person, a battery has occurred. The unlawful and nonconsensual touching of another person may be charged as a battery. The point about consent is important when you consider sporting contests such as football or boxing where the participants agree to be battered (enduring aggressive physical contact) within the rules of the respective sport. And while research does not generally support the efficacy of corporal punishment, a swat on a child's rear to gain compliance or redirect behavior is not likely to be pursued with a criminal charge unless used repeatedly or to excess, or with indications of intentional cruelty.

Aggravated Assault

If the person committing an assault or a battery does so with the intention to inflict severe injury or by using any item as a weapon to cause injury, this may be considered an **aggravated assault** or a battery. In tracking violent crime, and when considering homicide, it is important for police, policy makers, and researchers to examine rates, trends, and patterns of aggravated assault. While experiencing 20 victims shot but not killed in a community is tangibly "better" than 17 shot dead and 3 wounded, more accurate aim may have achieved the former. Tracking violent criminal acts is as important as their investigation. The FBI's Uniform Crime Report (UCR) tracks only aggravated assaults, not simple assaults. In 2016 an estimated 803,007 aggravated assaults represented a 5.1% increase over 2015, but a decrease of 7.3% from 2007 figures. From 2007 to 2016 there was a decrease in the rate of aggravated assaults of 13.5% (FBI, 2017). An attempted murder is considered an aggravated assault. This underscores the importance of competent investigation of the circumstances of such crimes (Tables 7.1 and 7.2).

Researchers and investigators know that victims of aggravated assault and homicide are similar. Victims and offenders are often acquainted and intimate partner violence (IPV) results in more female victims of aggravated assault, compared to homicide. Aggravated assaults are cleared at a relatively high rate in part because the victim and

TABLE 7.1 Aggravated Assault, 2-Year Comparison

YEAR	NUMBER OF OFFENSES	RATE PER 100,000
2015	764,057	238.1
2016	803,007	248.5
Percent change	+5.1	+4.4

Source: FBI, Crime in the United States, 2016.

TABLE 7.2 Aggravated Assault, Types of Weapons Used, and Percent Distribution by Region, 2017

REGION	TOTAL AGGRAVATED ASSAULTS	FIREARMS	KNIVES OR CUTTING INSTRUMENTS	OTHER WEAPONS (E.G., CLUBS, BLUNT OBJECTS)	PERSONAL WEAPONS (E.G., HANDS, FISTS, FEET)
Total	810,825				
Northeast		14.5	21.2	31.2	33.2
Midwest		30.3	14.7	27.3	27.7
South		32.1	17.1	32.2	18.7
West		20.7	17.1	31.5	30.8

Source: FBI, Crime in the United States, 2017.

offender do know one another. The victim is available, and hopefully willing, to bear witness against his or her attacker. The IPV incidents, discussed more in the next chapter, may have dynamics that complicate witness cooperation, but still allow police to identify the likely suspect. Such assaults may also be in the home, which limits the number of suspects. In aggravated assault cases that occur in a public place, there may be witnesses who can provide the identity of the assailant or furnish other information. Once again, given that many offenders and victims know one another, the dynamics of the relationship or acquaintance may impact the likelihood of cooperating with law enforcement in an arrest and prosecution. And, of course, this will subsequently affect the agency clearance rate but also help us understand the context of clearance as a measure of agency or criminal justice system effectiveness.

The first officer(s) on the scene must be cautious since the offender may still be present or the scene dynamic and potentially dangerous. A quick determination must be made as to whether victims are injured and medical assistance should be called to respond. If there is any possibility of injuries, law enforcement agency policies or practices will generally dictate calling for medical assistance as a precaution. If a victim is transported to the hospital an officer may accompany the person to gather information or possibly physical evidence that could be damaged or discarded during the trip to the hospital. Even the marks or injuries on a victim's body are evidence in the crime investigation. The psychological impact on a victim or their visible response should be observed and documented by the investigating officers. This includes recognition by the officers that a victim may be quite upset or in shock, and the officers must take care to be supportive and show understanding with a victim. This is not simply a professional duty in rendering informed aid to victims, but holds practical implications for the investigation as officers think ahead to winning or keeping the victim's cooperation.

Documentation

The primary officer assigned the call will, as in most other crime investigations, begin note-taking almost immediately and determine what steps need to be taken to protect the scene of the crime. This will include photographing the scene as well as collecting evidence. During this phase of an investigation the officer is collecting information to describe the event from beginning to end. This process will provide evidence about the intent of the offender, a needed element of the investigation. Witnesses may or may not be cooperative in furnishing information, but in the immediate aftermath of an aggravated assault or battery it is more likely people will talk. This includes the victim and suspect. The importance of obtaining witness and victim statements is underscored by research that supports a case moving forward with such testimony, while the same research shows that forensic is not always available, collected, or beneficial in robbery or assault cases (Baskin & Sommers, 2012). Officers must question each person who may have information and do this separately to the extent possible given conditions at the scene. The officer may need to issue a be-on-the-lookout (BOLO) through the communications center to alert other units or agencies of the flight of the perpetrator and possible mode of transportation.

How It's Done

DOCUMENTING THE CRIME SCENE

A recurring emphasis throughout the book is the need for officers and personnel to document. The National Forensic Science Technology Center (2013) notes in its section Document Actions and Observations, the following:

Principle: All activities conducted and observations made at the crime scene must be documented as soon as possible after the event to preserve information.

Policy: The initial responding officer(s) shall maintain documentation as a permanent record.

Procedure: The initial responding officer(s) should document:

a. Observations of the crime scene, including the location of persons and items within the crime scene and the appearance and condition of the scene upon arrival.

b. Conditions upon arrival (e.g., lights on/off; shades up/down, open/closed; doors and windows open/closed; smells; ice, liquids; movable furniture; weather; temperature; and personal items).

c. Personal information from witnesses, victims, suspects and any statements or comments made.

d. Their own actions and actions of others.

Summary: The initial responding officer(s) at the crime scene must produce clear, concise, documented information encompassing his or her observations and actions.

This documentation is vital in providing information to substantiate investigative considerations.

Source: *Crime Scene Investigation: A Guide for Law Enforcement,* National Forensic Science Technology Center (2013).

Of course, some witnesses may be frightened to speak with the police, fearing retaliation from the perpetrator. This may frustrate the investigation (and officers), but continued efforts may eventually yield witness or victim cooperation, so law enforcement personnel must be patient and professional in their dealings with everyone at the scene and be alert for indications of who may provide information. Investigators should check for any available video surveillance footage given the increased use by citizens and businesses alike. Some people may be willing to talk away from the scene at a time when it is less likely that others will know they are giving assistance to officers. Seeking witness information is one of the greatest and ongoing challenges to law enforcement in all manner of investigations, especially where there has been physical violence. The involvement of suspected gang members may scare potential witnesses, along with a neighborhood or cultural proscription against "snitching."

Confrontational Violence

When two individuals face off in an argument it can transition from a verbal disagreement to a physical fight. Throughout history people have confronted one another in anger over all manner of issues. The majority of reported argument-based fights involve males, although girls and women have increased in numbers of offenders. Often the fight will erupt after one individual has perceived some type of slight or insult to their honor by another person or the two both lay claim to something, such as the attention or affections of a third person. According to Hough and McCorkle (2017):

> The largest category of murder is that involving a conflict or confrontation between non-intimates. Almost half of all homicides are preceded by a fight or argument. Male-on-male homicides arising from such friction are the most common homicide situation. The lethal event may result from a brief, albeit emotionally charged, encounter or be a culmination of long-standing animosities. (p. 62)

Confrontational homicide One generally male individual who kills another in a conflict that likely escalated from an argument over what is perceived to belong to each person or a slight to honor; the most common form of criminal homicide.

While the FBI's uniform crime report (UCR) does not include a category of **confrontational homicide**, by looking at the circumstances we know that many began with an argument. The term *confrontational homicide* has been credited to researcher Kenneth Polk (1994), who cited a number of characteristics in many such murders. In addition to the perceived slight or insult, Polk observed that many confrontations were in a public location or in front of a number of other people, and often involved the use of alcohol. The insult suffered by one or both of the people involved may have been very minor indeed when examined after the fact but seemed tremendously important when one or both combatants believed they had to save face in front of others or when their inhibitions were muddled by alcohol. I investigated a murder case in which two friends and coworkers got into a Friday night argument over who would drink the last beer in the six-pack, which ended with one of the men shooting his friend in the chest with a small caliber handgun.

Victim-precipitated During the exchange the victim contributed in some significant way to the escalation of a situation.

Another important research effort, that of Marvin Wolfgang in 1958, examined homicide in Philadelphia between the years 1948 and 1952. The resulting book written by Wolfgang outlined a category of homicides that he referred to as **victim-precipitated**. This does not mean that the victim "had it coming," but rather that during the exchange the victim contributed in some significant way to the escalation that resulted in their death. Maybe the eventual victim shoved the other person, never thinking that their argument would or could lead to one of them dead. *Fighting words* is a term that describes comments that have the clear intention or effect of inciting a breach of the peace, and are typically words directed personally and specifically to someone. The U.S. Supreme Court, and the courts in other countries, has wrestled with free speech considerations against the danger of inciting violence when applying limitations to such words. Related to this may be how many people, and legislative bodies, view **stand-your-ground laws**. Such laws often establish legal protection for a person to use force or not retreat from a threat or perceived threat. There are limitations on such protections and the challenge for law enforcement and prosecutors to deal with a wide variety of assaults committed by people who claim they felt threatened, even in the absence of much evidence that a credible threat existed.

Stand-your-ground laws Such laws often establish legal protection for a person to use force or not retreat from a threat or perceived threat.

Victim blaming Suggests that the eventual victim is wholly responsible for whatever occurred.

Victim precipitation is a social science construct that helps us understand the interaction, but it should not be confused with the concept of **victim blaming**, which suggests that the eventual victim is wholly responsible for whatever occurred. All experienced law enforcement officers have faced many instances of two people pointing at the other and claiming that they started the fight. Witnesses, including friends and family members, may be hesitant to clarify who may have been most at fault in their opinion, and such cases are often declined for prosecution due to a lack of clear evidence one way or another.

Road rage may be considered a version of confrontational violence erupting between drivers or manifested by one driver acting aggressively toward someone else on the roadway who angered him. While younger drivers have been found more frequently to engage in aggressive driving behaviors, there are examples of drivers at most ages responding poorly to the frustration or hostility they feel. The American Automobile Association (AAA), the U.S. Department of Transportation (USDOT), and others have over the years linked aggressive driving and traffic accidents.

Workplace violence may come to our attention through media coverage of someone shooting and perhaps killing a number of coworkers or former coworkers. Yet most workplace violence falls into the category of an assault or a battery. Most workplace fatalities are accidental and not the result of criminal assault, but many nonlethal assaults do occur in the work setting. The risk of victimization varies across workplace occupations, as you might imagine. The National Institute for Occupational Safety and Health (NIOSH) estimated in 1996 that "Each week in the United States, an average of 20 workers are murdered and 18,000 are assaulted while at work" (Jenkins, 96–100). While the current number and rates of homicide are far less than they were in 1996, the reminder about the number of assaults is important and underscores not just physical injury and disruption in work settings, but economic losses to individual workers and businesses. NIOSH goes on to list previously established risk factors associated with assaults in the workplace and comment on the need for attention to environmental design of workplaces, administrative controls such as policies and staffing plans, and behavioral strategies that emphasize a need for ongoing employee training.

> **Road rage** A version of confrontational violence erupting between drivers or manifested by one driver acting aggressively toward someone else.

Robbery

Robbery is certainly one of the most frightening crimes that a citizen may encounter. The elements of the crime of robbery include taking something of value from another person by force or the threat of force, and putting the victim in fear. When initially reporting the crime of burglary, victims often mistakenly say they have been robbed. Communications personnel have to be sure to clarify as much as possible what crime someone is reporting so that the responding officers can act accordingly. There are several different types of robbery, but the common thread is the threat or actual use of violence to take property from another person. If the perpetrator uses his hand, fists, or feet to grab something from another person in a struggle, it is usually referred to as a **strong arm robbery**. A firearm or other clearly deadly weapon would generally be classified an *armed robbery*.

Robberies may occur in a variety of commercial establishments including banks, convenience stores, and liquor stores. But robberies also occur in people's homes, on the street, or even when driving a vehicle. This may be robbing a taxi driver, delivery driver, people stopped at a traffic light, or even an armored car transporting money. **Carjacking** involves robbing an individual of the vehicle they are driving. This became such a trend in the late 1980s and early 1990s that legislation was created making armed carjacking a federal crime. Solving these crimes can be challenging if, as with many other crimes, there is no physical evidence available at the scene to include specific identifying information about the robber

> **Strong arm robbery** Using violence or the threat of violence when taking something from a person.

> **Carjacking** Involves robbing an individual of the vehicle they are driving.

and possibly identifying information for a getaway vehicle. CCTV and various other video recordings when available are one of the most powerful tools in a robbery investigation and may be used in conjunction with eyewitness identifications by a victim or bystander.

Documentation

Patrol officers initially responding to a report of a robbery in progress must get there as quickly and safely as possible and assume they may arrive while the robber is still on the scene or fleeing. Officers are considering the time of day and the volume of traffic they encounter as they make their way to the reported robbery scene and whether or not to use their siren if they are attempting to catch the robber still at the scene. If multiple officers are available to respond, they will typically take a number of different routes to coordinate covering as many escape routes as possible knowing that the robber generally wants to complete his crime as quickly as possible and leave. The contemporary technology available through automatic license plate recognition systems can also assist in gathering license data for vehicles that the officers pass. It is also the case, however, that some robbers steal a vehicle in advance of the robbery to thwart any effort at their identification. One of the challenges that can occur when arriving on the scene of an active robbery is that of hostage-taking by the robber. The risk of injury to a victim increases in this event, and it may become more complicated if the robber or robbers barricade themselves at the location and there is a need for the assistance of a hostage negotiation team or SWAT. In the majority of robberies the report does not reach a 9-1-1 operator until the robber has already fled the scene. When the officers do arrive, and after ensuring that there is no active threat at the scene, they must try to obtain a description of the perpetrator and any vehicle involved to broadcast immediately to other officers.

As with most crimes, officers must be alert to what physical evidence may be present. If a note was used to demand money; if witnesses observed that the robber did not have his hands covered, leading to the possibility of fingerprints; or if there are areas in or near the crime scene where shoe impressions may have been left—these are among the types of physical evidence it is possible to gather. In addition, there may be video or photographic images available and what information can be furnished by the victim or witnesses. Victims and witnesses will likely be frightened, possibly in shock, or highly agitated, and officers must use their training and experience to manage and guide the information-gathering process with people in various states of mind. A canvass of the surrounding area or neighborhood is very often an important tool used by officers investigating a robbery. Speaking to people in the homes and businesses along the likely escape route will be the first priority; however, it is important to expand the canvass to speak with as many people in the area as is practical. Investigators are interested not only in whether witnesses observe the actual robbery, but whether they had observed a vehicle parked nearby that they thought to be out of place, or other information about the comings and goings in the local area that may tie into the robbery investigation.

Robbers' Methods of Operation

Modus operandi (MO) Method of operation.

Robbers also have different methods of operation, or **modus operandi (MO)**, including what type of businesses or people they target, what type of weapons or threats they use, and whether they have shown they will follow through on the use of violence, among other things. In advance of the crime, many robbers will conduct surveillance on the location and even perform dry runs to prepare. Some offenders have a distinctive way of moving through a robbery scene, specific phrases or ways of talking to victims and bystanders at a robbery, particular weapons they use, or types of disguises and specific valuables they demand. Robbers may establish a habit of threatening assault to gain the victim's compliance, or they may actually injure victims at a robbery scene. While not frequent, due to the amount of time it takes, some robbers may tie up victims or lock them into parts of a business while they complete the robbery. A series of robberies I once investigated involved a man and a woman who traveled around several adjacent counties robbing grocery stores and put baby powder in their hair to give the impression of gray or white streaks. Another robber placed an adhesive bandage on his face. In both these cases the individuals were creating a distinct and distracting feature that drew the attention of eyewitnesses and lessened the normal ability of a witness to take in other physical features of the robbers with any specificity.

When we think of the burglar who generally wants to use stealth to gain access to a place and steal goods without confronting anyone, we see a clear contrast with the robber who threatens harm in a direct confrontation with his victim. The consciously coercive nature of the robbery transaction holds real danger and might end in violence. The robber's motivation is generally to obtain money, although some who commit robbery, including juveniles, may also rob others for the thrill it provides. The 2016 estimate of average loss per robbery was $1,400, but the amount taken in bank robberies skews this number higher than the majority of other robberies of individuals or small businesses (FBI, 2017).

The fear and psychological distress, along with perhaps as many as one in three robberies resulting in a victim being injured, make this crime a high priority for law enforcement as well as in security efforts by businesses. Carelessness or lack of planning may increase the risk of injury for victims but may also provide clues for investigators. A bank robbery I investigated a number of years ago had the robber entering the branch bank with women's pantyhose pulled down over his face to obscure his features. When he arrived at the teller window and began the robbery he found that the hose impaired his vision too much, so he rolled them up to his forehead. The alert banking staff activated the bank cameras, which delivered several clear face photographs that I had on the evening newscast and in the next morning newspapers. By mid-morning two different people had already called in to identify the robber, and after securing an arrest warrant for the man and a search warrant for his house, we entered his residence shortly after noon and made the arrest. He still had most of the $450 he had managed to take from the bank.

Robberies are serious crimes that create real danger for members of the public and signal the potential for further danger or problems in an area of a community. Robbers may be amateurs who have only committed a handful of robberies, or professionals who have made the crime more of a vocation. Each type presents challenges for the investigators, though an amateur may not be as cautious about evidence or witnesses. Devoting resources to investigate the crime of robbery is a priority for any law enforcement agency, but as we have noted elsewhere, no agency has unlimited resources. If there is a pattern of robberies an agency may form a task force with other departments to work toward solving these crimes. Additionally, each robbery case that is not immediately solved by catching the perpetrator in the act may be assessed using solvability factors. This involves comparing available leads or clues against the statistical likelihood of solving a particular type of crime (Horvath, Meesig, & Lee, 2001). For example, if a criminal left behind fingerprints on a glass door as he pushed his way out of a bank and then fled in his own car with several people noting his tag number, that robbery will be solved quickly, with congratulations all around. If, on the other hand, the robber had on a mask, baggy clothing, left no physical evidence, and fled by an unknown method of transportation, the hopes of solving the case are quite a bit less, as you can imagine. But think about having a partial fingerprint, a partial tag number, and various other bits of information that investigators can follow-up on. This leads to a determination of what actions to take in a case, how many personnel to assign to the investigation, and how long to pursue the leads before suspending further action in the robbery investigation.

Robbery Statistics

The FBI's 2015 annual report on crime in the United States showed an estimated 319,826 robberies, a 1.5% increase over 2015 estimates. In 2015, 40.8% involved the use of a firearm (FBI, 2016). Again, the clearance rate for robbery reveals the challenges with the FBI's UCR program, reporting just 29.6% cleared in 2016 (FBI, 2017). Once again, an increase from one year to the next does not constitute a trend, and it is notable that the robbery rate in 2016 was 25.7% lower than 2007 estimates of robbery in the United States. As with most crimes it is also important to remember that the National Crime Victimization survey conducted for the 2016 year revealed that fewer than 42% of all violent victimizations were reported to law enforcement authorities (BJS, 2017). For robbery, it was estimated than 54% of robberies and 58% of aggravated assaults were reported, while rape or sexual assault was only reported in 23% of cases.

Victimology

Investigators consider the type of victim to help determine potential suspects or suspect types. The individual or team, who would enter a bank during daylight business hours with the institution's security features in place, is different from the robber or robbers who seek lone individuals in out-of-the-way places to "mug." Even in bank robberies, the more common method is the individual who does a "note pass" to a teller demanding money, as opposed to the dramatic—and more dangerous—group who takes over the bank until they get what they came for and flee. If the bank robber ends up with a dye pack that explodes to stain his clothing, law enforcement may end up with an important clue if the robber is captured. Studying the behavior of those who become victims, known as **victimology**, can provide information about why some people are chosen as victims. This, in turn, may allow for the dissemination of prevention literature to citizens so that they can be aware of how their daily habits may put them at greater risk of victimization. Where a person lives can be a component of risk, as well as where he or she works and during what hours. Reported home invasion robberies, for example, may involve victims who are known to the robbers as having valuables or perhaps something wanted by the robbers. In some home invasion incidents the victims may possess a quantity of drugs and the robbers not only want to take it, they may use violence with less restraint or calculate that the victims will not report the theft of their drugs to law enforcement.

Victimology The study of victims and their experiences, including interacting with offenders.

Sex Crimes

While sexual battery or rape is perhaps given the most attention among all sex crimes because of the impact on victims, there are a variety of crimes that have a sexual component central to the act. The term *sexual assault* may encompass a number of legally proscribed acts that do not include intercourse. The types and frequency of sex crimes must be understood within the context of the data challenges represented by underreporting and the particular circumstances that make victims hesitant to come forward. For example, in 2016 the crime of rape or sexual assault according to the National Crime Victimization Survey (NCVS) was only reported to law enforcement in approximately 22.9% of cases, while overall violent crime was reported in approximately 42% of cases (BJS, 2017). The NCVS data are gathered annually and complement the UCR reporting to provide a more complete picture of many crimes including that of sexual assault. More male victims are coming forward to report sexual assaults, and the CNVS and other surveys provide a clearer picture of female sex offenders than in years past.

Megan's Law

In the early 1990s few states had laws requiring sex offenders to register their domicile information subsequent to fulfilling their sentence. This changed in 1994 with the passage of the Jacob Wetterling Crimes Against Children and Sexually Violent Offender Registration Act. The act required local law enforcement agencies to maintain registries and for offenders to register when they move into a jurisdiction. Megan's Law, named for murder victim Megan Kanka, was a subsection of the broader act, and was directed at requiring that the sex offender registry information be made available to the public. Variations by state exist on how long registry is required (in some states, permanently), and for which sex crimes, including whether a minor was involved.

Rape and Sexual Assault Typologies

Crime in the United States, 2016, estimated there were more than 114,000 rapes committed (FBI, 2017). Again, these count only those known to police and are vastly underreported when compared to the NCVS. Most state statutes and the U.S. federal code addressing sexual assault recognize that offenders and victims may be male or female. We are also aware, however, that the majority of offenders are male, and the majority of victims of this crime are female.

As with many crimes, it can be helpful both to investigations and research to construct typologies to better understand variance within a crime category. Generally considered the most serious of sex cases, the crime of rape (sexual battery) is one in which the different typologies have been classified as based largely on the dynamic of the offender and the victim. In addition to the typologies that follow, detectives investigate rape cases involving varied victims such as spouses, the elderly, men, prison inmates, children, and others.

There is not a single motivation for rape; nor is there a single type of rapist. In fact, given the underreporting of the crime itself, we cannot be certain that we have a thorough understanding of all of the dynamics of this criminal behavior. Criminologists cannot point to one specific theory, and investigators cannot utilize one lone investigative approach to the various crimes falling under the title of rape or sexual battery. While sexual gratification is undoubtedly one component for many if not most rapists, power over another person and dominating their actions are clear aspects of the behaviors and motivations of many rapists. As noted by Alvarez and Bachman (2017), "To rapists, violence and sex are linked in a perverse way that allows them to project their insecurities, fantasies, and frustrations onto the bodies of their victims" (p. 205).

As with any framework, as a guide it can be helpful in understanding the broad characteristics of subgroups. At the same time, offenders can exhibit characteristics that blend the subtypes or not fit clearly into any one category alone. A framework of characteristics still in use today as a basis for understanding different types of rapists was developed and articulated by Groth (1979). Groth first divided his typology between "power" and "anger" rapists. The power category was then subdivided into the power-reassurance rapist and the power-assertive rapist. The anger category is subdivided into the anger-retaliatory rapist and the anger-excitation rapist. The FBI relabeled and expanded these categories, referring to the power-reassurance rapist as within their typology of *pseudo-selfish*. The power-assertive, anger-retaliatory, anger-excitement, opportunistic, and gang rapists fall under a second typology of *selfish*. The FBI developed the two additional rape classifications, the *opportunistic rapist* and the *gang rape*. The FBI describes a variety of characteristics under each typology that have been particularly useful in identifying repeat or serial rapists. Understanding or having insight to the offenders' motivations can help investigators identify specific behaviors as consistent with other observations about specific suspect behaviors.

As you might imply from the name, the **power-reassurance rapist** has a primary motivation of exhibiting or holding power over his victim. This will not necessarily involve a use of physical force, because this individual fantasizes a consensual sexual encounter. The offender is likely a stranger to the victim and will plan his assault to occur after dark in the victim's residence. He may apologize following his rape of the victim. The **power-assertive rapist** uses his assault to prove his masculinity as his primary motivation. He often may use strategies to con his victim into trusting him or going somewhere with him, only then to use force to rape his victim. The **anger-retaliatory rapist** is typically a man who is clearly hostile

Power-reassurance rapist Has the primary motivation of exhibiting or holding power over his victim.

Power-assertive rapist Uses his assault to prove his masculinity as his primary motivation.

Anger-retaliatory rapist Typically a man who is clearly hostile to women and commits his crime to intentionally demean and punish.

How It's Done

The National Protocol for Sexual Assault Medical Forensic Examinations (2013) was developed with input from experts across the country. A consistent procedure for the investigation of sexual assault involves the efforts of those in the medical field as well as law enforcement. Properly collected evidence will aid the case through victim-centered care and support and a coordinated approach (DOJ, 2013). In the Operational Issues section of the National Protocol, the following topics are addressed in detail:

- Sexual assault forensic examiners
- Facilities
- Equipment and supplies
- Sexual assault evidence collection kit
- Timing considerations for collecting evidence
- Evidence integrity

The section on the examination process contains the following topics:

- Initial contact
- Triage and intake
- Documentation by health care personnel
- The medical forensic history
- Photography
- Exam and evidence collection procedures
- Alcohol and drug-facilitated sexual assault
- Sexually Transmitted Infection (STI) evaluation and care
- Pregnancy risk evaluation and care
- Discharge and follow-up
- Examiner court appearances

Source: U.S. Department of Justice, Office on Violence Against Women.

to women and commits his crime to intentionally demean and punish. If the victim is not a known specific object of his hatred, she may resemble or remind him of someone he does hate. The **anger-excitation rapist** or **sexual sadist** wants to inflict pain and suffering on his victim to achieve his sexual satisfaction. This is typically the rapist who severely injures or kills his victim, often through means of torture.

Anger-excitation rapist or **sexual sadist** Wants to inflict pain and suffering on his victim to achieve his sexual satisfaction.

Investigating Sexual Battery/Rape

Although every victim shares the experience of having been violated by their attacker, they may differ in their reactions when investigators first encounter them. When a rape case is reported to law enforcement, a uniformed patrol officer may be the first official to interact with the victim. The primary responsibility for responding officers is to take care of the victim. The individual may be physically injured, in a state of shock, or exhibiting a range of emotions and behaviors that have to be viewed in context by the investigating officers. This is where it is important to say how critical it is that law enforcement and medical personnel are properly trained and supervised in not only how to identify and gather evidence, but more importantly, how to interact with and be supportive of the individual victim. For a crime that takes away all control from a victim, it is crucial that first responders understand the need to help the victim regain some sense of control. If officers or medical personnel act judgmental, skeptical, or even cold and "professional," they may complicate every step of the investigation that is to follow. If the victim has already been withdrawn, she may become even more so or defensive or even hostile if she perceives any treatment other than compassionately professional.

Officers may be called to meet with the victim in their home, at a hospital or doctor's office, or in some other location where the assault may have taken place. After ensuring that needed medical care of the victim is taking place, the officer must engage in rapport-building so that he or she can productively talk to the victim. Some of the initial interaction will be to explain the procedures needed to gather evidence. The victim will be able to furnish some or a good bit of information about her attacker, but she is also part of the crime scene based on the potential to recover biological evidence such as her attacker's semen or hairs, as well as any marks or injuries that he inflicted. The victim may be hesitant to talk, and so it is important to remember that physical evidence can be part of what speaks for the victim. If a suspect has been taken into custody, his clothing and physical being are also components of the overall crime scene, just as is the location where the assault took place. Gathering of any evidence must follow established procedures, including the choices a victim has prior to a sexual exam kit and maintaining the chain of custody. Investigators and medical providers must respect the choices made by victims regarding their treatment. A medical patient is not required to report to law enforcement, though employees of various government agencies may have a lawful duty to report a suspected crime victim to help ensure they receive assistance.

What are some important considerations to keep in mind when interviewing a victim of a sexual assault?

BSIP SA/Alamy Stock Photo

If the sexual assault has just occurred, then a preliminary interview of the victim may be in order before the arrival of the detective, if one is available at all. This initial interview will be to determine if it is likely that a rape actually took place and to gather information about the suspect so that attempts can be made to locate him. The interview with the victim has to establish that a sex act described by statute is alleged by the victim to have occurred. If so, any information that the victim can then offer about the identity of the suspect and his specific behaviors are the next priority. The victim may very likely be embarrassed, possibly in shock, fearful, very angry, or vacillating among several states and emotions. The increased focus on the seriousness and complications of sex crimes has led to specialized training and protocols within the emergency medicine profession. These include the use of sexual assault nurse examiners (SANE) and sexual assault response teams (SART). The support of a victim, while care is also taken with evidence collection, can be an invaluable resource to investigators.

As with any other investigation, documentation by officers should begin immediately. Anything said by the victim or witnesses should be recorded verbatim, if possible. Interviewing the victim should be done in a comfortable and private location to the extent possible. Documenting the victim's emotional state of mind is done by specifically describing how the victim is behaving in addition to her specific words. Photographing injuries and other physical evidence at the scene, whether done by the officer or crime scene technician, should also be explained to the victim so that she understands why she is being asked to do certain things such as revealing different parts of her body. While it is obviously important to note any preexisting relationship between the offender and the victim, it must be remembered that any relationship, including a prior sexual one, is not an excuse or explanation in a sexual assault case.

The initial investigating officer must also be very thorough in describing and documenting all available information regarding the suspect if he is still at the scene or if he has fled. As with the victim, the suspect's emotions should be described in detail based on observed behavior and his statements. If the preliminary interview is conducted by a patrol officer, or when a follow-up interview or interrogation is done by an investigator, the officer must also portray an attitude of nonjudgment with the hope that the suspect will talk. Any injuries must also be described and photographed. If the suspect is to be interviewed right away, and this is rarely the case for patrol officers on the scene, it should be done away from the victim or other witnesses.

The follow-up interview will typically be conducted by a detective and will be in great depth. Training for investigators in the various approaches and dynamics of the interview process are critical. There are differences not only by type of crime committed, but based on age and maturity of the victim. A misperception by many people is that only a female investigator is appropriate for conducting the interview with a female victim. Gender of the investigator is typically not relevant; what is important is that the officers are well trained and act in a professional manner. Investigators will again explain the need and importance of this interview and of the questions while showing support for the victim and an understanding of why the victim will likely be uncomfortable. Due to the nature of the crime and the questions that have to be asked of the victim, it is best that two officers carry out the interview both to support the victim and to preclude an allegation by a victim that questions asked were unprofessional. The victim's emotional state may very well hinder a full or completely accurate recitation of what occurred before, during, and after their assault. Effective interviewing in such cases often relies upon the investigator allowing the victim to narrate what happened with limited interruptions for questions.

While there is an ebb and flow to interviews and interrogations, the initial phase will be developing some rapport with the victim before transitioning to the stage of describing the event and answering specific questions about the event and the suspect. Each step of the way through an investigation, and certainly the interview, the investigators should explain to the victim why things are done and why she is asked the various types of questions. The investigator will ask the victim to describe graphic details of the assault and for a description of the suspect, including every detail she can remember about the assailant. If the attacker was a stranger, details about his physical appearance may be critical. This may include particular ways that he spoke, what he was wearing, his musculature or amount of hair on different parts of his body, any smell that she noted about his breath or body, and any scars, marks, or tattoos that she saw. The description of the attack itself will include any and all information about the interaction with the suspect before the assault. When the victim is asked to describe what happened during the assault itself, the investigator will want to know about each action the suspect took and how he interacted with the victim to include how he spoke to her and what he instructed her to do. At the conclusion of the assault, it is still important to know what the subject did, when and how he left, and whether and what he may have said to the victim.

Sexual Abuse and College Rape

Larry Nassar, a former Michigan State University sports clinic doctor and physician for USA Gymnastics, is accused of using his various professional roles to abuse as many as 265 girls and young women. Another contemporary example was the prosecution of a Penn State assistant football coach who was charged in 2011 on more than 50 counts of child molestation. Jerry Sandusky was convicted on 45 counts and sentenced to 30 to 60 years in prison. Other school officials were charged or fired for failing to properly report information and to protect children. Examples from other organizations such as the Catholic Church, Boy Scouts of America, and others are an unfortunate reminder of cases that highlight the failure of individuals and organizations in authority to act to protect potential victims through appropriate monitoring as well as acting on complaints that had been initiated by many of the victims.

College students face vulnerability to sex crimes, which led Congress to pass legislation in 1990 that required colleges and universities that received federal student aid to "prepare, publish, and distribute, through appropriate publications and mailings, to all current students and employees, and to any applicant for enrollment upon request, an annual security report." This 1990 act was revised in 1992 and in 1998 and renamed the Jeanne Clery Disclosure of Campus Security Policy and Campus Crime Statistics Act.

The act requires that institutions promulgate specific policies to promote awareness and prevention of sexual assault as well as to establish rights for sexual assault victims. Most victims of attempted rape or rape knew their assailant, with most being boyfriends, ex-boyfriends, classmates, or friends.

Criminal Deviant Sex Acts

There are a number of deviant acts that are sexually arousing to various individuals. These behaviors are known as **paraphilias**. Some of these are criminalized, are seen mainly in men, and likely began in adolescence. When such paraphilias victimize others, law enforcement may be called upon to investigate. Some of these behaviors coincide with sexual assault. **Exhibitionism** is obtaining sexual gratification from the exposure of genitals to strangers.

Paraphilia Deviant acts that are sexually arousing to various individuals.

Exhibitionism Obtaining sexual gratification from the exposure of genitals to strangers.

Fetishism A sexual attraction to various objects, very often articles of clothing.

Scatologia When a caller is sexually aroused by the fearful reaction of the person receiving the call.

Frotteurism Involves rubbing against another person who is not aware of the action, usually in a crowded public place.

Necrophilia An instance of someone being sexually aroused by a dead body.

Voyeurism Known to many as Peeping Toms, voyeurs are sexually aroused by secretly observing people in various stages of undress or having sex.

One such case I investigated involved an individual who would drive into fast food drive-through lanes with no pants on so as to induce a shocked reaction from the attendant. His gratification ended one day when an angry rather than shocked restaurant manager quickly dumped the drink she was holding into the man's lap, and as he tried to speed away she accurately wrote down his tag number, leading to his identity and arrest. **Fetishism** is considered a sexual attraction to various objects, very often articles of clothing. An individual may commit a crime to obtain such paraphilias items or take them as a trophy during another crime. Obscene telephone calls, known as **scatologia**, are when a caller is sexually aroused by the fearful reaction of the person receiving the call. **Frotteurism** involves rubbing against another person who is not aware of the action, usually in a crowded public place. Sometimes seen in murder cases, notably serial killer investigations, **necrophilia** is an instance of someone being sexually aroused by a dead body. As for **voyeurism**, voyeurs—known to many as Peeping Toms—are sexually aroused by secretly observing people in various stages of undress or having sex.

How It's Done
SEXUAL ASSAULT EXAMINATION

Criminal justice and health care system employees are uniquely situated to significantly impact a victim's overall recovery after being subjected to sexual assault. An important way that this is accomplished is through the use of a thorough and clear protocol for forensic examinations of rape victims coupled with a compassionate approach by law enforcement officers and medical personnel. Created by the U.S. Department of Justice, Office on Violence Against Women, and most recently updated in 2013, *A National Protocol for Sexual Assault Medical Forensic Examinations: Adults/Adolescents, Second Edition*, "provides detailed guidelines for criminal justice and health care practitioners in responding to the immediate needs of sexual assault victims" (iii). Within this document the steps recommended for initial law enforcement, 9-1-1, and EMS responders include:

- Assess victims' needs for immediate care for potentially life-threatening or serious injuries. Administer necessary first aid and request/obtain emergency medical assistance according to jurisdictional policy.

- Address safety needs of victims and others at the scene (e.g., offenders may be present), calling for assistance/backup if needed.

- Assess quickly the age, abilities, communication modality, and health condition of victims and tailor your response as appropriate (e.g., a qualified interpreter, assistive devices, or protective service worker may be needed).

- Respond to requests for victim assistance as quickly as possible. Understand that victims need immediate assistance for many reasons: They may not be safe, may be physically injured, and/or are experiencing trauma. Be aware that time delays in response can cause loss of evidence and increased trauma.

- If injuries do not appear serious, emphasize to victims the need for

- medical evaluation and address related health concerns. Also, explain the purpose of the exam and what happens during the exam process, keeping in mind that the amount of information that victims want at this time varies.

- Inform victims about exam facility options (if options exist) and seek their consent to transport them to the facility of their choice (if they had options) for treatment and/or medical/forensic evaluation.

- Encourage victims' interaction with advocates as soon as possible after disclosure of the assault, even if victims choose not to receive medical care and/or have the medical forensic exam. In a few jurisdictions, advocates may be dispatched directly to the scene to provide victim support and advocacy, if appropriate. Follow local procedures for activating an advocate.

- Ask victims if they would like family members or friends to be contacted.

- Explain options for interpretation and translation for victims who are not proficient in English or who may prefer to communicate in a non-English language.

- Take measures to preserve crime scene evidence, including evidence on the body and clothing of victims. Document victims' demeanor and statements related to the assault, according to jurisdictional policy.

- Explain to victims their reporting options. Keep in mind that the amount of information desired will vary per individual.

- Responding law enforcement officials should seek basic information from victims about the assault in order to apprehend suspects and facilitate crime scene preservation in a timely manner.

EXPLORE THIS

Navigate to the website of the U.S. Department of Justice, https://www.justice.gov/. Then go to the Publications tab and scroll down the page to the link to the Office of Sex Offender Sentencing, Monitoring, Apprehending, Registering and Tracking (SMART). Look around at the publications and resources available.

SUMMARY

While the investigation of all crimes is important to individuals and to a community, violent physical crimes committed against citizens receive a high priority from agencies. Throughout the chapter we examined several of the most common types of persons crimes including assault, battery, robbery, and sex crimes. A well-founded fear that someone's threatening behavior will immediately harm someone constitutes assault in most states. Actual unlawful physical contact by one person on another is generally considered battery. Either of these offenses may be part of a robbery, which is the threat or actual use of force to take something of value from another person. If an assault or a battery does take place during the commission of a robbery it may be considered a "lesser included" crime. This means that the robbery, as the most serious part of the incident, may be the primary charge made, although the other offenses could be charged as well. The decision for what charges to file or pursue is often based on the discretion of the prosecutor. Among the violent personal crimes, robbery is the one most likely committed by a stranger. This reality of the victim–offender dynamic in robberies is a barrier to an investigation, yet we have seen that many challenges also exist when victims and offenders in assaults and sex crimes are known to one another.

Among the various criminal acts that contain a sexual component, rape or sexual battery is the most serious. Many victims are hesitant to report the offense for a variety of reasons, but a serious challenge is when victims believe they cannot trust the criminal justice system employees to treat them professionally and with compassion in the wake of an intensely personal assault. This challenge has faced victims and law enforcement for a long time, and agencies, working in concert with medical and social service providers, must continually improve practices and outreach.

Persons crimes, absent a homicide, have the victim as at least one witness to the offense. In the case of an assault in a public place or the robbery of a business, there may be additional witnesses who can, if they are willing, provide information to investigators. Engagement with victims and witnesses is an essential part of how the people in the process—police and victims—interact and often determines not only the success of the investigation but how the victim is supported and able to deal with any number of psychological challenges they face as a result of the crime. If the information to begin an inquiry comes from a source other than the victim, it will be necessary for law enforcement officers to work hard and gather all available information to move an investigation forward. Once contacted, the victim may cooperate with investigators, though the time lapse from the actual crime may hamper the overall effort to find and hold the offender accountable.

Technology aids in the investigation of persons crimes in varying degrees based on the offense and the location where it took place. A fight that occurs at a nightspot may be recorded by video cameras. Similarly, many businesses and even taxi and service delivery drivers also maintain cameras to make recordings for their safety. Of course, crime scene processing is often critical, as in many crime investigations, and will rely on the training and care taken by technicians as well as investigators and prosecutors utilizing any evidence developed to help tell the story of what happened. The use of DNA in criminal proceedings has grown over the last three decades and has been critical both in prosecutions as well as the exoneration of suspects. When dealing with the victims of sexual assault in particular it is important to remember that the victim, the suspect, and the location of the assault are all potential crime scenes in the event and, if possible, evidence must be gathered from all three locations.

KEY TERMS

Aggravated assault or battery 146
Anger-excitation rapist or sexual sadist 156
Anger-retaliatory rapist 157
Assault 145
Battery 145
Carjacking 151
Confrontational homicide 150
Exhibitionism 159
Fetishism 160
Frotteurism 160
Modus operandi (MO) 152
Necrophilia 160
Paraphilia 159
Power-assertive rapist 156
Power-reassurance rapist 156
Rape or sexual battery 145
Road rage 151
Robbery 145
Scatologia 160
Stand-your-ground laws 150
Strong arm robbery 151
Victim blaming 150
Victimology 154
Victim-precipitated 150
Voyeurism 160

DISCUSSION QUESTIONS

1. How have the rates of intimate partner homicide changed in the last 25 years and why?

2. List and discuss several risk factors that appear to contribute to intimate partner homicide.

3. How does feminist theory contribute to our understanding of intimate partner homicide?

4. In what ways does social learning theory connect to intergenerational transmission of violent behavior?

5. How can agencies utilize lethality assessment to improve outcomes for battered persons?

6. What agencies and groups should be part of a fatality review team or process? Why?

SAGE edge™

- Get the tools you need to sharpen your study skills. SAGE edge offers a robust online environment featuring an impressive array of free tools and resources.
- Access practice quizzes, eFlashcards, video, and multimedia at **edge.sagepub.com/houghci**

©iStockphoto.com/sturti

8 Familial Crimes

RUNNING CASE STUDY: INTIMATE PARTNER VIOLENCE AND CHILD ABUSE

Introduction

Domestic Violence
    Investigating Domestic Violence

Intimate Partner Violence (IPV)

How It's Done: Determining Level of Risk for Intimate Partner Violence

    Changing Views on Intimate Partner Violence
    Correlation Between Intimate Partner Violence and Child Abuse
    Restraining Orders and Arrest of Intimate Partner Violence Offenders
    Responding Officers' Safety
    Stalking

How It's Done: Investigating Suspected Stalking Crimes

How It's Done: Law Enforcement Stalking Protocol
    LGBTQ Victims

Child Abuse
    Incidence of Child Abuse
    Risk Factors for Child Abuse
    Child Abuse Legislation
    Investigating Child Abuse
    Sexual Abuse of Minors
    Infrequent Situations of Sudden Death or Mistreatment

Elder Abuse

Explore This

Summary

Key Terms

Discussion Questions

LEARNING OBJECTIVES

After reading this chapter, students will be able to

8.1 Discuss the special considerations in investigating domestic violence.

8.2 Describe the dynamics of intimate partner violence.

8.3 Discuss the challenges of investigating crimes against children.

8.4 Explain motivations for the abuse of elderly people.

Running Case Study: Intimate Partner Violence and Child Abuse

Katie checked Christopher's eyes to make sure he was asleep, closed the *Batman* book she'd been reading, then stood up. The clock read 7:30 p.m.

Jeremy had said he was going "out," the way he had since he lost his job last month. Normally, she would study for her nursing degree, but the last time she was studying, Jeremy threatened to throw her computer out the window, so she'd dropped out.

Without the computer she'd have no way to chat with her mom, who lived out of state. She couldn't use her cell phone because Jeremy used the device he'd given her to track her every move. However, he didn't know how a computer worked.

Katie walked into the kitchen and started the dishes. She tried to remember when Jeremy had started hitting her one night, then apologizing and trying to make up the next. When had she started to feel trapped? Things had gotten worse when Jeremy lost his job, but they had been going downhill for months. She tried to remember Jeremy 18 months before, when he bought Christopher superhero books.

Katie undressed in the dark so she wouldn't see the bruises Jeremy had left under her clothes. She lay down in bed, but couldn't sleep. She didn't know when Jeremy would return and how he would act.

Her mother had e-mailed saying it was hard for women to leave men like Jeremy because they didn't have enough support in place before leaving. Katie had been saving the money that Jeremy "gave" her every week. She almost had enough for her and Christopher's bus tickets to her mom's. Once she was there, her mom would help her figure things out.

Even though Jeremy had lost his job, he seemed flush with cash. Most nights he returned to the apartment with a six-pack of beer and chocolate milk for Christopher. After a few beers, he'd get angry. One night, Christopher knocked over his milk, and the liquid spilled onto the carpet. Jeremy pinched the inside of the boy's arm until he cried. Katie checked her son every night for bruises. So far, he only had one. She had everything under control.

How did Jeremy have so much money? She figured it must be some kind of crime. He had been in trouble before, but he'd never talked about it. Was there anything she could do to help Jeremy change? If Jeremy was arrested, how would that affect Christopher? And if she ran to her mother, would Jeremy be able to find her? Would he retaliate?

At 11 p.m., Katie heard a car dropping Jeremy off. When he walked into the bedroom, she held perfectly still, pretending to be asleep so she wouldn't have to deal with him.

*

At midnight, Officer Carl Jayden knocked on the apartment door. Rayna's dispatch had relayed that a neighbor reported screaming from the apartment. She was sure the woman and young boy were in danger.

Officer Jayden heard no sounds. He knocked again. Eventually, a man wearing a pair of cotton pants appeared. At the sight of Officer Jayden, he ran out of view.

Officer Jayden was relieved that Rayna had called two backup units. He knew how heated the air in a small apartment could get when a couple was fighting. While he waited for the backup officers to arrive, he reviewed the facts.

Based on the screams and the belief that someone was in danger, he could enter the apartment without a warrant. He would have to be careful interviewing the boy, as the child would most likely be scared. The boy might also think that the way he and his mother were treated was normal and not have the maturity or insight to fully report what he'd seen and heard.

Officer Jayden would need to note every aspect of physical evidence, including bruises as well as the behavior of the mother and child in the presence of the suspect. That way if the mother did not want to cooperate, there were other ways to establish probable cause that the man had abused them.

Many people do not understand the complex dynamics in many abuse situations. What are some of the reasons why people remain for some period of time in an abusive relationship?

Introduction

Criminal investigation of physical harm done by one person to another is an important responsibility. Patrol officers and detectives routinely explore allegations and acts of such violence. Some of these incidents involve people who are or were in a familial or intimate relationship. We explored the topic of sexual assault in the previous chapter but must emphasize that family members can be the victim of such assaults by other family members. Similar to our need to separate topics so that we can devote specific attention to them, investigators may be assigned specific crime types to investigate so that they can develop expertise in the dynamics of the crime. While robbery, for example, is most often committed by someone not known to the victim, the offenses discussed in this chapter are almost exclusively perpetrated by those known, and likely trusted, by the victim. It is estimated that police spend up to 30% of their time on calls for service related to familial crimes.

Not all crimes considered in this chapter and investigated by police under this heading involve physical violence. Abuse in the form of emotional, psychological, financial, and other methods to inflict harm may form the basis of an investigation by government employees. Theft from a family member may accompany other domestic or **intimate partner violence (IPV)**. Child abuse through terrorizing a young child often happens in a home where intimate partner violence is also occurring.

Intimate partner violence (IPV) Various forms of abuse or violence committed by an individual with whom the victim is currently or had previously been in an intimate relationship.

Domestic violence (DV) Assault and violent crimes committed by family members against other family members.

Domestic Violence

Domestic violence (DV) includes assault and violent crimes committed by family members against other family members. An assault specifically committed against someone who is currently or was previously in an intimate relationship with the offender is referred to as intimate partner violence (IPV). We will take up that discussion in the next section of the chapter. As discussed in the homicide chapter, because of the body of the deceased, authorities generally are aware of murder victims, including those within the family. Far less known are instances of other acts of physical and psychological violence between family members. Victims within a family are quite often reluctant to discuss, let alone make a criminal complaint about, severe mistreatment at the hands of another family member. The stigma attached to family violence, along with personal embarrassment, is one reason why victims are hesitant to speak up. It is also the case that some family members think of the family interactions, including mistreatment, as normal or just the way it is in a family. The National Crime Victimization Survey (NCVS) report for 2016 indicated a rate of 3.9 per 1,000 of population for victims of domestic violence, which the survey defines to include "crimes committed by intimate partners (current or former spouses, boyfriends, or girlfriends) and family members" (BJS, 2018, p. 7).

Investigating Domestic Violence

Law enforcement officers as well as researchers recognize that the various acts of physical and emotional violence, abuse, and harassment may be part of an escalating pattern that can end in severe injury or death for a victim. Domestic violence has come to refer to a variety of offenses that may occur, such as battery, arson, robbery, and damaging property, that are elevated in level of charges and potentially at sentencing because the crime was done in the context of harming or intimidating those in a formerly intimate or family relationship. These circumstances were once frequently considered to be "family matters" and not vigorously investigated by police agencies, nor seen by political leaders or many in society as something appropriate for investigation by police. Times have changed. Society, legislators—and, by extension, law enforcement agencies and officers—have a deeper understanding of the innate wrongness of intimate partners and family members being victimized by other family members and the multiplier effect in communities as the result of such behavior. State statutes now clearly assert that domestic violence is to be treated like a crime and not a private family matter. Legislation has provided law enforcement agencies and prosecutors tools to help them investigate, arrest, and prosecute domestic violence. Legislation, both state and federal, has also given the courts the ability (and often the mandate) to impose enhanced sanctions and specific requirements such as offenders attending a *batterer intervention* program. Sadly, many judges routinely ignore such legislative mandates and fail to utilize the enhancements and requirements, sometimes substituting a shorter program such as *anger management* in place of batterer intervention, which is specific to the dynamics that need to be addressed.

One of the ongoing training aids developed by the International Association of Chiefs of Police (IACP) is a series of Report Review Checklists. The Domestic Violence report checklist poses the rhetorical question of whether the report includes "all needed information." Figure 8.1 includes a partial list of suggested questions that officers should address.

The checklist moves into questions about whether the suspect used or threatened to use a firearm, have access to a firearm, whether these facts frightened the victim, and whether the officer confiscated the weapons. The checklist asks about the officer assisting the victim with safety planning and, if both people used some type of force, if the officer took steps to determine if one acted in self-defense. The checklist includes more questions regarding whether an arrest was made and accompanying actions. It is important early in this part of the chapter to emphasize the necessity of a good investigation, especially in DV cases where we know that victims may not want to or be able to participate in the prosecution in a few months based on the unique dynamics that surround these types of crimes. These investigations are important and take time to do well.

■ **FIGURE 8.1** The Domestic Violence Report Checklist Developed by the International Association of Chiefs of Police (IACP)

☐ Is the time of the call recorded (including time of incident, time of dispatch, time of arrival)?
☐ Are the elements of the crime(s) articulated to meet state and/or federal laws that address domestic violence? Firearms?
☐ What were the observations upon approach?
☐ Is there a valid protection order in place?
☐ Is the scene concisely described/diagramed?
☐ Were photos taken and details recorded?
☐ Is the relationship of the parties identified?
☐ What is the history of the relationship? (include frequency of any violence, intimidation, and threats)
☐ Were all witnesses interviewed and documented?
☐ Were there children on the scene?
☐ Was information about previous incidents documented?
☐ Were weapons/objects used?
☐ What was the emotional state of the victim (what did they report they were thinking and feeling)?
☐ What evidence was collected?
☐ Is evidence of fear articulated in the report?
☐ Have all threats been clearly documented?
☐ Is the use of coercion and/or force articulated?
☐ Have all injuries (visible and non-visible) been documented? Were injuries existing or new?
☐ Was there any property damage? Theft? Burglary?
☐ Are stalking behaviors identified? (e.g., following, repeated calling, sending unwanted gifts)
☐ Did the officer inquire about possible strangulation (hands, ligature, etc.)?
☐ Did the victim report being strangled ("choked")? If so, was it described in detail?
☐ Did the victim request/need medical attention?
☐ Did the victim report sexual violence?
☐ Were all spontaneous statements captured?

Source: Adapted from IACP, Domestic Violence Report Review Checklist.

Intimate Partner Violence (IPV)

The majority of crime in the United States has continued to decline over the past 20 years. Statistics indicating rates of intimate partner violence and violence against women have matched these declines overall. We have already referred to the National Crime Victimization Survey (NCVS), which conducts interviews of household members age 12 or older in a sampling of approximately 135,000 households conducted annually by the Bureau of Justice Statistics (BJS) in the United States. The NCVS estimated in 2016 that the rate of intimate partner violence was 2.2 per 1,000. The 2016 survey estimated that 49.1% of domestic violence is reported to police and 46.9% of intimate partner violence (IPV; BJS, 2017). In other words, fewer than half of such crimes are reported to law enforcement.

Intimate partner violence includes a range of physical acts and also involves psychological and emotional abuse against someone with whom the abuser is in a relationship or was previously intimately involved. By the nature of the relationship, the suspect and victim are known, possibly for some period of time and possibly in a current marriage or other relationship. Various factors have been shown to be associated with greater risk of IPV, though it is important to remember that *correlation* of one or more factors does not equal *causation* for a specific act or crime. Some of the factors studied are significant age differences between partners, race, socioeconomic status, length of the relationship, and substance use and abuse (Breitman, Shackelford, & Block, 2004). Unsurprisingly, perhaps, jealousy and problems in a relationship, as well as a lack of social support, have been noted as well (O'Leary, Smith-Slep, & O'Leary, 2007). Increased risk has also been noted for pregnant women and their unborn child (McMahon & Armstrong, 2012). Information about such factors may be helpful for the law enforcement officer investigating a reported crime and for the counselor working to assist a victim, but again, no single piece of information can be conclusive.

How It's Done

DETERMINING LEVEL OF RISK FOR INTIMATE PARTNER VIOLENCE

Some questions suggested in the IACP protocol (and of interest in an interview or interrogation of an IPV suspect, too) to help determine level of risk are:

1. Present threats to kill the victim.
2. Past threats to kill this victim or other victims.
3. Use of weapons such as guns, knives, or other potentially lethal weapons.
4. Possession of lethal weapons.
5. Degree of obsession, possessiveness, and/or jealousy regarding the victim.
6. Violations of a restraining order with demonstration of little concern for the consequences of arrest and jail time.
7. Past incidents of violence against this victim and/or others.
8. Present or past threats of suicide.
9. Access to the victim and/or the victim's family.
10. Hostage taking.
11. Depression.
12. Other mental illness evidence or indicators regarding the stalker.
13. Drug or alcohol abuse of the stalker.
14. History of prior stalking of this victim or other victims. (2018, p. 41)

Changing Views on Intimate Partner Violence

Law enforcement officers and investigators in the various jurisdictions may reflect the priorities of their particular community. And so individual communities may not collectively see intimate partner violence as a priority. This may be a general ignorance of the effects such abuse has throughout their community and the larger society. A great deal of intimate partner violence is viewed through the lens of the *woman's* thought process or the collection of factors tabulated after an incident. It is simplistic to assert that women (the overwhelming victims of IPV) should simply stop seeing the abuser or move out if cohabitating. The dynamics of intimate partner violence are poorly understood by many, if not most, people in society, including law enforcement officers and the victims themselves. In the United States, social and criminal justice system progress was finally made in the investigation, prosecution, and sanctioning of impaired drivers. And while progress has been made regarding IPV, there is quite a way to go in addressing the use of power, dominance, and violence against those in vulnerable positions at the hands of those who should be most protective. IPV, long viewed in society as a "woman's problem," is statistically and responsibility-wise a man's problem, and one that is largely reduced by cooperative efforts to educate and speak out against mistreatment of girls and women. The collection of viewpoints for examining the victim analogy has included the cycle of violence, the so-called battered woman syndrome, **Stockholm Syndrome**, and some theoretical approaches including entrapment and traumatic bonding. Societal explanation, such as the treatment of women by men, arises from the feminist theoretical perspective. But society as well as families also faces the dynamics of conflict, a second theoretical orientation to explain domestic violence, and how people may use violence in resolving conflict. These predominantly focus on the mind-set of the victim while less research and training efforts for law enforcement focus on the behavior and motivation of the overwhelmingly male offenders. The abuser's behavior is described by many as ties to **social learning theories** that explain behavior arising from the observation and imitation of others.

Stockholm Syndrome A captive becoming psychologically dependent on or aligned with their captor.

Social learning theory Learning through observing and imitating others, especially admired individuals.

Correlation Between Intimate Partner Violence and Child Abuse

The correlation between IPV and child abuse is an important investigative consideration for law enforcement and child/family protective investigators. For law enforcement officers the use of tools such as the **Danger Assessment** may help reduce the potential for an escalating situation of violence. Still, law enforcement is typically called after the fact of crime, so what can they do but retroactive investigation? Importantly, fatality review teams and researchers note that in cases of intimate partner homicide (IPH), police have interacted previously with the victim and perpetrator. This fact alone does not allow law enforcement to put in place an iron-clad strategy to prevent further violence in a relationship, but it supports the use of risk assessments to inform victims, make referrals, and interact with prosecutors and probation officials regarding an offender.

Danger Assessment An instrument that helps to quantify and illustrate the level of danger posed by an intimate partner of killing the abused person.

Restraining Orders and Arrest of Intimate Partner Violence Offenders

A tool of the judicial system is the **restraining order,** or injunction issued by a court with the intention to force the batterer to stay away from the petitioner. Research regarding the effectiveness of such court orders is mixed with the reduced risk of violence shown in some cases but also some violations of these injunctions, many coming shortly after issuance of the restraining order. For some individuals, such injunctions impose the clarity about what consequences may come from their continued behavior, and that is a good thing. In some

Restraining order Court order intended to protect a person or organization from assault or harassment.

other instances this further blockage of access to the target of their control, dominance, and abuse enrages the abuser to a point of even lethal violence. Law enforcement officers can arrest batterers who violate the terms of the restraining order.

Important in the arsenal of the contemporary law enforcement officer are state statutes that allow and encourage the arrest of intimate partner violence offenders. Historically, the battery of one person by another has been classified as a misdemeanor crime and therefore required an officer to witness the crime to arrest the person without a warrant. Now, if the officer has probable cause to believe that the misdemeanor crime of battery in the context of domestic violence occurred, she may arrest the offender without a warrant. There are many examples of statutory exceptions to the warrant requirement, and battery in a domestic violence situation is one of those, as well as violating a domestic violence protective order injunction, violating a condition of pretrial release in a DV case, stalking, and crimes such as sexual cyberstalking (for example, Florida State Statute 784.049 [4][a]). If the officer has probable cause that a felony was or is being committed, she may make an arrest without a warrant as well.

Most states' statutes on intimate partner violence require that the officer attempt to determine the **primary aggressor**. It is important for law enforcement officers to receive specialized training in investigating domestic and intimate partner violence, as this training addresses best practice in determining the primary aggressor. A conscientious officer conducting an investigation of the crime will often be able to discern the responsible party through evidence and interviewing those knowledgeable of the past and present circumstances of the victim's relationship with the suspect. A poor and often damaging police practice may be dual arrest of people at a complaint of IPV or domestic violence. If both persons claim to have been battered by the other, and perhaps each has red marks from some physical contact, the law enforcement officer may arrest both and let the prosecutor "sort it out," claiming to have done his duty to keep the peace. Most states' statutes say that arrest in IPV cases is mandatory (if probable cause is established), arrest is preferred, or arrest is at the discretion of the officer. Individual agencies may be more stringent than the state law and perhaps make the arrest mandatory where the state law establishes that the arrest is *at least* preferred, for example. Given the hesitancy borne of fear and other reasons for some victims to cooperate, most states do not require the consent or insistence of the victim.

During any criminal investigation it is important for the officers to be thorough in notetaking and report writing as well as specific in the language they use. An important example of this is in the use of the term *strangulation*. In days past an officer may have written that a suspect was alleged to have *choked* a victim. I may choke if something is obstructing the passage of air in my throat. If a person is squeezing my throat to keep me from getting air, he is strangling me! Someone squeezing another person's throat shut does not mean her well. Recognizing the seriousness of the crime of strangulation, many states have now added strangulation as a unique crime or an aggravating factor raising the underlying crime of battery to a felony. For investigating officers, there may not be readily visible external indications of strangulation but the victim may have internal injuries. They may also observe redness, fingernail impressions or bruising, or petechial haemorrhaging—small, red spots in the eye caused by broken capillaries resulting from increased pressure in the head or neck and frequently seen in strangulation cases. If an officer suspects that a victim has been strangled, he should call for medical assistance.

Primary aggressor
An individual determined through investigation to pose the most threat or who has shown a history of abuse to the victim.

Responding Officers' Safety

As with many persons crimes that have just occurred, law enforcement officers should respond quickly and safely, be cautious when arriving at the reported location of the crime, determine if anyone is in need of medical care, and then separate the people involved and begin to gather information and evidence. The call for service of a domestic violence or intimate partner violence case is one that has long necessitated at least one backup officer being dispatched along with the initial officer. It is recognized that when emotions are running high and family and relationship dynamics are involved, there is a heightened potential for people to act out physically, including instances of one or both people involved assaulting the officer who has responded. Officers responding to a call of domestic violence cannot know everything there is to know about what has occurred and what is happening as he arrives at the scene. It is critical, however, that those communications personnel and the responding officers gather as much information as they can before the officers arrive and as those officers are on the scene. Whatever act of violence that a person will do to his own family member is important to put into the context of what that person might do to a law enforcement officer who is coming to try and help the family member.

Stalking

Although most people are aware of **stalking** when it involves a celebrity, the majority of such cases are incidents of people who are or were intimately involved. The U.S. Department of Justice Office on Violence Against Women partnered with the National Center for Victims of Crime (n.d.) to create the Stalking Resource Center (SRC) to provide information and resources to victims and to those who work with victims of stalking. One section of their website lists things people can do if they believe they are being stalked, including:

Stalking Following or monitoring someone's activities who interprets the stalking as harassment or intimidation.

- If you are in **immediate danger**, call **911**.
- Trust your **instincts**. Don't downplay the danger. If you feel you are unsafe, you probably are.
- Take **threats** seriously. Danger generally is higher when the stalker talks about suicide or murder, or when a victim tries to leave or end the relationship.
- Contact a crisis hotline, **victim services agency**, or a domestic violence or rape crisis program. They can help you devise a safety plan, give you information about local laws, weigh options such as seeking a protection order, and refer you to other services.
- Develop a safety plan, including things like changing your routine, arranging a place to stay, and having a friend or relative go places with you. Also, decide in advance what to do if the stalker shows up at your home, work, school, or somewhere else. Tell people how they can help you.
- **Don't communicate** with the stalker or respond to attempts to contact you.
- Keep **evidence** of the stalking. When the stalker follows you or contacts you, write down the time, date, and place. Keep emails, text messages, phone messages,

letters, or notes. Photograph anything of yours the stalker damages and any injuries the stalker causes. Ask witnesses to write down what they saw.
- Contact the **police**. Every state has stalking laws. The stalker may also have broken other laws by doing things like assaulting you or stealing or destroying your property.
- Consider getting a **court order** that tells the stalker to stay away from you.
- Tell **family, friends, roommates, and co-workers** about the stalking and seek their support.
- Tell security staff at your job or school. Ask them to help watch out for your safety.

The Centers for Disease Control and Preservation (CDC) published a summary report compiled from data gathered for the year 2011 on sexual violence, stalking, and intimate partner violence (Breiding, Smith, Basile, Walters, Chen, & Merrick, 2014). The summary reported that, "An estimated 15.2% of women and 5.7% of men have been a victim of stalking during their lifetimes. An estimated 4.2% of women and 2.1% of men were stalked in the 12 months preceding the survey" (p. 1). And in findings from the National Intimate Partner and Sexual Violence Survey (2010–2012 State Report) it was noted that "1 in 6 women and 1 in 19 men experienced stalking at some point during their lifetime" (Smith et al., 2017, p. 2).

These surveys reveal the seriousness of people experiencing the fear of stalking behavior, perhaps in conjunction with other actions by the stalker. The stalking behaviors measured by the Supplemental Victimization Survey of the NCVS listed the following "seven types of harassing or unwanted behaviors consistent with a course of conduct experienced by stalking victims," including:

- making unwanted phone calls
- sending unsolicited or unwanted letters or e-mails
- following or spying on the victim
- showing up at places without a legitimate reason
- waiting at places for the victim
- leaving unwanted items, presents, or flowers
- posting information or spreading rumors about the victim on the internet, in a public place, or by word of mouth (Baum, Catalano, Rand, & Rose, 2009).

The recommendations of the Stalking Resource Center listed above include notifying police. Documenting the occurrence of any of these behaviors with local law enforcement and the person's own records is important to provide information that may be useful in a criminal investigation. Stalking behaviors may be leading to more serious crime, so people should take such matters seriously and seek assistance. Law enforcement can use the documented and reported incidents to help triangulate the behavior or actions of a suspect if one is known, and track the behaviors to help develop a suspect if one is not already known. Stalking victims are typically women, and most frequently younger than age 24

(Baum, Catalano, Rand, & Rose, 2009). It is important that law enforcement training stresses the need for officers to recognize a pattern of actions and document them. In the absence of an as yet unidentified pattern, officers must realize that any single incident of a stalking behavior that someone reports may be the beginning or part of a larger pattern and so they should be prepared to write a report documenting the occurrence. Stalking cases can continue over a lengthy period of time, sometimes years, and behaviors may be reported to several different law enforcement agencies. Officers should ask victims about previous reports or the possibility of behaviors that only upon reflection the victim may realize could have been part of being stalked.

Stalking is an ongoing series of actions rather than a single offense. That a victim feels significant fear is a key element in legislation that criminalizes stalking behaviors. That the individual reasonably knows that his ongoing actions cause fear in his victim is also a key aspect of establishing the criminal pattern of behavior. The U.S. Department of Justice Office of Community Oriented Policing Services (COPS), in conjunction with the National Center for Victims of Crime, published a document summarizing research and recommendations for law enforcement agencies in dealing with suspected stalking cases (2002). In the document's model protocol, they offered some guidance on investigating these crimes and emphasizing that the investigator learns all he can about the stalker and his modus operandi. See How It's Done for three questions they suggest asking to establish a threat assessment.

How It's Done
INVESTIGATING SUSPECTED STALKING CRIMES

(1) Who is the suspect?

(2) What risks of violence does the suspect pose to the victim?

(3) How does the investigator manage the suspect and dangers posed to the victim? (p. 34)

The section on evidence collection notes that the investigating officer should look for information about the suspect's identity such as:

1) Name
2) Description
3) Personal information
4) Residence
5) Place of work
6) Mode of transportation
7) Vehicle make and model
8) License plate number (p. 36)

From the victim, the protocol suggests gathering evidence that "corroborates the stalking behavior." These include letters or notes the victim received, phone records or recordings, social media or Internet messages, and journals or logs that the victim has compiled. In addition, investigators should determine if the victim has changed her residence, telephone number, habits, or possibly obtained a weapon due to her fear of the stalker.

Law enforcement officers may be able to obtain search warrants for the car, residence, or workplace of the suspect. If officers obtain search warrants they may look for things identifying the victim or her movements, photos or video of the victim, telephone records or an actual cellular phone, or other items that show attention focused on the victim. With the omnipresent computer, digital, and social media aspects of contemporary life, cyberstalking behaviors may be significant in an investigation, with more than one in four victims reporting e-mail or instant messaging being used (BJS, 2009). The investigation will examine the suspect's background and movements, and if investigators are able to question the suspect they will ask about a great many things. These are only some of the considerations during the conduct of stalking investigations. The protocol from the National Center and others like it is more extensive than this brief example and is the basis for training officers and others who might investigate crimes of violence.

How It's Done
LAW ENFORCEMENT STALKING PROTOCOL

The National Center for Victims of Crime received funding from the U.S. Department of Justice Office of Community Oriented Policing Services in 2002 to research and develop a protocol to assist police in responding to stalking cases.

This protocol does the following:

- Promotes a strategic approach that encourages early intervention.
- Broadly defines the roles of officers in functional areas including patrol, 9-1-1 operators, detectives, and community relations.
- Presents guidelines for developing and participating in a coordinated community response stalking.
- Encourages the use of collaborative problem-solving techniques.
- Defines the need for centralized, preferably computerized, case management.
- Describes appropriate threat assessment techniques.

LGBTQ Victims

Victims of intimate partner violence span every lifestyle, race, and gender. Research varies as to the rates of such abuse within same-sex relationships as compared to heterosexual partners, but the violence is present and the challenges to investigators can be significant. Already marginalized by many in society, a victim with a lifestyle not considered mainstream in their particular community may be fearful or hesitant to report their victimization to authorities. In addition, the victim may believe that people will not believe them based on the perceptions people have about who can be a victim and who can be an offender. For police agencies, this requires outreach efforts in the community, and for officers and investigators the implication is that they must use their skills in rapport-building to reassure victims that their crimes will be investigated with the same level of professionalism as any other.

Child Abuse

The most extreme form of child abuse is when a parent or caregiver kills the child. There are many other forms of child abuse, including physical abuse, sexual abuse, emotional and psychological mistreatment to the point of abuse, and the harmful neglect of children under a person's care. Estimating overall numbers of child abuse is complicated by the relationship dynamics within the family, societal expectations about what is discussed or reported, and the fear of embarrassment, not being believed, or further pain at the hands of the family member if a child does report the abuse.

Children being victimized is tragic, and something that contributes to how their victimization can happen is their vulnerability both physically and psychologically. Given the challenges with discovering or reporting the types of crimes against children it should be recognized that the numbers that are gathered likely underreport the overall level of victimization. One of those challenges to accurate accounting is the fact that a majority of these crimes occur in the household and behind closed doors without the presence of witnesses. Because the idea of harming children is a foreign thought to most people there may be a tendency to mistake discussion of these crimes for different ways that families interact and how parents choose to raise their children. This presents a clear challenge for investigators as they come up against strongly held beliefs about directing or disciplining someone's child or children. A generational condition of neglect within a family can also perpetuate such maltreatment as children grow up and sometimes neglect the well-being of their own children.

Incidence of Child Abuse

The abuse or neglect of children often results in significant physical harm and sometimes death. The risk and incidence of abuse and neglect decrease with the age of the child, and victims are both girls and boys. It is also important to consider the emotional and psychological harm done to juveniles, which they may carry forward into their adult life, causing a ripple effect into the future impacting others. Those abused may become abusers, thus continuing a generational cycle. Children who suffered abuse may act out as adults against family members or nonfamily individuals.

Here are some of the key findings from the U.S. Department of Health and Human Services 2016 annual Child Maltreatment Report (2018):

- The national estimate of children who received a child protective services investigation response or alternative response increased 9.5% from 2012 (3,172,000) to 2016 (3,472,000).
- The number and rate of victims have fluctuated during the past 5 years. Comparing the national rounded number of victims from 2012 (656,000) to the national estimate of victims in 2016 (676,000) shows an increase of 3.0%.
- Three-quarters (74.8%) of victims were neglected, 18.2% were physically abused, and 8.5% were sexually abused.
- For 2016, a nationally estimated 1,750 children died of abuse and neglect at a rate of 2.36 per 100,000 children in the national population.

If effective prevention occurs, the need to investigate crime is always less. In cases of crimes against children this may be especially true. Given the generally expanded involvement

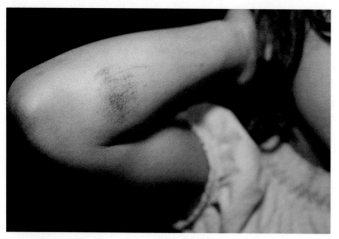

What are some of the key risk factors for child abuse?

Charles Gullung/Science Source

of family members during a young person's upbringing, the interaction with school teachers and staff members, children interacting with their friends and peers, and the generally benevolent view of society toward children and their vulnerable status, it seems possible to coordinate more support around issues of child abuse and neglect.

Risk Factors for Child Abuse

There are causes of abuse as well as factors that have been identified as correlated with neglect and physical mistreatment that give communities a starting point on preventive strategies. Children raised in poverty are overrepresented in reported abuse and neglect cases, and children living in a household marked by violence between the adult caregivers may become victims of violence themselves. Children who live with one parent who has a live-in partner face the highest rates of abuse.

In addition to the environment of the home and family, a number of risk factors associated with abuse and neglect focus on the individual aspects of the child. Some of these characteristics such as premature birth or being born to teenaged parents can set the stage for stress that many parents or families are ill-equipped to deal with. If the newborn or infant suffers from medical and physical challenges, the result may be an extended hospital stay that separates them from frequent contact with parents, interfering with the important initial bonding. This disruption in bonding may have consequences that manifest in how the child is cared for after going home. If these or any other conditions lead to the abuse of a child, then delinquency may darken the short-term future of that child and possibly farther into adulthood, with the individual engaging in violent or other crime.

Child Abuse Legislation

Both the federal government and each state have created laws to address child abuse and neglect. An important component of such legislation, and a necessary and helpful part of how investigations come about, is the requirement for various individuals to report suspected cases of abuse or neglect. The first report of suspected abuse or neglect often comes through a teacher, neighbor, doctor or nurse, or even the child's grandparents. Most such investigations proceed under state statutes that utilize the child protective services of a state agency established to address the welfare of children. Most state statutes or rules require the responsible agency to actively investigate within a period of days. These services are rooted in the view that government entities should represent the interests of the child when the family does not. Child protective investigators, certain health care employees, social workers, and law enforcement personnel intersect in many cases of child abuse as well as instances of neglect. Each of these professionals is motivated to and mandated by protection of the child. Legal authority generally exists to remove a child from their immediate setting if it appears there is a continued danger to the juvenile. This authority to take custody is not unlimited and typically requires a court hearing within a few days. This time allows workers to prepare a preliminary report for the court that provides as much information as possible about the individual circumstances of the case.

Investigating Child Abuse

In cases that involve potential criminal charges, law enforcement officers may be called to investigate. As with many other types of reported crimes, one of the first responsibilities of a law enforcement officer is to try and determine the likelihood that a law has been violated. This is more challenging than in most criminal investigations due to the specific aspects of child victimization. The officer, if presented with a physical injury, must work in concert with medical personnel to assess whether the injury was accidental, as the family may claim, or the result of an assault on the child. Investigators are faced with potential signs of abuse that include bruises and lacerations, as well as physical damage not immediately discovered such as broken bones, brain damage, and other internal injuries.

The investigation of abuse or neglect will involve the victim, the location or locations where victimization is likely to have occurred, any suspected family members or other suspects, and any other individuals who may have knowledge about the circumstances of the case. If the child is too young to adequately articulate what happened, then the officer may be largely left with the medical assessment to begin with. Interviews with family members, reporting parties, and others identified during an initial investigation may lead to further evidence of abuse or being reasonably certain the injury was the result of an accident. Accurate recording of all information from the medical examination is critical if the investigator will try to determine through the examination or records whether there have been any other reported injuries to the child. The investigator may be able to determine through the medical records whether the infant or child was premature or faced any physical or psychological disability. And the investigator and medical personnel should be alert to whether the injuries were reported in a timely way and by the parents or someone else.

If the child is old enough to talk about what happened to him the officer and child protective investigator face the task of establishing a rapport with the child and making him feel comfortable talking, to the extent possible. A parent or other person responsible for the abuse may have told the child not to talk with anyone about what happened or to make up a story about it. Sometimes the interview will take place in a setting like a child advocacy center designed with children in mind that also allows remote viewing of the interview by one or more members of an interdisciplinary team investigating potential abuse so the child is not intimidated by the presence of too many people asking him too many questions. The investigator must also evaluate the child's credibility in regards to specific statements while understanding that inconsistencies in the child's narrative may indicate his hesitancy to discuss what happened to him rather than represent an intentionally false statement about a caregiver. Individual children may speak relatively freely or need prompting in all or different phases of an interview. The child may fear what his parents will do if he speaks about the abuse or he may view anything done to him as something he deserved because of his behavior. The investigator is also working to assess whether the child grasps the difference between lying and truthfulness and the level of the child's maturity. In considering sexual abuse, for example, this may include what level of knowledge the child has about specific acts or behaviors that are beyond the typical knowledge of children of the same age.

At the suspected location of child abuse, investigating personnel will pay close attention to the living conditions. The behavior of the child's parents and any siblings or extended family members may give the investigator indications of circumstances in the home. Examining others in the family can be important to an investigation since other children or a spouse may also have been physically abused. The examination of the home setting and interviews and interaction with parents and family members are likely to mix together at certain points in the investigation, which provides the investigators the opportunity to make observations, including how family members interact, and ask questions of family members. The parents may contradict one another in their statements, and they may exhibit harsh discipline. Emotional abuse may also appear alongside physical abuse, though it may stand alone as well. While child protective investigators may become involved in complaints of psychological or emotional abuse, law enforcement is less likely to be called initially for such investigations. The longer-term nature of such abuses may manifest physically in the child as slower physical development and with conduct disorders and awkward or stunted interactions socially.

Sexual Abuse of Minors

Instances of the sexual abuse of a minor, if discovered or suspected, will involve law enforcement authorities in the investigation. There may be physical indications on the victim including bruising or tearing of genital and rectal areas, and there may be behavioral indications possibly noticed by teachers (if the child is not homeschooled) including withdrawal, sexual knowledge or behavior beyond what is normal for the child's age, delinquency, and other behaviors. Sexual abuse of a minor is most often committed by someone known to the child, and when a parent, a biological parent. When the sexual abuser is not the child's parent, it may be another juvenile or individuals with the ability to physically isolate the victim and have some position of trust, including faith community members, school personnel, or babysitters, among others. Of those who seek to habitually sexually abuse children, the pedophile is most known. This is usually an adult male who targets boys or girls within a certain age range. Less frequently uncovered at the local level are Internet-based or large-scale commercial sexual exploitation rings that may operate across state lines or internationally.

Infrequent Situations of Sudden Death or Child Mistreatment

Several other statistically infrequent situations may present themselves to investigators. Particularly difficult investigations for officers include Sudden Infant Death Syndrome (SIDS) and Sudden Unexplained Infant Death (SUID). SUID, the death of children younger than 18 months, can pose challenges for the medical examiner or coroner seeking the cause or manner of death. The investigation may determine a number of physical causes of the death including infections, accidental suffocation, or previously unknown diagnosed heart defects. If no cause can be determined for the death, the medical examiner will usually consider the case SIDS.

An infrequent criminal mistreatment of a child is called Munchausen Syndrome by Proxy (MSBP). This form of abuse involves a parent or caregiver harming, sometimes by poisoning, a child along with furnishing false medical information to authorities so that the person can gain the sympathy of others who believe the individual is caring for their seriously ill child. Munchausen Syndrome is when the individual induces symptoms in themselves with a similar name of gaining attention and sympathy.

Elder Abuse

Another segment of the population that may be vulnerable to criminal victimization includes elderly members of society. The U.S. Census Bureau has provided mid- and long-range projections that show the number of those over age 65 in America growing to around 83.7 million by the year 2050 (U.S. Census Bureau, 2014). These numbers represent a significant percentage of the U.S. population and clearly show the importance of devoting resources to crime prevention as well as investigation of criminal abuse of older individuals. In part because of the demographics of projected aging and due to the vulnerability often accompanying senior years, the U.S. Administration on Aging (AoA) 30 years ago established the National Center on Elder Abuse (NCEA) as a resource on research, training, and education on the challenges and appropriate responses to elder crime. The NCEA, through its initiatives, is a resource used by many law enforcement agencies to help in training officers and to develop department policies on handling elder abuse investigations.

What are some of the challenges presented when investigating a case of elder abuse?

fototext/Alamy Stock Photo

Con artists and family members may take advantage of an older person through fraud and financial trickery or by physical or emotional mistreatment using the abuser's control over someone. "Elder abuse perpetrators are non-family members as well as family members, may be in a position of control or authority, or may be anyone who establishes a relationship of trust with an older person" (Daly, 2010, p. 153). Disturbing reports may put a spotlight on how some people in long-term care or retirement facilities are battered or have their medicines intentionally manipulated. Psychological and emotional mistreatment by scaring, humiliating, or screaming at older members of society is an unknown figure. An elderly person may also be neglected by withholding food, medicine, or clothing. And investigations have also revealed forced sexual activity against the elderly or those medically confined to a facility and unable to object to their treatment and abuse.

And so abuse of the elderly may be physical, psychological, or financial. Investigators may face similar difficulties to investigations involving other vulnerable victims. These crimes occur in the household without other witnesses. The elderly person may not be able to provide clear information or they may be too afraid to speak out against their abuser. Similar to some domestic abuse and intimate partner violence, the victim may continue to wish for the abuse simply to stop without seeing their family member criminally charged. For the investigator, looking into and interviewing those closest to the victim may provide clues to move the investigation forward. Those individuals the elderly person is most dependent on will receive the attention of the investigator. These are often the spouse or adult children of the victim, though it may also be a person who has ingratiated themselves to the victim and gradually began to take advantage of the victim's age and physical or mental decline. A social service agency may receive a referral to assist the elderly person in a number of ways that may also serve to minimize their vulnerability to being victimized. Elder abuse and neglect investigations require a multidisciplinary approach and highlight the need for police agencies to partner with other government, nonprofit, and health care organizations to hold abusers accountable and craft a plan to help ensure the safety of the victim.

> **EXPLORE THIS**
>
> **LETHALITY AND RISK ASSESSMENT**
>
> Intimate partner violence (IPV) has historically been a challenge in most countries and continues to be a problem faced by every community in the United States. Jacqueline Campbell, a faculty member at Johns Hopkins School of Nursing, developed the first version of the Danger Assessment with the hope of assisting survivors of abuse in part by raising their awareness of the danger they are in (2004). The Danger Assessment is generally completed by law enforcement officers, social work advocates, and health care personnel, but anyone can access the assessment and other documents online. This process is considered to be an important part of the investigative process in intimate partner crime as it can also help law enforcement and prosecutors determine which crimes may escalate to homicide.
>
> Go to the website at https://www.dangerassessment.org/. Examine the information gathered from the victim and then reflected back by a trained person to the victim in a discussion about risk.

SUMMARY

While each state and the federal government have created laws prohibiting the persons crimes described in this chapter, investigation and enforcement remain challenging. Part of the difficulty in such cases lies in the dynamic of fear of a known offender, mixed emotions over how best to deal with offenders who are family members, and a tendency for individuals to underestimate the danger they are in, to name a few. When law enforcement does become involved, officers must be diligent in gathering information about a suspected crime and about the individuals involved, and should be trained in the dynamics of domestic and interpersonal violence. The procedures and protocols that have been developed reflect the collected experience of professionals and the research insights of accumulated outcomes about family and relationship crimes over the last several decades. This information should guide the policies put in place by agencies and the training devised for their officers.

The topics of our discussion in this chapter included the abuse of children, intimate partners, and older family members. Each of these areas has moved slowly into the larger social conscience as representing serious societal problems. The consequences of child abuse and neglect include transmission of abusive behavior into the adult patterns of many as they grow past adolescence. Intimate partner violence has seen some reductions possibly owing to lower marriage rates and the ability to leave marriage that was once not so easy. The number of Americans living longer and increasing the percentage of citizens over the age of 65 (and more) focuses attention on protecting them from intentional harm and the need to aggressively investigate crimes against the elderly.

Investigative effectiveness involves competence, persistence, thoroughness in documentation of evidence and actions in the case, and the ability to build rapport—with victims, witnesses, and suspects. But effective methods also include openness to exploring new strategies and technologies, leadership and commitment by the agency, working with partners in the health care and advocacy fields, and a personal pursuit of investigative excellence by the people in the process—officers and investigators.

KEY TERMS

Danger Assessment 171
Domestic violence (DV) 167
Intimate partner violence (IPV) 167
Primary aggressor 172
Restraining order 171
Social learning theory 171
Stalking 173
Stockholm Syndrome 171

DISCUSSION QUESTIONS

1. How have the rates of intimate partner violence changed in the last 25 years and why?
2. List and discuss several risk factors that appear to contribute to intimate partner violence.
3. Why does child abuse happen and how frequently does it occur?
4. In what ways does child abuse connect to intergenerational transmission of violent behavior?
5. How can agencies utilize lethality and danger assessment to improve outcomes for battered persons?
6. What agencies and groups are involved in combating family violence? What role does each play?

SAGE edge™

- Get the tools you need to sharpen your study skills. SAGE edge offers a robust online environment featuring an impressive array of free tools and resources.
- Access practice quizzes, eFlashcards, video, and multimedia at **edge.sagepub.com/houghci**

©iStockphoto.com/stevecoleimages

9 Burglary, Theft, White-Collar Crime, and Cybercrime

RUNNING CASE STUDY: MULTIPLE-CAR BURGLARY ON HALLOWEEN NIGHT

Introduction

Burglary
- Types of Burglaries
- Preliminary Investigation of Burglary
- Types of Burglars

How It's Done: Burglary Investigation

Theft
- Preliminary Investigation of Theft

How It's Done: Organized Retail Crime (ORC)
- Employee Theft
- Motor Vehicle Theft
    - *Investigating Motor Vehicle Theft*

White-Collar Crime
- Fraud
- Embezzlement

Identity Theft
- Warning Signs of Identity Theft
- Reporting Identity Theft

Cybercrime
- Types of Cybercrime
- Investigating Cybercrime
- Cybercrime Targeting in the United States

Explore This

Summary

Key Terms

Discussion Questions

LEARNING OBJECTIVES

After reading this chapter, students will be able to

- 9.1 Describe evidence sought in burglaries.
- 9.2 Explain theft motivations.
- 9.3 Explain how white-collar crime differs from street crime.
- 9.4 Describe the challenges with identity theft.
- 9.5 Explain the steps in investigating cybercrime.

Running Case Study: Multiple-Car Burglary on Halloween Night

Will had started casing Summit Street on Halloween, when he and Jeremy took Christopher trick-or-treating.

"See how close the houses are," Jeremy said. "People park in the street, close together. In December these cars will be full of money and packages."

A few days later, Christopher and his mom disappeared, and Jeremy went to jail. Will was on his own.

Now it was early December, and though brisker than Halloween night, Will was sweating under his jacket. Shortly after midnight, he walked up Summit, remembering the quiet street clogged with costumed kids.

Will liked working on his own, making his own rules, but he missed Jeremy's tone of reason, his reminders to stay calm, to keep an eye on his surroundings.

The first car he tried, a 90s Subaru, was unlocked. Inside, Will found a stash of grape bubblegum, and in the glove compartment, a sheaf of paperwork. *Maybe he could learn how to steal someone's identity.* He grabbed the papers, and stuffing the gum in his pocket, Will slid the door shut.

He was on the lookout for electronics. In a white Jeep he saw a black bag sitting on the passenger seat. *There had to be something good in that bag.* Will tried the door. *Locked. Yeah, definitely something good.* Will pulled out the Slim Jim he carried under his jacket. With this long stiff piece of wire, he pushed between the glass and window frame until he could reach down to the door lock button and jiggle it open. Will grabbed the heavy bag, then swiped some spearmint gum. Will was excited, getting that twitchy feeling, but he knew not to open the bag on the street. He would search a few more cars before returning to his VW parked down the hill.

Because the burglar had attempted to break into 30 cars, crime scene supervisor Charla Lynne, CSI Chip Traci, and Detective Wes Thompson were all on the scene. It was 1:30 a.m. on a clear December night.

Chip Traci and Sam Ryan photographed tool mark evidence left from the burglar breaking into the cars, including a scale to indicate accurate measurements of the marks. They gathered impression evidence of a shoe mark in the soft mud alongside a Land Rover with casting material that would dry to a hard finish and could later be compared to the footwear of the suspect.

Detective Wes Thompson interviewed residents who reported stolen change and electronics. The suspect had also taken chewing gum from many cars and discarded pieces on the street.

CSI Chip Traci processed the discarded gum for evidence of fingerprints.

The middle-aged woman who'd called the police stood outside with her dog, her hands trembling from shock that someone had been inside her car. "He stole Ogar's favorite blanket!" she screamed. "I can't believe he was in my car."

This woman called the police twice a week, for nuisances, but tonight she had reported a real crime. Her terror at having her car broken into was heightened by her delight at having made the call that led to the entire street being searched.

"What time did you see the suspect in your car?" Wes Thompson asked.

"Five minutes ago. You don't even have your sirens on!"

Charla Lynne smiled to herself. Because they suspected that the potential burglary was still in progress, the team had not used their sirens as they approached so as not to alert the burglar.

Officer Carl Jayden searched the area for a potential getaway car. A 20-year-old VW Bug, parked with the right front tire over the curb, looked out of place. Carl Jayden parked two blocks from the crime scene and began walking toward Summit. He saw a young man hurrying toward him carrying a computer bag over his shoulder. The man had several smaller bags in his hand and appeared to be concealing something under his jacket. He was chewing gum and had pieces of purple Bubblicious falling out of his pocket.

"Good evening," Officer Jayden said.

The man turned and began to run in the opposite direction, gum tumbling from his pocket, along with a Slim Jim.

Officer Jayden relayed his location to the backup officers and began to pursue the suspect.

Why do law enforcement officers and investigators have a difficult time solving burglaries? What technologies may begin to help locate evidence?

Introduction

We have spent time looking at crimes committed against a person and how they are investigated. We turn our attention now to crimes that are committed against property. Stealing things that belong to someone else is understood by almost everyone as theft. If this theft is accomplished by sneaking into your home and stealing while you are asleep or away, we label this a **burglary**. If your money is taken from you by someone in a position of trust at a business or organization without the use of violence we consider it to be **white-collar crime**. And with the omnipresent technology of the global society any number of people may find ways to benefit monetarily by victimizing us through the use of computers and the Internet; we examine **cybercrime** here and comment about it again in a later chapter. Most of these crimes are committed so that someone may gain at the expense of someone else.

Investigating crimes against property can be a daunting task. Often, there is little or no physical evidence to process at a laboratory, or the identity of a specific individual who perpetrated a financial or cybercrime may be "virtually" impossible to determine. While property crime rates have declined generally, they are still of great concern to citizens and they exact a significant monetary tool, often on those who can least afford a loss.

The property crimes recorded and tracked by the FBI in its Uniform Crime Reporting (UCR) program include burglary, **larceny/theft**, motor vehicle theft, and arson. Property

Burglary Entry into a building or conveyance or remaining within it with the intent to commit a crime.

White-collar crime Money is taken by someone in a position of trust at a business or organization without the use of violence.

Cybercrime Benefit monetarily by victimizing others through the use of computers and the Internet.

Larceny/theft Unlawful taking of property.

crimes are those where no force or threat of force is used. The 2016 Crime in the United States report from the UCR provides the following overview:

- In 2016 there were an estimated 7,919,035 property crime offenses in the nation. The 2-year trend showed that property crime offenses declined 1.3% in 2016 when compared with the 2015 estimate. The 10-year trend showed that property crime offenses decreased 19.9% in 2016 when compared with the 2007 estimate.

- In 2016 the rate of property crime was estimated at 2,450.7 per 100,000 inhabitants, a 2.0% decrease when compared with the 2015 estimated rate. The 2016 property crime rate was 14.5% less than the 2012 estimate and 25.2% less than the 2007 estimate.

- Larceny/theft accounted for 71.2% of all property crimes in 2016. Burglary accounted for 19.1%, and motor vehicle theft for 9.7%.

- Property crimes in 2016 resulted in losses estimated at $15.6 billion.

The 2016 report also notes that 18.3% of property crimes were cleared and 45.6% of violent crimes (https://ucr.fbi.gov/crime-in-the-u.s/2016/crime-in-the-u.s.-2016/topic-pages/property-crime) (Figure 9.1).

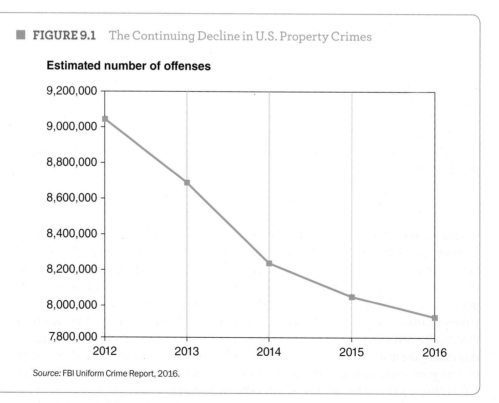

■ **FIGURE 9.1** The Continuing Decline in U.S. Property Crimes

Source: FBI Uniform Crime Report, 2016.

Burglary

People are robbed, buildings are burglarized. A victim understandably may confuse the two crimes when calling their local law enforcement agency in an emotional state following the discovery that their home or business has been entered and things of value taken and possibly the structure damaged. Recall from our discussion of persons' crimes (including robbery) that they involve the actual physical person as suffering a physical assault, either by threat or battery. The elements of the crime of burglary are entering or remaining in a building or vehicle with the intent to commit a crime. A residential burglary can happen when a house is under construction, when a home is unoccupied, or when the residents are present and perhaps asleep. If a business is broken into or entered, this would be classified a commercial burglary. In most states burglary is also committed if someone enters or breaks into a car or other **conveyance** and steals something from the vehicle. The charge of burglary is a felony crime and may have increasingly serious degrees depending on a variety of *aggravating* factors such as the structure being occupied, if the burglar is armed or arms himself while there, or if the criminal assaults someone.

Conveyance A vehicle used to transport people.

While it is obvious that the value of stolen items in a burglary may be upsetting to the residents, there is also an emotional component to realizing that someone has been in your family's home, maybe even while you were there. As officers and detectives investigate, they should be mindful of the shock and trauma that such a personal invasion of privacy may have on people. Let alone if the victim became aware of the intruder while they were present. The victim may have been confronted and assaulted, as in some rape cases, or the person may have entered unlawfully and then taken something by force from the victim—in which case the crime transitions to a home-invasion robbery and ceases to be a simple burglary. Each crime must be charged and proven separately. Two other aspects of the investigation that will bear on the charges filed by the prosecutor and reported by the law enforcement agency are largely unknown to victims: (1) The prosecutor may "bargain" some of the charges with the accused to secure a plea of guilty on the main or strongest charge in the case, and (2) the manner that multiple crimes in one incident are reported to the FBI's Uniform Crime Reports (UCR) program is to record only the most serious crime. In other words, for the UCR reporting, if a person was raped during a burglary, only the rape would be tabulated in the UCR figures, though the prosecutor would charge as many of the crimes as she deemed worthy and the police agency would still maintain reports of each crime and investigate them.

Types of Burglaries

Burglary is one of the Part I, or more serious, crimes tabulated by the FBI in the annual Uniform Crime Report (UCR) figures. The FBI (2017) estimated that in 2016, 1,515,096 burglaries occurred at a rate of 468.9 per 100,000 of population, which continued a decline in the rate of burglaries extending back more than 20 years. The estimated average value of property taken in the burglaries was $2,361. Law enforcement cleared 13.1% of these crimes by arrest or exceptional clearance within the same reporting period. The 2016 UCR report (FBI, 2017) shows residential burglaries to be just under 70% of all burglaries, occurring during the day (35.9%), at night (20.6%), and at unknown times (13.1%). Commercial buildings or property (nonresidence) can be targeted by criminals as well, and in 2016 accounted for 30.5% of reported burglaries (FBI, 2017).

Criminal mischief/vandalism Unlawful damaging of another's property.

But someone may also *trespass* onto private or commercial property with the intent to commit an additional crime such as theft, **criminal mischief**, or in the attempt to commit a burglary. Trespass is generally understood as being on someone's property without permission. Law enforcement may be called either to arrest the individual or to issue a *trespass warning* documenting that the person has been told not to return or face arrest or prosecution. In most states the crime of trespass is classified a misdemeanor, but if the trespass occurs on a construction site or possibly government property, the charge may be elevated to a felony. A person may intentionally damage another's property, which is charged as *criminal mischief,* or in some states referred to as **vandalism**. The dollar value of the damage done will determine whether the crime is considered a misdemeanor or a felony. Once again, thorough report-writing is demanded in property crimes incidents to ensure appropriate charges may be filed by the prosecutor in the case. While trespassing or preparing to commit a burglary, an individual may also be discovered by police and found to be in possession of tools that indicate the further intent of the person to commit a burglary. The individual may be charged with the crime of possession of burglary tools in some cases, even without being charged with a burglary.

Preliminary Investigation of Burglary

The preliminary investigation of a reported burglary is (as in many crimes) critical in determining the chances of solving the crime. With the previously mentioned dismal clearance rate for burglaries, each action taken by the initial patrol officer and every effort at crime scene work, victim interviewing, and neighborhood canvassing takes on great significance. If patrol officers are responding to a potential burglary in progress, they will be alert to people in vehicles or on foot leaving the area of the reported burglary. They will typically not use their siren as they get close to the location so as not to alert the burglar and in hopes of catching him or her in the act. The burglar may have parked any vehicle they used at some distance from the actual place of their burglary, and officers will be on the lookout for any vehicles that may appear out of place as the officers themselves park just out of the sight of the home or building so they can approach in a stealthy manner. Officers must be trained in these issues of approach the same way they must learn the specifics of processing a crime scene. The importance of where to park is illustrated in one of my own cases as a patrol supervisor responding to a burglary in progress with two of my officers. Each officer parked almost in front of the house on either side of a double driveway and went around behind the two-story building looking for the burglar. I parked several houses away, and as I walked up along the darkened street and reached the driveway a rear door opened on one of the three cars parked there. What emerged at this point was an individual dressed from head to toe as a ninja and—yes—he was armed with a sword, which he brandished when I spoke to him. I convinced him to put down his weapon and whistled for my officers, who probably to this day talk about the night Sarge caught a ninja. Because he was carrying a pack of cigarettes and had to use a ladder to reach the second floor of the home that he had burglarized I feel confident assuring you he was not an actual ninja.

At the point of entry, officers or crime scene technicians will try to locate tool marks from whatever was used to break into a property as well as potential trace evidence or fingerprints. Tool mark evidence must be photographed, including a scale to indicate accurate measurements of the marks. These marks may also be documented by taking impressions using a compound that dries to a hard finish. Other impression evidence may be tire tracks or shoe impressions that may also be gathered by using a compound or casting material for the print, which may later be compared to the tires or footwear of a suspect.

Types of Burglars

The way the criminal gained entry to a home or building may show a pattern over several crimes. Whether the individual looks for unlocked doors and windows or whether they use some type of tool to break through a lock, or even smash out a window that is not easily viewed from the street or a neighbor's house, can provide a clue that a group of burglaries may be a series of crimes committed by the same person.

Burglars are generally considered either professional or amateur. The professional may move around the region or country, sometimes even moving with the seasons, and essentially make their living by committing burglaries. In contrast, the amateur may commit a burglary as but one of many crimes they commit as the opportunity presents itself. The amateur may also be motivated to secure money to buy drugs. Additionally, and as you might imagine, even amateur burglars vary in how careful they are in selecting a target and how skilled they are in carrying out the burglary. In one business burglary I investigated, the burglar had climbed in through a window above a doorway and when jumping down to the floor his wallet came out of his pants unnoticed. When I arrived to meet with the owner of the burglarized garage, it was not a sophisticated investigation (though steps had to be taken) that revealed the suspect; rather, I opened the wallet and looked at his driver's license. On the other hand, another case I handled involved the former employee of a large grocery store chain who knew of a specific roof access in the design of all of the grocery stores. He exploited this vulnerability to enter grocery stores after closing hours in several adjacent counties and then broke into the safe in the manager's office.

The unskilled or semiskilled appearance of a burglary is one aspect of the crime that the investigator will consider. Whether or not the building was a residence or business, whether they enter at night or on a weekend as opposed to during a weekday, the way the building was entered, and the type of property stolen are also aspects of the modus operandi of the burglar. Some criminals who burglarize do so with one or more others, and some commit the crime alone. Each of these factors can help the investigator in determining whether the crime is linked to other burglaries. For example, juveniles have often been known to commit daytime residential burglaries during the summer months when they are out of school but when typical adult residents are at work.

Routine activities theory is the criminological concept that explains that if a motivated offender finds a suitable target, in the absence of a capable guardian a crime may very well take place. Law enforcement officers are very familiar with this dynamic and how it is often seen through the crime of burglary. The component of a capable guardian does not mean necessarily that a person is present to protect the property; it can also refer to a burglar alarm or surveillance technology monitoring the property as well as nearby businesses or residences that have clear sightlines to a potential target. The burglar will generally consider these factors and move on to pick a target that has a lower chance of discovery. This explains why police routinely advise home and business owners to install good locks, adequate lighting, and ensure that landscaping does not interfere with being able to view potential entry points to property.

> ## How It's Done
> **BURGLARY INVESTIGATION**
>
> - Discovery or report of burglary.
> - Officer examines scene for point of entry and exit, determine what was stolen, or other acts committed.
> - The report will move to a detective for follow-up on any leads in the case.
> - If a suspect is identified and arrested the investigator may obtain a search warrant in an attempt to recover stolen property.
> - The investigator will meet with a prosecutor to determine appropriate charges by reviewing the evidence at the scene, recovered from the suspect, and provided by the victim or any witnesses.

Theft

The formerly common-law crime of larceny is often referred to in state statutes as theft and is defined as the unlawful stealing or taking of something of value without violence. And compared to burglary, theft does involve being in a building unlawfully or taking something with force. Reported larceny/theft is tracked in the Uniform Crime Reports (UCR) each year. In 2016 there were 4,971,925 reported theft cases (not including motor vehicles) in the United States. The types of theft and circumstances vary. Most larceny/thefts (45.4%) were estimated at a value of no more than $200 (FBI, 2017). Someone who commits a theft might be in your home as a guest and think to steal something when you are out of the room. They may also come in the night and steal the car from your driveway. The criminal may also find their way by virtual means into your bank account or steal your identity and profit by masquerading as you in subsequent purchases by setting up fraudulent credit accounts. The value of what has been stolen will generally categorize the crime as either grand theft or petty theft. The dollar-value threshold between the misdemeanor crime of petty theft and the felony crime of grand theft varies from state to state but may be between $500 and $1,000.

Preliminary Investigation of Theft

Most people discover they are the victims of a theft only after it has occurred. Police are then notified and respond to the location of the crime and are often able to do little more than write a report of the loss. While the initial investigation of a theft case is much like a burglary investigation, there is often little or no physical evidence with which to work. Sometimes, however, a theft such as shoplifting may be stopped in the act, allowing store personnel and police to apprehend and arrest a thief. These shoplifting thefts may go undetected or without a suspect, but if the perpetrator continues stealing from similar businesses in the same area there may be the chance of store video surveillance documenting the same person in multiple locations where property has been stolen. While obviously a thief might try to avoid being caught on camera, the reduced cost of various technologies to capture images has increased the ability for even small-business owners to provide some level of deterrence or detection. One of my sons, who owned a skateboard shop at the time, suffered a theft from an individual who had been acting suspiciously inside the store. Playback of the in-store video produced a clear image of the thief taking the property. My son notified the local police department to report the theft and provided photographs of the individual but also, using the power of contemporary social networking, put the image onto a social media platform. Two people contacted him with the identity of the thief, which he passed on to law enforcement, who subsequently arrested the individual.

In many instances there is neither technological monitoring nor an eyewitness to theft. This fact goes a long way in explaining the very low clearance rate for the crime. It is also why law enforcement officers always recommend that if you have valuables with serial numbers, record those numbers. If you are able to inscribe or attach serial numbers to other items, do so; also, to photograph for your own records items that might be stolen. One way that law enforcement can sometimes solve a theft case is if the thief sells the item to a pawn shop whose records may be examined later by law enforcement where the item may be matched to a report of stolen goods. Many jurisdictions have an ordinance that directs pawn shops to maintain copies for inspection or provide duplicate copies or data access for law enforcement to examine records since the sale of stolen goods remains a brisk business. Most of what is stolen are items of property as opposed to cash. So the challenge for thieves as well as burglars is how to dispose of the stolen property. Aside from some pawn shops, individuals may sell goods at flea markets, trade items for drugs, or sell them to someone at what is clearly an unreasonably low price. The person receiving the stolen property may be a **fence**, who makes a living or routinely receives stolen goods for resale, or an individual trying to benefit from the low-cost item.

Fence Person who makes a living or routinely receives stolen goods for resale, or an individual trying to benefit from the low-cost item.

Shoplifting by customers or those who come into a retail location is certainly a challenge for businesses. Shoplifting as a form of theft may be addressed through visible signs of surveillance such as closed-circuit television (CCTV), electronic article surveillance (EAS) that alerts to tagged merchandise, or the presence of security personnel or attentive employees greeting each customer and following up to give the customer attention—and thus heighten the risk of being caught if attempting to shoplift. What many do not realize is that employee theft in many stores is responsible for more dollar-value loss than theft by customers.

Each form of theft can have specific considerations as well as avenues of investigation. Consider mail theft that may involve U.S. Postal Service investigators working with local or state law enforcement. Theft of farm and agricultural equipment may take resources to steal large items, and require officers to coordinate via computer and Internet to search for serialized items being sold elsewhere. The theft of art, currency, and coins—even animals—points out the enormity of the seemingly straightforward crime of larceny. This is certainly true when the item stolen can often be driven away, as in motor vehicle theft.

How It's Done

ORGANIZED RETAIL CRIME (ORC)

Within the large problem of retail theft is a particularly challenging approach to crime: organized retail theft. ORC is defined by the National Retail Federation as "groups, gangs, and sometimes individuals who are engaged in illegally obtaining [substantial quantities of] retail merchandise through theft and fraud as part of a criminal enterprise." This approach allows criminals to steal large quantities of merchandise using the organized method of a group. In addition to shoplifting, ORC groups may be engaged in identity theft, cargo theft, and various fraud schemes. In shoplifting, experienced shoplifters such as those working as part of an ORC group will use techniques to defeat electronic article surveillance (EAS) that may be based on radio frequency tags, magnetic devices, and microwave technology. Something using aluminum foil, such as a shopping bag, is a common way to block signals from various EAS, and allows the thief to take items past the sensors. This is also why many retail establishments have a policy of receipt-checking to further deter theft.

Employee Theft

Shrinkage Theft of inventory by employees of a business.

Referred to as **shrinkage**, theft of inventory by employees of a business results in huge losses each year, losses often passed on to customers through increased prices. Some employee theft may be ongoing for a lengthy period of time and only is discovered inadvertently as the result of an inventory check or audit. Some may be discovered after an employee has left the business or company. Incidental theft is when an employee takes items home from work such as business supplies, or consumes something while at work such as food in a convenience or grocery store. Opportunistic theft by employees, referred to as situational theft, may arise in a number of circumstances where the employee views the opportunity and seizes the moment. If there has been a burglary and the employee is called to the store to secure it after meeting with law enforcement, he may take several more items and rationalize that insurance will reimburse the owner for the items. Route delivery drivers may have items apparently unaccounted for after their deliveries and take advantage of the faulty tracking so as to steal packages. One of the technological advancements that has helped companies track their inventory, including identifying what has been stolen, is the use of *radio frequency identification* (RFID), which uses small tracking devices in or on each item of inventory that can be scanned or identified by the inventory management system so as to keep up with ordering or unusual patterns of inventory movement.

Employee theft is a problem for many businesses. What are some strategies employers can use to deter this behavior?

©iStockphoto.com/Nicki1982

Many employers choose to fire employees caught stealing rather than involve law enforcement or prosecution. This can have larger implications, as the former employee moves on to a new employer and perhaps a continuation of their stealing. Some companies use internal security personnel (asset protection) to monitor inventories and identify employees potentially stealing. The use of the polygraph to question employees may occur by company personnel or during a criminal investigation by law enforcement. As always, the polygraph will not be admitted into court but can be of great assistance in narrowing down suspects or shifting focus from apparently uninvolved employees. Investigators will also ask potential employee suspects a variety of specific questions, which may elicit an incriminating statement. If the days and times of thefts are known, questioning will begin with employees that the company or investigator determines may have been working or had access to the items.

Motor Vehicle Theft

Motor vehicle theft, sometimes referred to as grand theft auto, refers to the stealing of cars, trucks, motorcycles, or other vehicles. The crime, or the variation unauthorized use of a motor vehicle, is charged when a person drives or takes a motor vehicle without the owner or agent's permission. The thief or thieves will take the vehicle for one of a number of reasons. Juveniles or others may take a car for joyriding or for the thrill of driving around in something that does not belong to them, and possibly without benefit of a driver's license. Following the excursion, they will abandon the vehicle somewhere. Criminals will also steal vehicles as an instrument to commit another crime such as robbery. That way, if a witness or victim describes the getaway car it will not immediately benefit law enforcement, as they learn the vehicle was stolen from yet another victim. Sometimes, however, when the stolen vehicle is recovered, evidence inside may lead to the identity of the robber as well. More organized groups or individuals stealing cars as a profession may immediately take the vehicle to a chop shop, where the car is dismantled and the parts sold off individually. They may also sell the entire car in another area, state, or even outside of the country. This may involve removing or changing the vehicle identification numbers from the car where they have been placed by the manufacturer in several locations on the engine and the chassis. Yet another motivation for vehicle theft is the owner arranging for the car's theft or destruction so as to defraud their insurance company. The UCR estimate for 2016 was 700,143 motor vehicle thefts (FBI, 2017).

Motor vehicle theft Stealing or attempting to steal a vehicle.

A variation of auto theft that became more frequent in the 1980s was the threat or actual use of force or weapons to *carjack* a vehicle that was occupied. The criminals would typically confront the driver with a weapon and then drive the car away. The investigative steps for most vehicle thefts include taking the initial report once the theft is known, determining whether the vehicle was repossessed by someone with a financial claim to it rather than stolen, gathering the identifying information about the vehicle and having the communications center broadcast an alert for patrol officers to watch for the vehicle, input the information to the state and national computer databases, and then forwarding the initial report to investigators. If the vehicle is recovered it becomes a crime scene of its own and will be processed for fingerprints and the collection of any other evidence inside or on the vehicle.

A person may take possession of a motor vehicle with permission, initially. At some point the person who rented, leased, borrowed, or otherwise originally was authorized to have the vehicle exceeds the conditions under which he had the vehicle. This may be viewed as embezzlement or another crime distinguished from motor vehicle theft. Often, the person taking a vehicle in this way (or those who commit motor vehicle theft) will take the car or truck across state lines. Federal legislation known as the National Motor Vehicle Theft Act, or Dyer Act, was enacted 100 years ago to make a federal crime of the interstate transportation of a stolen vehicle.

INVESTIGATING MOTOR VEHICLE THEFT

When an officer begins the investigation of a motor vehicle theft, he will try to gain as many details about the status of the vehicle as he can. Had the car been loaned to a friend? Had it been repossessed and not actually stolen? There are various ways that a thief may steal cars, even those equipped with electronic keys or ignitions, GPS for locating the vehicle, etc. The officer will certainly get a detailed description of the vehicle, along with information about the license and registration of the vehicle. A staple of the investigation of vehicle theft is obtaining the vehicle identification number, or VIN. This number is unique to the vehicle so that investigators can be certain when they recover all or part of the vehicle.

Investigators and agencies coordinate with state, federal, and private industry groups such as the National Insurance Crime Bureau to track groups committing high-volume or specialized vehicle thefts (even boats, motor scooters, and airplanes). Most stolen vehicles are recovered locally and in a fairly short period of time. Awareness campaigns about motor vehicle theft can be effective in communities as well as proactive methods such as sting operations. Automotive industry antitheft devices have continued to improve, though criminals continue their efforts to defeat technology in vehicles.

White-Collar Crime

The FBI defines white-collar crime this way:

> Reportedly coined in 1939, the term white-collar crime is now synonymous with the full range of frauds committed by business and government professionals. These crimes are characterized by deceit, concealment, or violation of trust and are not dependent on the application or threat of physical force or violence. The motivation behind these crimes is financial—to obtain or avoid losing money, property, or services or to secure a personal or business advantage. (FBI.gov, *What We Investigate, White-Collar Crime*)

Many white-collar crimes are also examples of large-scale theft, and some of these receive media attention based on the dollar amount lost by victims. The industry group the American Society of Industrial Security (now known as ASIS International) has devoted resources and training to assist businesses and financial institutions in preventing or discovering incidents of white-collar crime. Interestingly, just as with many victims of much smaller theft cases, embarrassment keeps organizations and top executives from sometimes reporting being victims of white-collar theft.

Classified as Part II crimes and tracked by the FBI's Uniform Crime Report (UCR), forgery, fraud, and embezzlement cause financial losses to victims that far exceed violent street crimes such as robbery.

Investigating white-collar crime often takes time. This is one of the many limitations already noted to the average law enforcement agency conducting these types of investigations. If the case has only state jurisdiction, the department or investigator may have little to no resources to draw from at the federal level. The gathering of information and documents is painstaking, and the analysis of the information gained can be complicated, to say the least. Investigators must often prepare comprehensive search warrant applications to gain access to all manner of financial records, credit documents, real estate, taxes, and more. Victims of such crimes may themselves not have well-organized records, only knowing that they have been victimized.

Fraud

Fraud involves some element of deception or trickery to gain access to the money or resources of the victim. The offender often has a position of trust, possibly a family member or someone who has ingratiated themselves to their intended victim. The victim may have believed they were helping the offender by providing money, only to learn they had been fooled. Perpetrators also take advantage of greed with some victims by convincing the target they will get a significant amount of money by participating in the financial scheme. Some people are defrauded by carelessness such as not monitoring services they have been signed up for without their permission.

> **Fraud** Involves some element of deception or trickery to gain access to the money or resources of the victim.

Corporate fraud is a category of white-collar crime typically investigated by the FBI's financial crimes section. Often these cases revolve around companies altering their financial records or involve insider trading, or kickbacks paid to individuals or companies to illegally act to the benefit person or company paying the kickback. Investigation can include a component to make a charge of obstruction of justice when individuals of a company withhold or conceal information from investigators.

One of the many variations of white-collar crime is what is known as *affinity fraud,* where the criminal takes advantage of his social circle such as fellow churchgoers or faith community members. People committing crimes of fraud often prey upon the emotions or trust of others using their skill as manipulators to convince victims to give up their money or valued items. With many of these various forms of fraud the investigator's focus is primarily *how* the crime was carried out, since *who* did it is generally known. To accomplish this, circumstantial evidence is often sought to augment tangible records of the use or conversion of the victim's money, property, or other resources. If the suspect went through extensive means to conceal their actions, this is indicative of an intention to keep others from learning of their activities.

In 2018 the U.S. Justice Department coordinated an effort aimed at perpetrators of fraud that largely targeted older Americans through various "fraud schemes, ranging from mass mailing, telemarketing and investment frauds to individual incidences of identity theft and theft by guardians" (DOJ, Office of Public Affairs, February 22, 2018). In this effort coordinated with the Federal Trade Commission (FTC), the U.S. Postal Service (USPS), and state attorneys general, more than 250 defendants are alleged to have affected more than one million Americans, costing them up to more than a half-billion dollars in financial exploitations including:

- "Lottery phone scams," in which callers convince seniors that a large fee or taxes must be paid before one can receive lottery winnings;
- "Grandparent scams," which convince seniors that their grandchildren have been arrested and need bail money;

- "Romance scams," which lull victims to believe that their online paramour needs funds for a U.S. visit or some other purpose;
- "IRS imposter schemes," which defraud victims by posing as IRS agents and claiming that victims owe back taxes;
- "Guardianship schemes," which siphon seniors' financial resources into the bank accounts of deceitful relatives or guardians.

Many of these cases illustrate how an elderly American can lose his or her life savings to a duplicitous relative, guardian, or stranger who gains the victim's trust. The devastating effects these cases have on victims and their families, both financially and psychologically, make prosecuting elder fraud a key Justice Department priority (DOJ, Office of Public Affairs, February 22, 2018, press release number 18–225).

Embezzlement

Embezzlement Theft or conversion of funds under a person's care that was entrusted to them by another.

Embezzlement is a financial crime that is committed by someone entrusted with the property or finances of another. Once the criminal has control over the victim's resources he uses them in a manner not approved by the owner. Banks are often the victim of embezzlement by employees, and the losses can be significant. The key distinction here between theft, other types of fraud, and embezzlement is that with the latter the offender had lawful control or possession of the property or resources when he used or converted it to his own purpose or gain. Company CEOs, comptrollers, nonprofit employees, banking industry employees, and others are just a few examples of people in positions of fiduciary trust who may commit this type of crime.

From the standpoint of the financial crimes investigator, they must work to discover evidence that shows the suspect had the *intent* to deprive the victim of property. The investigator will likely inquire about certain warning signs of embezzlement that may have been noticed within the organization such as:

> (1) an employee who suddenly has an improved standard of living with no extra income to explain it, (2) sudden or protracted decrease in company's profits or holdings, (3) disorganization of financial records, (4) unexplained changes in revenue or expenses or increased write-offs, and (5) extremely loyal employees who never take vacations so that only they manage the books with sufficient auditing practices in place. (Dimarino, 2015, p. 120)

A resource available to financial crimes investigators is the Financial Crimes Enforcement Network (FinCEN). FinCEN is housed within the U.S. Department of the Treasury, and its mission is "to safeguard the financial system from illicit use and combat money laundering and promote national security through the collection, analysis, and dissemination of financial intelligence and strategic use of financial authorities" (https://www.fincen.gov/about/mission). While the network is most recently focused on money laundering and national security threats via the financial system, its other current strategic goal is the sharing of financial intelligence with government and private industry partners (FinCEN Strategic Plan 2014–2018). Resources are available for law enforcement as well as for financial institutions and include advisories and frequent bulletins in addition to aiding law enforcement

CHAPTER 9 Burglary, Theft, White-Collar Crime, and Cybercrime

efforts to fight money laundering through the analysis of banking information as required by the Bank Secrecy Act (BSA).

Identity Theft

A crime of alarming frequency and far-reaching impact is identity theft. USA.gov on its identity theft page defines it as "a crime where a thief steals your personal information, such as your full name or Social Security number, to commit fraud" (https://www.usa.gov/identity-theft). The identity thief gains access to personal information about an individual or the person's existing credit card accounts and uses them without permission.

The Federal Trade Commission (FTC) began to track commission of this crime in 1998, when identity theft was deemed a federal law violation, by passage of the Identity Theft and Assumption Deterrence Act. Identity theft has been labeled the fastest growing crime in the United States, and confusion over jurisdiction leaves many citizens frustrated in reporting being a victim to authorities. Another challenge to the brief definition offered above is the number of crimes that may involve identity theft, such as the use of credit or debit cards, forgery or counterfeiting, checking account fraud, the use of false or stolen identities for illegal immigration, use by terrorists, and others.

Those who seek to profit from crimes such as identity theft are many and occupy positions throughout society, not just on its fringe or seated in front of a computer in a dark room somewhere. In 2018 the former chief counsel of the U.S. Immigration and Customs Enforcement's (ICE) Office of the Principal Legal Advisor, Raphael Sanchez, pleaded guilty to a scheme involving his use of the personal identification information of seven undocumented people being removed from the United States. Sanchez opened bank and credit card accounts in the names of those he targeted and acquired loans in their names as well as using their credit accounts to acquire goods and funds for himself (DOJ, Office of Public Affairs, February 22, 2018, press release number 18–189). Later in 2018, a federal grand jury returned an indictment against a postal employee for his part in a refund fraud scheme using stolen identities (DOJ, Office of Public Affairs, August 29, 2018, press release number 18–1126).

Identity theft is a major crime problem. What should you do if you have been a victim of identity theft?

©iStockphoto.com/MotoEd

Warning Signs of Identity Theft

The FTC provides the below information on the warning signs of identity theft:

What Do Thieves Do With Your Information?

Once identity thieves have your personal information, they can drain your bank account, run up charges on your credit cards, open new utility accounts, or get medical treatment on your health insurance. An identity thief can file a tax refund in your name and get your refund. In some extreme cases, a thief might even give your name to the police during an arrest.

Clues That Someone Has Stolen Your Information

- You see withdrawals from your bank account that you can't explain.
- You don't get your bills or other mail.
- Merchants refuse your checks.
- Debt collectors call you about debts that aren't yours.
- You find unfamiliar accounts or charges on your credit report.
- Medical providers bill you for services you didn't use.
- Your health plan rejects your legitimate medical claim because the records show you've reached your benefits limit.
- A health plan won't cover you because your medical records show a condition you don't have.
- The IRS notifies you that more than one tax return was filed in your name, or that you have income from an employer you don't work for.
- You get notice that your information was compromised by a data breach at a company where you do business or have an account.

If your wallet, Social Security number, or other personal information are lost or stolen, there are steps you can take to help protect yourself from identity theft (https://www.identitytheft.gov/Warning-Signs-of-Identity-Theft).

Reporting Identity Theft

Societies around the globe have crossed over into an era of virtual commerce and communication that likely will continue to grow. Unfortunately, identity theft has grown as well and stands as a dark symbol of how criminals adapt to new circumstances to commit old crimes. If a person discovers that someone has misused their personal information it is important to report the identity theft and to follow the steps recommended by the FTC. Federal legislation such as the Fair and Accurate Credit Transactions Act (FACT) of 2003 states the rights and remedies for victims of identity theft. FACT established the ability for consumers to receive free credit reports to help combat identity theft and at the same time increase the ability to build or repair a consumer's credit. This act required a police report and encouraged law enforcement agencies to train personnel in the specifics of identity theft crimes.

Cybercrime

The FBI, which takes the federal lead in the investigation of cybercrime, clearly expresses the scope of the threat this way:

> Our nation's critical infrastructure, including both private and public sector networks, are targeted by adversaries. American companies are targeted for trade

secrets and other sensitive corporate data, and universities for their cutting-edge research and development. Citizens are targeted by fraudsters and identity thieves, and children are targeted by online predators. (FBI.gov, *What We Investigate, Cyber Crime*)

A challenge that law enforcement agencies faced after the advent of the automobile was the transitory nature of crime and criminals as offenders moved from one jurisdiction to the next. With jurisdictional boundaries and a less than effective way of sharing information in the early years of organized law enforcement, many criminals made their livelihood facilitated by the lack of available technology and intelligence capabilities by law enforcement. The power of databases such as the National Crime Information Center (NCIC) and the state equivalents of NCIC was a tremendous boon to law enforcement nationwide as well as through partnering with other countries through Interpol. The Internet opened possibilities for near-instantaneous sharing of greater volumes of data and intelligence information to aid agencies and their investigation of crimes, especially perhaps cybercrimes involving the use of computers and other devices to steal or manipulate the data of victims.

As I sit and type on my keyboard, I am also connected via the Internet to networks of computers across the globe. This represents how so many people in modern life have access to huge volumes of information in what we conceptualize as a worldwide web of data. My connection to this sprawling and somewhat abstract well of information is also a vulnerability. The computer is at once a potential tool or instrument to commit crime while also being the target of a computer used by someone else as a tool. Relatively few general-service law enforcement agencies have the trained personnel or resources to investigate cybercrimes. Many victims do not know they have been victimized, decide not to report the crime, or simply believe they have faulty equipment or software issues.

Types of Cybercrime

Crimes committed using a computer include infecting the computers or computer components of equipment of others through viruses and worms, hacking into a computer or network of computers to acquire information that can be used to the hacker's advantage, spreading unlawful materials such as child pornography, or launching denial of service attacks to slow or shut down an entity's web address. Cyberterrorists can use the Internet to incapacitate public and private sector computer-dependent equipment within the infrastructure of every sector of the country. Efforts by government agencies and private companies work constantly to develop ways to effectively block such attacks. As mentioned, the average local law enforcement agency is ill-equipped to deal with this thoroughly modern method of crime.

Investigating Cybercrime

When officers and investigators do seize a computer that has been compromised, they will isolate the computer if it is part of a network. Evidence on a computer is often fragile in the sense that many actions taken to view or acquire it may cause it to be deleted or corrupted. Officers or technicians will work to identify any useful information that may allow tracing the origin of the attack, or at least eliminating the weakness that made the attack possible.

This preliminary investigative work will include documenting the computer, system, and any peripheral devices linked to the computer. Documentation involves photographing the scene and devices, including anything displayed on the screen, and the front, sides, and back of the computer where connections are made. As with any evidence, the chain of custody must be maintained; this evidence is likely heading to a lab for analysis by trained computer technicians.

Cybercrime Targeting the United States

In 2018 a grand jury in the District of Columbia indicted "thirteen Russian nationals and three Russian companies for committing federal crimes while seeking to interfere in the United States political system, including the 2016 Presidential election." Within the indictment, the Special Counsel's Office alleged that the defendants conducted "information warfare against the United States," with the stated goal of "spread[ing] distrust towards the candidates and the political system in general."

From the press release by the Department of Justice:

> According to the allegations in the indictment, twelve of the individual defendants worked at various times for Internet Research Agency LLC, a Russian company based in St. Petersburg, Russia. The other individual defendant, Yevgeniy Viktorovich Prigozhin, funded the conspiracy through companies known as Concord Management and Consulting LLC, Concord Catering, and many subsidiaries and affiliates.

> Internet Research Agency allegedly operated through Russian shell companies. It employed hundreds of persons for its online operations, ranging from creators of fictitious personas to technical and administrative support, with an annual budget of millions of dollars. Internet Research Agency was a structured organization headed by a management group and arranged in departments, including graphics, search-engine optimization, information technology, and finance. In 2014 the agency established a "translator project" to focus on the U.S. population. In July 2016, more than 80 employees were assigned to the translator project.

> To hide the Russian origin of their activities, the defendants allegedly purchased space on computer servers located within the United States in order to set up a virtual private network. The defendants allegedly used that infrastructure to establish hundreds of accounts on social media networks such as Facebook, Instagram, and Twitter, making it appear that the accounts were controlled by persons within the United States. They used stolen or fictitious American identities, fraudulent bank accounts, and false identification documents. The defendants posed as politically and socially active Americans, advocating for and against particular political candidates. They established social media pages and groups to communicate with unwitting Americans. They also purchased political advertisements on social media.

> The Russians also recruited and paid real Americans to engage in political activities, promote political campaigns, and stage political rallies. The defendants and their co-conspirators pretended to be grassroots activists. According to the indictment, the Americans did not know that they were communicating with Russians.

> After the election, the defendants allegedly staged rallies to support the president-elect while simultaneously staging rallies to protest his election. For example, the defendants

organized one rally to support the president-elect and another to oppose him—both in New York, on the same day.

On September 13, 2017, soon after the news media reported that the Special Counsel's Office was investigating evidence that Russian operatives had used social media to interfere in the 2016 election, one defendant allegedly wrote, "We had a slight crisis here at work: the FBI busted our activity.... So, I got preoccupied with covering tracks together with my colleagues."

EXPLORE THIS

Visit the consumer information page about identity theft on the website of the Federal Trade Commission at https://www.consumer.ftc.gov/features/feature-0014-identity-theft. Go to the section on protecting your identity to read about what you can do to keep your personal information secure from the prying eyes of criminals. Read over the section for law enforcement to see the recommendations available. Look at the free resources section and read some of the information compiled there. Share it with others!

SUMMARY

Property crimes come in a variety of forms, from snatching a purse sitting next to a victim to driving away in a car whose owner left the keys in the ignition, to bilking a bank out of millions using the Internet as the tool to commit the theft. If there is a physical crime scene, as occurs with a burglary, or if there is the possibility of photographic or video evidence, or a witness is located through the proven method of a neighborhood canvass, officers or crime scene technicians have important work to do first with notetaking, then with photography and possibly lifting prints or impressions, securing witness statements, and finally writing competent reports that reflect everything that occurred, including the actions of all criminal justice personnel.

Many forms of theft leave little if any physical evidence. Some, such as white-collar crime, may leave a paper trail that investigators must follow—increasingly through computer files and networks. One of the most prolific crimes perpetrated largely through online means is identity theft. Not only can criminals obtain money and services illegally by stealing a person's identity, but the challenges and frustrations of overcoming such a crime are significant for the victim. Trying to investigate such crimes may rely on sophisticated computer tracking, informants, or increased educational efforts with the public.

KEY TERMS

Burglary 187
Conveyance 189
Criminal mischief/vandalism 190
Cybercrime 187

Embezzlement 198
Fence 193
Fraud 197
Larceny/theft 187

Motor vehicle theft 195
Shrinkage 194
White-collar crime 187

DISCUSSION QUESTIONS

1. How have the rates of property crimes changed in the last 20 years and why?
2. What are the elements of burglary and how should an officer incorporate them into his incident report?
3. How does routine activities theory contribute to our understanding of property crimes?
4. What can people do to reduce the likelihood of becoming victims of identity theft?
5. How do agencies investigate fraud cases?
6. Why are many cyberattacks against businesses not reported?

$SAGE edge™

- Get the tools you need to sharpen your study skills. SAGE edge offers a robust online environment featuring an impressive array of free tools and resources.
- Access practice quizzes, eFlashcards, video, and multimedia at **edge.sagepub.com/houghci**

PRACTICE AND APPLY WHAT YOU'VE LEARNED

▶ edge.sagepub.com/houghci

WANT A BETTER GRADE ON YOUR NEXT TEST?

Head to the study site where you'll find:

- **eFlashcards** to strengthen your understanding of key terms.
- **Practice quizzes** to test your comprehension of key concepts.
- **Videos and multimedia content** to enhance your exploration of key topics.

©iStockphoto.com/gorodenkoff

10 Drug Crimes, Organized Crime, and Gangs

RUNNING CASE STUDY: UNDERCOVER WORK—REGIONAL GANG SURVEILLANCE AND INVESTIGATING A POTENTIAL GROW HOUSE

Introduction

Drug Crimes
- History of the U.S. Illegal Drug Problem
- Defining and Classifying Controlled Substances
- Scope of the Problem
- Fentanyl: A Synthetic Opioid

Investigating Drug Crimes

How It's Done: Handling Narcotics Evidence
- Undercover Operations

How It's Done: Undercover Investigations
- Grow Houses and Labs
- Knock and Talk
- Asset Forfeiture

Organized Crime
- Extortion
- Investigating Organized Crime

Gangs
- Identifying Gangs
- Combating Gang Crime
- Public Response

Explore This

Summary

Key Terms

Discussion Questions

LEARNING OBJECTIVES

After reading this chapter, students will be able to

10.1 Discuss aspects of the illegal drug problem.

10.2 Explain the approach to drug investigations.

10.3 Discuss the crimes common to organized crime groups.

10.4 Describe different types of gangs.

Running Case: Undercover Work—Regional Gang Surveillance and Investigating a Potential Grow House

Assigned to the regional gang task force 6 months ago, Sergeant Kevin Lloyd's main assignment had been surveillance of the Hellions' "clubhouse," a warehouse at 4120 Tasker Street. The Hellions had formed 9 years ago and begun dealing drugs across the southwest section of the city, pushing out the Dusters, a low-profile gang dominating that area since the 1950s. The Hellions used violence to scare away competitors, raise their own profile, and intimidate witnesses. Last month, Dusters had retaliated against the Hellions by shooting two of their street-level dealers.

As an undercover agent, Kevin Lloyd drove past the clubhouse, tailed gang members, and tried to infiltrate the gang via the Internet. At the park near the clubhouse, he'd talked with juvenile drug dealers, both Hellions and Dusters, and given them his undercover phone number, in case they ever wanted to make a big trade. On the Internet, he'd also made contacts, but it was hard to know who he was talking to as many hid behind anonymous faces and fake names. So far, none of these contacts trusted him enough to reveal anything. He was tired of driving around and achieving nothing or sitting slumped in his chair in front of the computer screen.

Sergeant Lloyd was calculating how long he had until he could rotate off the narcotics unit. The demanding schedule of constantly being on call was taking its toll on his body and mind. Hanging out with people involved in the drug trade—informants, dealers, and users—and seeing only the gritty parts of citizen behavior, as well as the harm caused by drugs, was making him cynical.

Eighteen months tops.

His undercover phone bleeped with a text from an unknown number: "Big delivery at 1919 Whelan Street."

Sergeant Lloyd looked up the address and saw it was the local office of a national shipping company. He suspected the tip came from a Duster. He looked up the phone number and saw it was registered to Janelle White, who lived on a side street near the clubhouse. Sergeant Lloyd knew enough to be wary.

He called Detective Bradley Macon, now assigned to the task force, and they drove to the shipping office. After talking to the manager, they called for a K-9 unit to conduct a sniff lineup of several boxes. Canine Officer Simone Grace responded with K-9 "RJ." RJ alerted them to the suspicious box, which allowed Detective Macon and Sergeant Lloyd to obtain a search warrant.

The box was addressed to the Hellions' clubhouse. Inside they found 30 pounds of marijuana.

"Arrange a controlled delivery," Sergeant Lloyd said.

Bradley Macon nodded and ran through the procedure. He would ensure that the delivery was monitored in real time. Once the box reached its location, he would make sure it did not leave. If the receiver tried to leave, Detective Macon would intercept the package. When serving the search warrant, he would isolate the occupant from any of the evidence, document the evidence in the building, use appropriate safety and health measures to collect and package suspected drugs or paraphernalia, and complete evidence forms to transfer the drugs to the storage facility or laboratory for testing.

Both Kevin and Bradley knew that if the controlled delivery went well, they could serve an "anticipatory search warrant" by stating drugs were at the clubhouse, and search the entire location for evidence of trafficking drugs.

Then Sergeant Lloyd received another text from "Janelle White." The informant wrote, "Indoor grow house 4120 Tasker."

Who was providing this intelligence? Kevin Lloyd knew they'd been lucky with the box, but what would be awaiting them at the Hellions' clubhouse? Grow house risks included fights with gang members, pressurized growth-speeding canisters of carbon dioxide that could explode, booby traps to secure the grow house from unwanted visitors, and electrical hazards from using direct power feeds, slipshod wiring, and high-voltage lighting sources.

"Look for holes in the warehouse floor. Nails protruding from surfaces. Gang members may be ready to fight," he told the crime scene officers.

A less obvious, and often invisible, hazard to officers and K-9 dogs were toxins and spores from plants and exposure to tetrahydrocannabinol.

"Suit up," Kevin Lloyd said.

They stepped into their hazmat suits.

Aside from proving the commission of a drug crime itself, what challenges do officers face in confronting drug dealers? Is it likely that organized crime groups would be involved in drug crimes? Why?

Introduction

One aspect of the crimes discussed in this chapter that binds all of them together is that in their most damaging forms they typically involve the coordinated efforts of a number of people rather than the independent actions of a single criminal such as a burglar or even a lone robber. While not exclusively involved in the drug trade, many organized crime groups and gangs involve themselves in some way in illicit drug sales, transportation, or associated crimes.

How, then, to approach crimes that are at once individual, such as an illegal drug transaction, and collective, as a supply-chain operation whose goal is to generate profit for a group, or maintain a level of control over parts of a community through acts of intimidation or behavior that affects the quality of life for residents? Most law enforcement agencies because of their relatively small size deal in the enforcement of the micro-event, sometimes joining with other local departments to form task forces to more effectively leverage resources in a pooled effort to address crime that extends beyond one jurisdiction or that requires an approach that uses more resources than are possessed by a single agency.

Some state-level agencies lead efforts to combat drug crimes and gang activity and provide an information network among investigators and departments. The federal government also has the ability to coordinate and provide legal authority across jurisdictions, supply equipment and access to technology not available to smaller organizations, and commit agents and officers with highly specialized training to help make impacts that single, or even several departments cannot effect working alone.

Drug Crimes

Taking the issue of illegal drugs or drug use first, we realize that many people believe that the use of various **controlled substances**, just as drinking alcohol, they contend, should only be the business of the individual involved. A challenge to this position is that acquiring the drugs that are often addictive and certainly the basis of a habit for many people who choose to use them, costs money that the user may acquire through criminal activity. Cocaine usage has decreased in recent years, but marijuana use has jumped significantly (Caulkins, Kilmer, Reuter, & Midgette, 2015; Salas-Wright et al., 2017). The social assessment of individual rights to use or abuse substances stops when such exercise of free will infringes on the health or rights of others.

Controlled substances Drugs or chemicals produced in a manner regulated by government.

History of the U.S. Illegal Drug Problem

Historically, drug usage within the United States is neither a new phenomenon nor a new problem. In fact, religious, regal, and governmental authorities have railed against various substances including tobacco, coffee, alcohol, coca, and more for hundreds of years in the United States and abroad. Initially lauded as a miracle cure in the early 1880s, cocaine became available only by prescription in some states, but also appeared as an ingredient in Coca-Cola in the mid-1890s, which was marketed first as a medicine before being advertised as a refreshing beverage. Around the same time heroin was sold as a cough suppressant.

The U.S. Congress passed the Pure Food and Drug Act in 1906 to require producers to accurately label foods and medicines. Not quite a decade later, in 1915 the state of California banned cannabis for other than medical applications. In 1919 the 18th Amendment banned the manufacture, sale, or transportation of liquor. This brought on the Prohibition era, during which it was unlawful for citizens to drink alcohol. It took nearly a decade and a half for the official recognition of Prohibition's failure when the 21st Amendment repealed Prohibition at the end of 1933.

In 1937 Congress passed the Marijuana Tax Act, which furthered the criminalization of the substance. The 1960s saw significant increases in recreational drug usage in the United States among various demographics. At the end of the decade, in 1969 the federal government launched Operation Intercept, with the main objective of blocking marijuana from entering the United States. One year later, in 1970 Congress passed the Comprehensive Drug Abuse Prevention and Control Act, which provided law enforcement agencies with expanded authority for lawful searches related to drug investigations. And then, in 1971 President Richard Nixon employed the metaphor of a "War on Drugs" to rally political and public support in combating drug crimes. A significant step in the federally led approach was the establishment in 1973 of the U.S. Drug Enforcement Administration.

While the 1960s and 1970s saw increases in the use of illegal drugs in America, the 1980s became known for the challenges of imported cocaine and other drugs, much of them from

Mexico and Central and South America. With renewed emphasis by the federal government on the threat of drugs, President Ronald Reagan launched a number of initiatives to combat illicit drugs. First Lady Nancy Reagan led efforts against drug abuse and popularized the slogan "Just Say No." Prison populations also rose during the 1980s as mandatory sentences were enacted for various drug crimes at the state and federal levels. A blanket approach to drug manufacture, distribution, or use of drugs shown to be ineffective. And a result of treating powdered and crack cocaine differently was the disproportionate punishment of African Americans.

Defining and Classifying Controlled Substances

The Controlled Substances Act (CSA) within United States Code has evolved since being enacted in 1970 by President Nixon. The CSA addresses the control of specific substances designated within five **schedules** based on their medical use, safety from a psychological or physical dependence standpoint, and the potential for abuse of the substance. The top two schedules, Schedule I and Schedule II, have the highest potential for abuse and also carry the highest potential penalties for their use in a criminal violation. These five schedules are ordered by dangerousness with Schedule I as the highest level of danger. The U.S. Drug Enforcement Administration (DEA) oversees the implementation and enforcement of the CSA.

Schedules Categories of various substances, chemicals, and drugs that are used to make other drugs. Many classified drugs have a medical use but may also have the potential to be abused.

The possession of controlled substances is in flux to some extent when the average person considers the debate over the use of marijuana and its varying legal status in different states as being used medicinally by some people and even permitted under some state laws for recreational use. While medical reviews regarding any benefit coming from the cannabinoids in marijuana have increased, the use of the actual marijuana plant has not received adequate support of the benefits outweighing the risks to achieve approval by the U.S. Food and Drug Administration (FDA). Of far greater concern to most citizens, law enforcement professionals, and legislators is the illegal use of prescription and nonprescription drugs and narcotics, including opioids, that lead to a great many deaths in contemporary America. Methamphetamine availability has increased in the United States due to both production within the country and transportation in from Mexico and other locations.

Illegal-drug investigations cover many substances.

©iStockphoto.com/Stas_V

As mentioned, the federal government developed the system of schedules that separate drugs into five categories. While illegal narcotics such as cocaine and heroin are familiar to readers as dangerous and addictive, prescription medications have long been abused but have garnered renewed attention because of the unprecedented number of overdose deaths and other problems associated with addiction to, notably, opioids and painkillers. As reflected in Figure 10.1, overdose deaths from opioids, including heroin and synthetics, from 2002 to 2017 increased fourfold. Figure 10.2 represents the CDC data showing that more than 72,000 Americans died from overdose in 2017 alone.

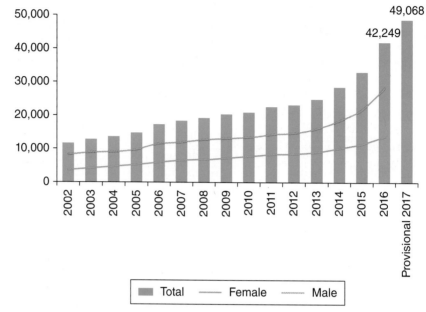

FIGURE 10.1 U.S. Overdose Deaths Involving Opioids

Source: National Center for Health Statistics, CDC.

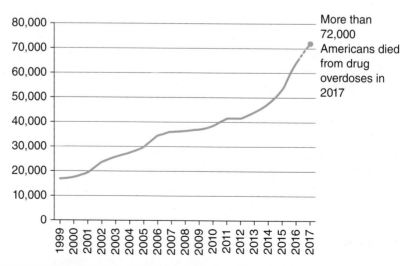

FIGURE 10.2 Total U.S. Overdose Deaths

More than 72,000 Americans died from drug overdoses in 2017

Source: National Institutes of Health, U.S. Department of Health and Human Services.

Scope of the Problem

Although the presence of illicit drugs and their various impacts on society change over time, the challenges are always real and affect individuals, neighborhoods, and communities. Some indication of the relative rates of drug-related crime is revealed by examining statistics gathered by the National Institute of Justice's drug abuse monitoring program. The FBI's Uniform Crime Report also offers numbers of how many people are arrested for drug offenses. But trying to accurately account for all human and economic impacts of drug abuse, drug crimes, and drug-influenced crime is a daunting task, to say the least. Tracked or perceived increases may result in greater focus on enforcement of crime statutes, education efforts on the danger and damage of drugs, and in some cases and places, efforts to treat those with addictions. Health and treatment cannot be artificially separated from unlawful behavior. While society vacillates over time about what behaviors to punish, and to what extent, research shows the efficacy of treating addictions as one tool to address the social harms of many crimes.

The National Institute on Drug Abuse (NIDA, 2014) notes the following about the interplay of drug abuse and crime:

> Drug abuse is implicated in at least three types of drug-related offenses: (1) offenses defined by drug possession or sales, (2) offenses directly related to drug abuse (e.g., stealing to get money for drugs), and (3) offenses related to a lifestyle that predisposes the drug abuser to engage in illegal activity, for example, through association with other offenders or with illicit markets. Individuals who use illicit drugs are more likely to commit crimes, and it is common for many offenses, including violent crimes, to be committed by individuals who had used drugs or alcohol prior to committing the crime, or who were using at the time of the offense.

The findings from research over several decades have established that alcohol and drugs are involved in one way or another with a majority of crimes that lead to arrest and incarceration. In particular, violent crimes including, assault, rape, and murder have been shown to be associated with alcohol in particular. While the majority of government expenditures related to illegal drugs goes to enforcement (Hughes et al., 2018), the correlations of substance abuse and criminal behavior have social implications far beyond the investigative efforts of law enforcement agencies.

Fentanyl: A Synthetic Opioid

One synthetic opioid is fentanyl. This is a Schedule II drug that produces the effect of morphine in the human body—but at 50 to 100 times the potency. The potency has led to significant increases in overdoses. In the 2017 Drug Enforcement Administration's Fentanyl Briefing Guide, this warning is given to first responders:

> WARNING
>
> There is a significant threat to law enforcement personnel, and other first responders, who may come in contact with fentanyl and other fentanyl-related

substances through routine law enforcement, emergency or life-saving activities. Since fentanyl can be ingested orally, inhaled through the nose or mouth, or absorbed through the skin or eyes, any substance suspected to contain fentanyl should be treated with extreme caution as exposure to a small amount can lead to significant health related complications, respiratory depression, or death.

Investigating Drug Crimes

The investigation of drug crimes spans everything from a street officer making an arrest for possession to surveillance of a local or regional drug house producing illegal products to multinational law enforcement operations over a period of years to uncover or infiltrate drug networks. The range of offenses also calls for a variety of investigative skills to address them. Officers receive training and information in their recruit academy and during in-service training about drugs and their effects, addiction and its manifestations, drug users and dealers and their behaviors and crimes, as well as how to investigate the many activities associated with illegal drug use and sale.

Patrol officers make arrests for possession of drugs when they come upon persons using drugs, or subsequent to arrest for another charge and the officer finds drugs on the person when they are searched. This may include someone possessing paraphernalia such as materials to smoke, inject, or ingest a drug. Unfortunately, officers may encounter a person affected by drugs when they respond to a call involving an individual acting in a bizarre or violent fashion, as the drug affects the individual. Law enforcement efforts also involve those who manufacture illicit drugs, and the illegal distribution of drugs or alcohol.

Most people are familiar with the use of K-9s in detecting the presence of various drugs. Trained police dogs are used at traffic stops, airports, on school grounds, and other locations for drug detection. Use of narcotics dogs at schools can also enhance security and provide educational opportunities (Reynolds, 2007). Where police have drug dogs present can have implications for whether or not privacy expectations exist and therefore whether the dog sniffing around constitutes a search (Chase, 2015). In addition, K-9s are used for locating live and deceased victims, security at events, tracking suspects, explosives detection, and more.

Small and mid-sized dealers of illicit drugs get their product from wholesalers or distributors, who in turn may receive the initial bulk product from an importer. The person selling drugs at the street level may be in a stationary location where customers come to them, or they may be mobile. They may also set up in various locations, using lookouts and runners to watch for police or potential rip-offs, and to take the drugs from a designated location (stash) to the person buying the drug. The variations of this way of doing business aim to keep the upper-level people from the risk of being caught with any drugs in their own possession. Runners may be juveniles who stand to face less criminal sanctions if caught with drugs or drug money. The training and experience of officers and investigators allow them to make street-level cases and often arrest someone at a low level and convince them to cooperate (turn or flip) in the investigation of the people higher up.

How It's Done

HANDLING NARCOTICS EVIDENCE

- Collect evidence following standard procedures; take care against exposure.
- Photograph, weigh, and count narcotics evidence as appropriate.
- Describe the evidence thoroughly and complete all chain of custody forms.
- Package items with consideration for whether evidence will be affected by environmental conditions in the evidence storage area or on the way to a laboratory; different types of drug evidence need different types of containers.
- When evidence is returned from laboratory testing, another evidence check must be completed; if the evidence goes to court and back to storage this must be repeated.
- When the case is concluded, the agency will typically request a court order to authorize destruction of the narcotics.

Large-scale or ongoing illegal drug operations have also been addressed in federal statutes through the Continuing Criminal Enterprise Statute. Also known as the Kingpin Statute, this law is used to prosecute five or more people who conspire to commit violations of the Comprehensive Drug Abuse Prevention and Control Act of 1970. The complexity and time length of such investigations are why it is generally necessary for local law enforcement agencies to collaborate with state and federal agencies in a task force approach that leverages the resources available from the larger, specialized enforcement agencies. Given the nature of the illegal drug crime process, which involves many steps and individual components as a product reaches an individual and money changes hands at various levels, there is cause to continue seeking ways to link the investigative process together in a more comprehensive fashion (Chistova, 2017).

Undercover Operations

Law enforcement agencies at the local, regional, state, and federal levels operating independently in some cases, and in coordination for larger operations, make efforts to stem the production and distribution of drugs to communities. Street-level sales can be disrupted through surveillance of known drug-dealing areas or the use of undercover officers or informants to actually participate in the purchase and receiving of illegal drugs. This type of evidence is difficult to fight for the person arrested after selling illegal drugs to an officer or agent of law enforcement. These arrests can yield benefits of expanding an investigation to the next person up the chain who has provided the drugs for sale on the street. Officers also use sting operations and reverse buys to sell confiscated drugs to others engaged in distribution before arresting them and reseizing the drugs. These operations are often lengthy and complex, but can also come up in a short span of time when an opportunity presents itself.

Undercover operations can be dangerous for officers or informants. For officers in undercover operations, this can include psychological issues including dysfunctional behavior on the job (Krause, 2009). The FBI and other agencies have developed programs and protocols that seek to mitigate the effects of working undercover, especially for lengthy periods of time. Even with potential physical and psychological dangers, it is recognized that an undercover operation may be the only viable way to get at the significant criminals in a drug enterprise (Bonney, 2015).

The investigation of drugs lends itself to undercover techniques that put the officer in direct contact with criminal suppliers and sellers. Agents may also use informants to conduct **controlled buys** that are monitored in real time and use specific procedures to ensure legal requirements are followed for the gathering of evidence (Darst, 2013). Agents may also pose as higher-up members of a criminal organization to entice lower-level dealers to reveal information by promising to provide the dealers access to bigger suppliers (Puddister & Riddell, 2012).

Undercover investigations conducted on the Internet are also a standard tool of law enforcement agencies. A major challenge in these types of investigations is determining where a supplier of illegal drugs is located and then working with other agencies to coordinate efforts to arrest the person. The double-edged sword of the Internet provides commission and facilitation opportunities across a number of crimes while also equipping law enforcement with capabilities to combat and investigate crimes. Regardless of the initial way an undercover investigation is begun, or what technology or techniques are used, personnel must be well trained and psychologically prepared (Schreiber, 2013), and the monetary resources to conduct the operation must be available (Logue, 2008).

> **Controlled buy** The use of a police officer or informant, under strict conditions, to purchase illegal drugs or other contraband from an offender.

How It's Done

UNDERCOVER INVESTIGATIONS

Television and movies have dramatized the actions of undercover police operations. Below are some issues highlighted by the United Nations Office on Drugs and Crime (UNODC):

Undercover operations occur where investigators infiltrate criminal networks or pose as offenders to uncover organized crime activity. These operations occur in many countries with different types of oversight. Major issues faced by jurisdictions are listed below:

- In what kind of cases and in which format are undercover operations allowed?
- Are there limits on the type of undercover operations permitted?
- What are the preconditions for conducting undercover operations?
- Is authorization from a judicial or other independent source required?
- Are there guidelines for the appropriate use of undercover officers?

In most jurisdictions, undercover officers are not permitted to encourage suspects to commit crimes they would not ordinarily commit, either as an agent provocateur or through entrapment (i.e., a situation in which an agent or official originated the idea of the crime and induced the accused to engage in it; in some jurisdictions, it is used as a defence to criminal charges). Their role is usually to become part of an existing criminal enterprise. Jurisdictions vary in the nature of the restrictions they place on undercover operations, with most focusing solely on prohibiting undercover agents from providing opportunities to commit crime, and committing crimes themselves.

Grow Houses and Labs

In communities across the country and around the world, drug suppliers produce or prepare drugs for sale in small to large physical operations. Some of these are clandestine **drug laboratories** and cannabis **grow houses** (CGH), also known as indoor marijuana growing operations (IMGO). **Clandestine labs** may be more familiar to many because of the news stories they can generate when officers and lab personnel have to enter the premises in full hazmat suits.

The grow house also poses many risks to first responders and evidence technicians. This begins with the hazards resulting from the way a grow house is powered. To mask the significant amount of electricity required to maintain equipment and conditions for growing, growers will often bypass the power company's meter and rig a direct power feed that can be quite dangerous. In addition to electrical hazards from slipshod wiring and high-voltage lighting sources, there may be pressurized canisters of carbon dioxide to speed growth that may explode, and outright booby traps to secure the grow house from unwanted visitors, such as police and firefighters (Gustin, 2010).

Drug laboratory Government or private laboratory that conducts analysis of substances suspected to be illegal drugs.

Grow house Generally, the use of a residential house to cultivate marijuana indoors.

Clandestine lab A laboratory set up secretly to illegally produce controlled substances.

A less obvious (and often invisible) hazard to officers and others comes in the form of toxins and spores from plants, and exposure to tetrahydrocannabinol (THC). The removal of plants from the grow houses will often increase the volume of fungal spores in the air, which can be a risk to personnel (Martyny, Serrano, Schaeffer, & Van Dyke, 2013). Exposure to THC in the quantities found on surfaces or in the air at a grow house has not been shown to be as much of a risk as the fungal spores, chemicals for fertilization of plants, carbon dioxide, or electrical risks, though the THC has been identified as toxic to police drug dogs in some concentrations (Doran, Deans, De Filippis, Kostakis, & Howitt, 2017).

Even though marijuana has been legalized in several states, unlicensed growing operations are still illegal. What are some of the other impacts of legalization on drug investigations?

©iStockphoto.com/contrastaddict

Clandestine laboratories in the United States are often used for the production of methamphetamine, or meth. As early as 1990, the Drug Enforcement Administration (DEA) and the Environmental Protection Agency (EPA) published a set of guidelines for law enforcement on the clean up of clandestine drug laboratories. In addition to the criminals present at clandestine labs when they are raided, there are other physical dangers. The fumes created by the mixing of various compounds during the process of making meth can be toxic (Michael, 2008). Residue of the methamphetamine itself is also a health hazard that has affected many first responders. Security measures may include holes cut in the floors and concealed boards with projecting nails to step on, which are also used in many outdoor marijuana-growing areas.

Thousands of such clandestine labs are discovered in the United States each year, with the Drug Enforcement Administration (DEA) reporting more than 85,000 illicit meth lab incidents between 2004 and 2011 (Green, Kuk, & Wagner, 2017). After shutting down the location, and hopefully arresting the operator, authorities must also arrange for "clean up" of the site. The EPA and local officials have worked together to mitigate the hazards to residents, but clean up can be expensive and time-consuming (Mirel, 2011).

Knock and Talk

An investigative technique used in a large number of suspected drug operations is the "**knock and talk**." Essentially, the courts have long held that voluntary interactions between police and the public do not present a legal problem for evidence seized in a search consented to by the citizen. The reality of many nascent criminal investigations is that there is little or no solid evidence of a criminal act. At times, however, there are tips and reports of criminal wrongdoing, but ones that fall short of the probable cause needed to acquire a search warrant from the court. These may be an anonymous community crime line tip that says, "The people at 123 Elm Street are growing marijuana in their house." Armed with a specific location, but no evidence or direct testimony of a person with knowledge of the drugs, law enforcement is left with little to go on. Setting up surveillance of houses or other buildings are labor-intensive and potentially long-term efforts with no clear indication of likely success. So what to do? Knock on the door and ask the person there if they are growing marijuana in their home. Before dismissing this approach as absurd or unlikely to yield results, consider the status of statements given voluntarily to police as well as other searches, such as cars, when an officer simply asks if she may search the driver's vehicle. People often say yes—even if they *know* they have some illegal substance or the fruits or instrumentalities of a crime in their car.

People submit their property or person to search and speak against their own interests quite often. As noted, the courts monitor the manner in which this comes about, but they have also held that if it is clear that there was no coercion, the actions by law enforcement are legal. Harken back to the dismal rate of solving or clearing crimes. If there was no opportunity to pursue a criminal case with the voluntary statements and consent searches of suspects, would you imagine that such case resolutions would be even lower? For many people, the notion of officers using information obtained in this way, or even through deception, creates a feeling of unease. Even with the scrutiny afforded by the legal system and courts, many feel that police should not even be able to ask individuals for the permission to search.

As for the grow houses, the first police department I worked for was a small agency with only one or two patrol officers working at a time. During the shift, the officer on duty might meet with the officer on duty with the adjacent small department. There was a specific motel parking lot at the city limit where we would sit and chat for a few minutes before returning to our patrolling. One day, my counterpart officer and I were talking when he pointed out the end unit of the sleepy little beach motel. We saw that the windows were all covered in aluminum foil from the inside, and a hose was run inside one of the windows. We were both aware that these were possible signs of illegal marijuana cultivation. We walked to the motel room door and knocked. The fellow who came to the door seemed initially surprised, but then hung his head. We asked him if he was growing marijuana. Seemingly disappointed with himself, he replied that yes he was, and he asked us if we wanted to come in and see the plants. We did. He went to jail.

Asset Forfeiture

An effective tool used in drug investigations is that of **asset forfeiture**. When a law enforcement investigation shows that a vehicle, property, cash, or other items were used in a drug crime (and other felonies), the agency may initiate a civil proceeding to seize the items. At times this may be in the form of a proceeding initiated against a sum of money, if no one

---

Knock and talk Police investigative method of asking a resident if the police may come in and search the premises.

Asset forfeiture Seizure of property or funds used in furtherance of a crime.

claimed the money because they believed it would show their involvement in the crime. Instead of the criminal courts' burden of showing "clear and convincing" evidence of something tangible being used in the crime, the U.S. Congress lowered the requirement to "a preponderance of the evidence," seen in civil cases. Thus, a person may be found not guilty, or perhaps not even be criminally charged, but still lose possession of property because the government showed that the property was used in a drug (or other) crime. This specific point has led to some concern that police may pursue assets at times more aggressively than the suspect because it may prove easier to link the money or property to a crime (Kelly & Kole, 2016). During my investigation into two drug dealers fighting inside the car owned by one, one dealer bit a portion of the car owner's ear. After I concluded my investigation, the piece of his ear was reattached in the emergency room, but I seized his car as having been used in a drug transaction.

Organized Crime

The concept of organized crime (OC) is fairly straightforward: multiple people working together, with different or overlapping roles, reporting in a structure to someone in charge, to commit crimes. And, of course, this simple description can expand to thousands of members of a criminal organization located around the world but working toward common objectives set by the group's leaders. In different countries around the globe, and in different locations throughout the United States, groups organized to commit mostly profit-making crimes may benefit from the lack of resources available to address their criminal acts, or because the groups may enjoy some level of "protection." We have noted throughout the book that given the relatively small size of most law enforcement agencies, they simply cannot devote much in the way of resources to something as well coordinated as organized crime. Smaller agencies may work together in a task force arrangement and coordinate with state and federal agencies for assistance.

Some of the crimes seen most often in connection to organized crime groups are those involving *vice*, or public morals. Crimes such as prostitution, gambling, and illegal or unregulated loans (loan sharking), often associated with gambling, are examples. The individuals participating are often willing, though many sex workers are coerced into the work they do to make money for others. This includes sex trafficking that involves kidnapping or coercion of women and children into prostitution. If the crimes are not particularly visible to the public, there may be little push to address such crimes.

Extortion

Organized crime groups do not exist only in the so-called victimless crime space. Organized crime groups, such as the Mafia and a number of gangs, may also make money by extorting businesses for protection from others or against having the group itself do damage or harm to the business or owners. Crimes of violence are used to intimidate those forced or choosing to do business with a group, as well as to send warnings to competitors. Money is central to much of organized crime, and so trafficking in drugs or weapons, sales of contraband or stolen items on the black market, and bribery of officials to gain money directly or indirectly through business arrangements, are also common. Because of the profit motive of most organized crime, the groups are generally working to expand their reach into every

Launder Filtering illegally obtained money through a legitimate business to obscure its source.

aspect of money-making surrounding a criminal activity. The organized crime group may then **launder** the money it makes by funneling it through otherwise legitimate businesses it owns or controls. Some of the legislation that has come about to address this aspect of organized crime makes it illegal to buy or gain control of a business with monies made through unlawful activities.

One of the most effective federal laws, used for nearly five decades, is the Racketeer Influenced and Corrupt Organizations Act, or RICO. The act provides penalties and civil recourse against organizations and their leadership for acts committed as part of an *ongoing* criminal organization. Used extensively by the FBI in its efforts against the American Mafia, this important tool was relatively recently applied in prosecuting members of the MS-13 criminal gang. RICO and other laws authorize assets forfeiture that allows law enforcement to seize monies and items used in the ongoing criminal enterprise.

Investigating Organized Crime

With the many ways that organized crime groups may pursue crimes as varied as retail theft, to extortion, to government corruption, the law enforcement responses must vary as well. As noted, effective countermeasures often involve multiple agencies working in concert to leverage resources. Federal law enforcement agencies often lead the fight against organized crime due to their access to greater resources and expanded jurisdictional authority. Local and state law enforcement, when resources and political will permit, can form specialized units or task forces to combat organized crime. These groups will work closely with prosecutors to ensure effective case preparation and comprehensive efforts to prosecute cases as well as seize assets. Collecting information now spans the cybersphere in addition to a continued focus on identifying involved criminals and seeking information through arrest and subsequent plea-bargaining for further information from an offender.

Gangs

A quick examination of gangs will give the impression that they are simply another form of organized crime. While this is true in certain ways, given that gangs operate locally, regionally, nationally, or internationally, there is also an argument that many such groups are "disorganized." Street gangs such as outlaw motorcycle gangs (OMG) or "biker" gangs have involvement in a variety of crimes. Some gangs may engage in low-level crimes and remain in a specific area of a community. Others may link with other gangs to collaborate or compete and fight with other gangs over the territory of a community with the winner being the one who commits more crime in the previously contested area. The use of violence by gang members is often for similar reasons as organized crime groups: intimidate or eliminate competition. Police are called upon to suppress gang activity and make neighborhoods safe. And corrections officials must track and address the problems caused by prison gangs and other **security threat groups (STG)**.

Security threat group (STG) Group or gang within a jail or prison setting that has a pattern of violent or disruptive behavior.

The rate and number of gangs and gang members have decreased in the United States, but the numbers are still large (Reynolds & Carlson, 2018) and the problems caused by gang activities remain. Gangs can exist in virtually any community, sometimes keeping a low profile to be able to operate without law enforcement attention. While most gangs were historically exclusive either by gender or ethnicity, hybrid gangs have formed as the business calculation of making money may involve, for example, a local drug gang teaming up with a more mobile biker gang to transport or distribute the illegal drug products. The credo of most gangs revolves heavily around demanding respect from others. Not being shown this respect will most likely result in a violent act against the offending person or group.

Identifying Gangs

Many gangs choose specific colors or symbols to wear that signify their membership in the group. Gangs typically have hand signals, and often tattoos, that are also unique to the group. Many gangs initially formed for self-protection from other groups, and this became the origin of neighborhood-based groups that defended their "turf" or geographic area. Gangs will often mark the area with **graffiti** painted on buildings or other objects (called tagging). Law enforcement officers can often identify gang members or associates through these colors, clothing, tattoos, and locations. Most states have statutory language that defines gangs and helps agencies classify gang members and associates. An example is the Florida State Statute definition of a *criminal street gang*:

What might the officers have used to identify these two men as gang members?

South_agency/E+/Getty Images

Graffiti Words or symbols drawn or painted onto buildings and objects with the intention to mark the territory of a gang or group or to convey a message to others.

> (1) "Criminal gang" means a formal or informal ongoing organization, association, or group that has as one of its primary activities the commission of criminal or delinquent acts, and that consists of three or more persons who have a common name or common identifying signs, colors, or symbols, including, but not limited to, terrorist organizations and hate groups. (according to s. 874.03, F.S.)

If a crime is shown to have an element of gang involvement, prosecutors may present a case for an enhanced penalty, if the perpetrator is convicted. Again, Florida statutes provide an example of how gang members and associates may be identified:

A ***criminal street gang member*** is defined as a person who meets two or more of the following criteria as per s. 874.03 (3), F.S.

A ***criminal street gang associate*** is defined as a person who meets one of the following criteria as per s. 874.03 (3), F.S.:

a) Admits to criminal gang membership.
b) Is identified as a criminal gang member by a parent or guardian.
c) Is identified as a criminal gang member by a documented reliable informant.
d) Adopts the style of dress of a criminal gang.
e) Adopts the use of a hand sign identified as used by a criminal gang.
f) Has a tattoo identified as used by a criminal gang.

g) Associates with one or more known criminal gang members.

h) Is identified as a criminal gang member by an informant of previously untested reliability and such identification is corroborated by independent information.

i) Is identified as a criminal gang member by physical evidence.

j) Has been observed in the company of one or more known criminal gang members four or more times. Observation in a custodial setting requires a willful association. It is the intent of the Florida Legislature to allow this criterion to be used to identify gang members who recruit and organize in jails, prisons, and other detention settings.

k) Has authored any communication indicating responsibility for the commission of any crime by the criminal gang.

Combating Gang Crime

Street officers interact with possible gang members or gang associates and generate field information cards or intelligence reports to try and track members and their activity and share this information with other officers, agency units, or other departments. Documenting the signs, symbols, and individuals involved in gangs is a critical investigative practice for officers. Proactive and hot-spot police patrol is one tactic to address gang activity. But in the contemporary era of intelligence-led policing, an increased emphasis on gathering data on gang members and activity, and changing conditions within communities, is geared to work smarter on intervening in the right places, at the right time, with the right people. This often involves the task force approach again, and using both technology and a cooperative approach to share information with other agencies. Agency tracking systems can be basic or quite advanced, depending on the available resources. The Internet and secured communication platforms can facilitate intelligence-sharing among law enforcement agencies at all levels.

Law enforcement agencies cannot change the culture or conditions that lead to much of the gang activity in a city or neighborhood. And when interacting with possible gang members, officers must be extremely cautious and aware of safety concerns. A concerted effort is needed that involves schools, businesses, the faith community, social and medical agencies, and other parts of a community. Community-oriented policing (COP) is generally seen as being more conducive to reducing the influence of gangs because the focus of COP is the ongoing and close interactions with the community, rather than simply responding to calls for service.

Public Response

Some states and communities have used antiloitering laws that address some of the visible aspects of potential gang activity. It is important to remember that much activity, such as drug dealing, is not entirely conducted out in the open, nor completely covertly. This reality of gang activity calls for the varied enforcement approaches of law enforcement and the educational efforts of other agencies of the community. Focused deterrence strategies have also include the so-called "pulling levers" strategy of coordinating as much of the authority of government as possible at gang members who commit violent crime. Addressing gang crime, heavily involving juveniles, is an enduring challenge that is not solely one to be addressed by law enforcement.

EXPLORE THIS

View the agency website of the Drug Enforcement Administration (DEA) at https://www.dea.gov/. Locate the Resources section and examine what is available to the public (and students!). Discuss in class how the various listed sources can directly benefit communities and citizens. Are you aware of strategies and resources being used in your community?

SUMMARY

Drug crimes affect the entire U.S. society in various ways: health, economics, crime, and quality of life are some of these. The use of illicit drugs or the abuse of legally obtained drugs are contributors to crime and present challenges for law enforcement officers and the entire criminal justice system. Although drug crime is not at the rate it once was, the human toll and tragedy of opioid overdoses is an example of how the problems associated with drug use and abuse are not strictly law enforcement matters.

Investigating drug crimes is handled at every level of law enforcement, which then passes suspects on to the courts and, for many, on to the corrections system. Investigative techniques and legislative tools combine to help officers identify importers, distributors, dealers, and users, and hold them accountable for their actions.

Organized crime has been known around the country and around the world. Groups participate in a wide variety of crimes aimed at profiting through provision of illicit services, items, and substances; extorting businesses and individuals; and harming those who oppose them. Since 9/11, funding and priorities of the FBI, which had major responsibility for investigating organized crime, have shifted to the fight against terrorism. This split focus has likely allowed a resurgence of organized crime. Some organized crime also is employed to provide funding for terror groups.

The fear and damage created in communities in the United States from gang activity are still significant. Even with some reduction in gangs and gang membership, various individuals turn to gang participation to have money or material things, seek the protection of a larger group, do something they consider exciting, or adhere to a subcultural lifestyle that satisfies other psychological issues. Many gangs have an organized crime approach, and some form business partnerships with other, formerly rival, gangs to profit from crime.

KEY TERMS

Asset forfeiture 218
Clandestine lab 217
Controlled buy 216
Controlled substances 210
Drug laboratory 217
Graffiti 221
Grow house 217
"Knock and talk" 218
Launder 220
Schedules 211
Security threat group (STG) 220

DISCUSSION QUESTIONS

1. What harms come from illegal drug crimes? How should agencies address the social aspects?
2. How do law enforcement agencies investigate those who deal drugs? What are the challenges?
3. Describe and discuss the crimes of organized crime groups.
4. Why is it that people join gangs?
5. What methods do street-level officers use in response to gang crimes?

- Get the tools you need to sharpen your study skills. SAGE edge offers a robust online environment featuring an impressive array of free tools and resources.
- Access practice quizzes, eFlashcards, video, and multimedia at **edge.sagepub.com/houghci**

PRACTICE AND APPLY WHAT YOU'VE LEARNED

▶ edge.sagepub.com/houghci

SAGE edge™

WANT A BETTER GRADE ON YOUR NEXT TEST?

Head to the study site where you'll find:

- **eFlashcards** to strengthen your understanding of key terms.
- **Practice quizzes** to test your comprehension of key concepts.
- **Videos and multimedia content** to enhance your exploration of key topics.

©iStockphoto.com/400tmax

11 Terrorism and Homeland Security

RUNNING CASE STUDY: HATE CRIME–MOTIVATED MASS SHOOTING PLANNED AT SULLIVAN'S SPA

Introduction

Terrorism
- Background
- Domestic Terrorism
  - *Recruiting Homegrown Violent Extremists*
  - *Left- and Right-Wing Groups*
  - *Militia Groups*
- International Terrorism
- Suicide Bombings

How It's Done: Bomb Threat Checklist
- Lone Wolf
- Hate Crimes
- Role of the Media

Homeland Security
- State and Local Level
- Federal Level
- Fusion Centers

Explore This

Summary

Key Terms

Discussion Questions

LEARNING OBJECTIVES

After reading this chapter, students will be able to

- **11.1** Discuss dangers posed by terrorism.
- **11.2** Describe the types of terrorist groups.
- **11.3** Explain the role of local law enforcement in terror investigation.
- **11.4** Explain the difference between international and domestic terrorism.
- **11.5** Describe the functions of homeland security.

Running Case Study: Hate Crime–Motivated Mass Shooting Planned at Sullivan's Spa

Criminal Investigations Division (CID) Commander Kimberly Kellan and Assistant State Attorney (ASA) Corbin Kelsea sat together for the quarterly Homeland Security Task Force. On the agenda was the attack on a Jewish cemetery in a neighboring city. Gravestones had been overturned and swastikas spray-painted on them. Because the act had been carried out by a man in alignment with neo-Nazis, it was classified as domestic terrorism. Because his actions were racially motivated, it was also classified as a hate crime.

"These types of crimes are so hard to intercept," Kimberly said.

Corbin nodded. "The lone wolf doesn't tell anyone his plans. There's no advance intelligence."

"One way is to infiltrate their forums," Kimberly offered. "Sometimes they'll give clues online. I've been thinking we need to assign Bradley Macon to follow the AN, the KKK, and all the other hate groups." Kimberly made a list: male supremacists, white supremacists, neo-Nazis, skinheads, white nationalists, anti-LGBTQ, anti-Muslim, neo-Confederates.

"The trouble with that," Corbin said, "is how toxic it is for an officer to be exposed to those extreme views all day."

"So, we'll switch them over, the way we do with gang surveillance," Kimberly said. "We have to track these groups at the local level."

James Sullivan walked through the darkened neighborhood. He was nocturnal, sleeping only every other day, which gave him the advantage of planning at night. He laughed to himself, thinking that the assistant state attorney lived next door. If only the man who mowed his grass every Sunday knew what James was planning.

He'd been lucky to get the waterfront house away from his ex-wife. But that wasn't enough. He hated women. He hated all five of his ex-wives and their attempts to get him to pay child support. Even though he hadn't seen her in years, he hated his teenage daughter. Misogyny wasn't strong enough a word. Woman-hater was so much better. There were thousands of men online who agreed with him.

He hated the IRS and its attempts to get him to pay his back taxes. So what if he never got Social Security? He hated the whole country of America. He especially hated the loose morals of American women. They didn't know enough to be subservient to men like their counterparts in South America. He was going to make American women pay.

So what if he didn't have a car? An angry man with a plan could walk miles in one night.

On the other side of town, he walked past the spa his wife ran. Yeah, she'd managed to keep the spa, but now that would change. After he was through, she'd be lucky to be alive. He walked past the spa over and over, trying to decide into which entrance to break. If he entered the main door, he could shoot all the women in the reception area before storming the yoga room, then move on to the private massage rooms. But if he broke in the back, fewer people would see him coming, and the cops might not arrive before he'd killed all the women.

He knew from Sandra's stories that the spa was busiest Saturday mornings. There would be dozens of women inside, relaxing. Getting their nails done, getting massages.

Just what the world didn't need: relaxed, empowered women. There would less of them by the end of next weekend. If the cops got there before he killed them all, he would turn the gun on himself. He had nothing more to live for, and he definitely didn't want another stint in jail.

What he wanted more than anything was for the story to appear on the news, national television, the Internet. He envisioned the headline: "Man Takes out 36 Women at Sullivan's Spa: Sandra Sullivan Included." He wanted other men to see how easy it was. Just pick up a gun and show women who's boss. Women wouldn't listen any other way. Men had to be violent. What he wanted more than anything was for his 20-year-old son—living on the West Coast with ex-wife number 3—to see the headlines and know what a great man his father had been. Maybe the kid would be inspired to follow in his footsteps.

Think about the challenge of stopping the intended attack of a lone wolf individual. Why is the community important in such cases? How can citizens help law enforcement fight terrorism?

Introduction

I attended my first terrorism seminar in the early 1980s. Like most people, I failed to understand the intertwined nature of acts typically understood as terrorism and the country's infrastructure as both a practical and symbolic target of attack. Law enforcement agencies and local police, in particular, are tasked with the investigation of criminal events that may be linked directly or indirectly to terrorism. Local law enforcement are also the first responders to in-progress or just-occurred acts of terrorism. Yet the dramatic and terrifying act of terror by bombing a place or building, with or without many civilians present, is not the only crime that a terrorist group is likely to commit. While the motivation for crimes may differ, the basic investigative methodology remains the same. And so, just like 40 years ago as I sat in class, today's law enforcement must continually learn about the methods of criminal and terror groups and how to investigate them.

For investigators and those tasked with trying to prevent acts of violence or determine after the fact who committed such acts, it is important to understand the motivations and objectives of those committing crimes considered terrorist. Regardless of whether the crime is classified as a domestic or international terrorist act, it is aimed at affecting how people think as opposed to a crime of profit or an expression of violence for its own sake. It is difficult for the average citizen in most countries to understand how a terror organization or even an individual believes they will affect a broad change of government policy or gain the sympathies of the public through an act of destruction or harm to civilians.

Terrorist actions often serve several purposes. A group certainly behaves in ways to affect what it perceives as the audience of the public, but it also seeks to recruit new adherents and generate funding. In a number of countries, the efforts of government groups and others are geared toward not only preventing acts of terror but working to resolve social issues that may give rise to such acts. Americans have become more knowledgeable in

Genocide Killing a large group of nationally or ethnically similar people.

Hate crime Nature of such crimes goes beyond the single offense to symbolically target whole demographics of people.

International terrorism Variously defined, but generally including violent or dangerous acts intended to frighten a civilian population and influence government action; primarily occurs outside of the United States.

general about terrorist activity over the last two decades and are rightfully concerned when criminal homicides caused by terror hit home inside U.S. borders.

It is beyond the scope of the book to catalog and explore each terror or extremist group or the ideology followed by every such group, let alone examine the large-scale killing of an ethnic group as in **genocide**. Killing people in acts of terror or hate crime are appropriately considered for the student who seeks to understand homicide. Yet it is important to understand some of the general categories that allow us to grasp the orientation of various groups. Understanding the typical motivation of people within these groups can provide investigators clues or avenues to explore as they seek criminals who are also classified as terrorists. Within the United States crime committed by terrorists is a small percentage of the overall number of crimes, yet the larger goals of terrorists brings significance and a sense of urgency to the investigation of these crimes.

Investigating **hate crimes** is also addressed here because the motivation of such crimes goes beyond the single offense to symbolically target whole demographics of people. Such bias in committing crime is in part defined by the U.S. Congress: "because of the actual or perceived religion, national origin, gender, sexual orientation, gender identity, or disability of any person. . . ." In such cases, if there is a potential violation of civil rights the FBI may be the investigating agency to work in conjunction with a local or state agency.

Terrorism

We begin with the definition used by the FBI for terrorism:

18 U.S.C. § 2331 defines "**international terrorism**" and "domestic terrorism" for purposes of Chapter 113B of the Code, entitled "Terrorism":

"International terrorism" means activities with the following three characteristics:

- Involve violent acts or acts dangerous to human life that violate federal or state law;
- Appear to be intended (i) to intimidate or coerce a civilian population; (ii) to influence the policy of a government by intimidation or coercion; or (iii) to affect the conduct of a government by mass destruction, assassination, or kidnapping; and
- Occur primarily outside the territorial jurisdiction of the U.S., or transcend national boundaries in terms of the means by which they are accomplished, the persons they appear intended to intimidate or coerce, or the locale in which their perpetrators operate or seek asylum.*

*FISA defines "international terrorism" in a nearly identical way, replacing "primarily" outside the U.S. with "totally" outside the U.S. 50 U.S.C. § 1801(c).

"Domestic terrorism" means activities with the following three characteristics:

- Involve acts dangerous to human life that violate federal or state law;
- Appear intended (i) to intimidate or coerce a civilian population; (ii) to influence the policy of a government by intimidation or coercion; or (iii) to

affect the conduct of a government by mass destruction, assassination. or kidnapping; and

- Occur primarily within the territorial jurisdiction of the U.S.

18 U.S.C. § 2332b defines the term federal crime of terrorism as an offense that:

- Is calculated to influence or affect the conduct of government by intimidation or coercion, or to retaliate against government conduct; and
- Is a violation of one of several listed statutes, including § 930(c) (relating to killing or attempted killing during an attack on a federal facility with a dangerous weapon); and § 1114 (relating to killing or attempted killing of officers and employees of the U.S.).

There are a variety of stereotypes about terrorists, but as noted by Laqueur (1999) and others, no one profile of specific demographics or other characteristics exists to encompass all terrorists and their motivations. At any point in history, and continuing through today, one group may use similar methods as another yet be motivated by completely different objectives and goals. Most groups and individuals would completely reject being labeled a terrorist. Their view of themselves would be as "freedom fighter, a guerrilla, a militant, an insurgent, a rebel, a revolutionary—anything but a terrorist, a killer of random innocents" (Laqueur, 2006, para. 22). There is, sadly, nothing new in how extremist ideology and violent behavior of a group comes about. A group may seek to maintain the status quo, push a nationalistic political agenda, change labor or individual economic conditions, or persecute others based upon a twisted view of religion.

Background

Sadly, the American public has become familiar with many aspects of terrorism. While the types of terror attacks that employ suicide bombers or many attackers working together in a single attack seem like distant crimes, single-person attacks in the United States are also familiar. There is the possibility that any country's population may be subjected to terror-based crimes or attacks. No doubt you yourself are aware of a crime (probably violent) that was recently committed that was classified as an act of terror, domestic or international. Such crimes in any country may come from groups with a worldview or ideology that runs counter to the majority of society. Subnational political groups or those supported by a sovereign country may be responsible. The crimes target civilians, not military facilities, although a group may also carry out a parallel attack against the government. Again, the aim is to create fear in the populace and engender feelings of insecurity.

Terror tactics have been used throughout much of history and around the globe. Two thousand years ago the Sicarii Zealots resisted the oppressive control of Roman rulers in the first century A.D. The Sicarii members would attack prominent Romans or moderate Jewish leaders while they were in public to achieve this same goal of making people know that an attack could come anywhere and at any time. The Roman military and other military forces throughout history have been used to suppress what have often been peaceful demonstrations. This has also often led to a subsequent use of violence in response to crackdowns.

While the group may be known, and thus provide a place for an investigation to begin, multiple groups may claim "credit" for the same attack or crime.

With the reality of more than one group claiming responsibility for property damage or physical attacks, it falls to government agencies and the media to help inform citizens of the motivations and differences among the groups. An individual who may or may not be affiliated with a terror group may still claim such an association because they support the group's cause or to rationalize their own behavior. In a later section we will look at this more closely. Some academic and professional examination of terrorists points to the inherent psychological aspects of who we are, which underlie the search for identity, sometimes through group affiliation, and through our individual acts including murderous terror.

Many crimes are opportunistic, or committed because the criminal has an immediate need or impulse. An ideologically motivated bomber or assailant may plan far in advance. The domestic or international terrorist may also be intent more on using his crime to send a message than on the immediate financial gain or property damage caused in a single crime. The terrorist may have no expectation of surviving his attack. While the question of who is responsible may thus be known, the connections and actions of the larger group will likely remain to be investigated. The intended audience is the entire society; the act is one of drama as well as harm. As Poland (2011) put it, "Today, terrorism is a global theater" (p. 9).

Domestic Terrorism

Both the FBI and the USA Patriot Act, define a domestic terrorist crime as one that occurs within the territorial jurisdiction of the United States. The geographic distinction of a crime occurring inside U.S. borders may facilitate an investigation in some ways, but not always supply the immediate motive for the terrorist acts. Most terrorism is domestic, and the offenders and victims share a nationality. The majority of Americans are not consciously aware of domestic terrorists, those "people who commit crimes within the homeland and draw inspiration from U.S.-based extremist ideologies and movements" (Bjelopera, 2017, p. 2). The primer report by the Congressional Reference Service (CRS) goes on to note that

> The Department of Justice (DOJ) and the Federal Bureau of Investigation (FBI) do not officially designate domestic terrorist organizations, but they have openly delineated domestic terrorist "threats." These include individuals who commit crimes in the name of ideologies supporting animal rights, environmental rights, anarchism, white supremacy, anti-government ideals, black separatism, and beliefs about abortion. (2017, p. 2)

Domestic terrorism U.S.-based group of extremists who perpetrate crimes based on, generally, a single issue or a belief system.

The semantic overlap of **domestic terrorism** and hate crimes can be confusing and result in misunderstanding of the actual volume of the two types of criminal acts. The challenges to investigating such groups and the methods used run along similar lines. Legally tracking the movements of suspected or identified members, devoting resources to monitor social media of the group and its members, and acquiring reliable informants from inside or on the fringe of these groups are ongoing efforts. Norris and Grol-Prokopczyk (2018) looked at the issue of entrapment claims made against U.S. agencies in their investigative efforts. While the authors found variation among right-wing cases, neojihadi cases, left-wing cases, and combinations of these groups, the practical implication for investigators is less clear.

Domestic terrorist groups have a variety of motivations, each group often having a **single issue** they are concerned with and act upon. Groups or individuals have also been referred to as extremists—those with a specific ideology and who commit criminal acts to further it. Also, if there are terrorists in the United States who are motivated by international terrorist groups, the FBI and DHS refer to them as homegrown violent extremists (HVE). In the 2018 report of the Office of the Inspector General (OIG) of the Department of Justice (DOJ), combating domestic as well as foreign terrorism was listed as one of the department's top management and performance challenges (USDOJ, 2018). The OIG noted that in fiscal year 2017, the FBI dedicated more than 5,000 personnel and in excess of $1 billion to this priority. While the DHS and FBI share responsibility for countering violent extremism (CVE), the OIG report noted that more needs to be done in the area of assessing whether the CVE Task Force is adequately meeting its goals.

Single-issue terrorism Violent acts carried out focused on one issue, such as animal or environmental rights.

RECRUITING HOMEGROWN VIOLENT EXTREMISTS

The Awareness Brief, Homegrown Violent Extremism (HVE), from the International Association of Chiefs of Police (IACP), notes an important use of social media to assist HVE recruiting.

Radical recruiters use a combination of mainstream and specialized social media sites and websites to encourage individuals and groups already in the West to rise up and use the "blend factor" to their advantage. The "blend factor" refers to the notion that there are no visual or physical cues, such as a dress code or location, that automatically identify someone as a violent extremist to law enforcement, making it easier for homegrown violent extremists to blend into the general population in disengaged communities.

LEFT- AND RIGHT-WING GROUPS

Two movements considered **left-wing groups** are animal rights and eco-terrorism. The Animal Liberation Front (ALF) and People for the Ethical Treatment of Animals (PETA) work against those they believe are harming animals and are considered animal rights extremist groups. The Earth Liberation Front (ELF) and Greenpeace, through their focus on the environment, perpetrate crimes against entities they believe are causing harm. Both of these movements have tended to use the destruction of property as a tactic, causing a great deal of damage to equipment and property, but they have generally avoided hurting people.

Left-wing group Communist- or socialist-influenced group that seeks to overthrow capitalist societies or systems.

Right-wing groups may espouse extremist religious ideologies and claim that there are national and international conspiracies involving people controlling the economy. Twisted views of various religions can attract individuals adrift and seeking someone or some view to guide them. Such people may be all too easily molded or their actions channeled by manipulative groups. Right-wing groups include the Army of God (AOG), which commits crimes against individuals and physical facilities offering abortion counseling or services. Aryan Nations (AN) is a white supremacist extremist organization acting on an ideology of whites being superior to all others. Problematic is the fact that neither the FBI nor DHS maintains an official list of groups actually categorized as domestic terrorists. The reason this inhibits aspects of investigation or prevention is that much of the hateful or inflammatory speech and activities of such groups, up to the point of harming someone, is protected speech. Those white supremacists who have murdered African Americans, Jews, and others are classified as right-wing terrorists as well as having committed hate crimes.

Right-wing group Far-right groups such as neo-Nazi and white nationalists that seek to overthrow government and replace it with a system of their own ideologies.

MILITIA GROUPS

Militia groups are another challenge to law enforcement and represent a threat of violence. Such groups reject central government and believe that federal authorities have an agenda to take guns away from citizens. They will often train in paramilitary tactics and weapons use, while stockpiling firearms and even explosives. Several of these groups have initiated confrontations with authorities and have shown a willingness to use violence against law enforcement officers. An antigovernment group known as "sovereign citizens" considers themselves apart from or "sovereign," though they live in the United States. In opposing government, these individuals reject paying taxes, including driver licenses or vehicle license fees, and they do not recognize court authority. Sovereign citizens may engage in what is known as paper terrorism, wherein they file false liens and frivolous lawsuits to harass government officials and others.

International Terrorism

U.S. law enforcement efforts today have focused much in international organizations such as Al-Qaeda. Al-Qaeda conducted the 9/11 attacks on Washington, D.C., and New York and has waged a lengthy campaign in various countries. Governments have targeted the leaders of groups such as Al-Qaeda, which met with some success but also caused the groups to alter their top strategy to one of smaller, autonomous groups or so-called **lone wolf terrorists**. Terrorist attacks outside the United States have also resulted in the deaths of American citizens. An objective of attacking Americans or American interests is the symbolic value. Neumayer and Plümper (2011) assert that terrorist attacks on Americans by groups in conflict with their own government may be seen by the groups as benefiting the terrorists to influence politics in their country.

Lone wolf terrorist May act consistent with the ideology of a terror group but without interacting with or being directed by that group.

Because the investigation of terrorist acts may be aided by understanding the motivation of the terrorists, typologies such as Martin's (2003) framework can provide governments added focus. Martin's five motivation categories are:

(1) State-sponsored terrorism, which occurs with the direction or funding of a government;

(2) Dissident terrorism by those attempting to overcome their own government;

(3) Terrorists from the left and right, usually in a country divided by politics with the group in power acting in oppressive ways;

(4) Religious terrorism, seen most frequently; and

(5) Criminal terrorism, such as the drug cartels in Mexico.

Alvarez and Bachman (2016) cited how the drug cartel has used assassinations and bombings to frustrate the efforts of the Mexican government and as a way to protect the cartel's profits.

So we see each group or group type willing to use violence when they believe it furthers their ideological effort. The brutality of an attack is often the hallmark of terror organizations. The ability to strike fear into civilians is accomplished to greater and lesser extent based on the actual vulnerability of the society.

If the country does not have a strong governmental structure or stable political and economic system, and citizens feel disenfranchised on the whole, terror attacks may indeed have a demoralizing effect and lead people to believe they may benefit by not resisting terrorists. If the protection of citizens is generally sound, and the government is viewed as legitimate and relatively competent, there is far less likelihood that terrorists would achieve significant or any progress with their agenda (Hough & McCorkle, 2017).

To this point, killing civilians often increases solidarity in a populace rather than demoralize or fragment citizens. The 9/11 attacks and the Boston Marathon bombing, for example, did not have the effect intended by the terrorists. Within a short period of time after each of these events, people did not simply shelter at home or slowly try to get back to "business as usual." One World Trade Center was built in the same location as the destroyed Trade Center. "Boston Strong" became the motto as runners from across the country signed up en masse to participate in the following year's Boston Marathon to show support to Bostonians.

The specific targeting is based, in large part, on the ability to attract media. Whether a kidnapping or hostage-taking, a ransom may be secondary to increasing the profile of the group. Financing group operations is partly achieved through these acts as well as various other schemes. Al-Qaeda is but one of many groups to use credit card fraud to finance its operations. Kidnapping for ransom is often used by Boko Haram in Africa. Al-Qaeda and its affiliates, and many others, conduct kidnapping for ransom. Many activities to raise funds for terrorists are not new.

Suicide Bombings

Media coverage is intense after such attacks, but the use of social media can be helpful to investigators following the event. Retracing the steps of attackers can reveal a great deal, and average citizens and businesses contributing their video clips may be indispensable. Explosives can be concealed in many ways, and compounds and devices have increased their lethal effectiveness. Governments have added legislation to track a variety of compounds and items used in the construction of explosive devices that can alert law enforcement to potential developing plots of destruction. Coordinating the influx of potential leads and evidence after an event as well as before a crime calls for the assistance of many investigators, analysts, and others. The burden of many investigations and task forces is that of coordinating a multijurisdictional effort and joint investigative teams such as the FBI's Joint Terrorism Task Force (JTTF) approach. Terrorists also may work in "teams" to carry out one or a group of coordinated crimes or attacks. Understanding team dynamics and the psychology of group behavior can aid investigators (Spitzmuller & Park, 2018). Targets are also selected by single individuals, or "lone wolf" actors.

First responders are called on during and after an attack. How has the U.S. response to terrorism changed over time?

AP Photo/The Boston Globe, John Tlumacki

> ## How It's Done
> **BOMB THREAT CHECKLIST**
>
> Suicide bombers pose a unique challenge to law enforcement. One aspect of facing the prospect of a suicide bomber is the information-gathering by communications personnel when someone calls law enforcement about a possible offender. In the Model Policy on Suicide Bombings developed by the International Association of Chiefs of Police (IACP, 2008), one portion of the Bomb Threat Checklist suggests:
>
> Communications personnel receiving threats or warnings from callers concerning suicide bombers shall, to the degree possible:
>
> a. Keep the reporting party on the line;
> b. Alert the appropriate supervisor;
> c. Identify the location of the suicide bomber with as much precision as possible;
> d. Collect any information on possible targets; and
> e. Attempt to determine:
>
> A description of the bomber;
> The type of explosive device;
> What the device looks like and how it is being worn or carried;
> What type of bomb it is;
> What will cause it to detonate;
> The caller's name;
> The basis for the caller's knowledge or suspicion; and
> The individual or group responsible for the threat.

Lone Wolf

Whether using an explosive device or another weapon, a distinguishing feature of the so-called lone wolf terrorist individual is that he acts without the direction of a terrorist organization. He is outside the structure of a larger group or even the small-cell format used by many groups. Where it may be possible to intercept or learn of a group's communications, the challenge is far greater with a single person because there are no communications to intercept. Seemingly attacking without warning is clearly difficult to defend against because the individual may have virtually no suspicious two-way communication with anyone. Keep in mind, they may have become "radicalized" online or through reading or listening to various sources; they just have not interacted personally with anyone else.

It is absolutely desirable in any terrorism investigation, team or sole individual, to obtain access to computing or communications devices used by those involved. The recovery of digital information or evidence from mobile devices can illuminate a number of activities of terrorists/criminals. Cahyani, Rahman, Glisson, and Choo (2017) recount and discuss four generally acknowledged uses of information communication technology (ICT) as "(1) information propagation, (2) information concealment, (3) fund raising, and (4) recruitment and training" (p. 2). ICT represents the double-edged sword of utility for both terrorist and investigator alike.

The motivations of lone wolf attackers are not always clear. Some individuals may be looking for a cause or group with whom to claim association so they can rationalize their acts. And some of those groups may be all too pleased to take responsibility for an act carried out by the person. Many people will not understand the need to determine whether the person was acting in concert with the goals of a group or truly be themselves, but it is important to understand motivations as well as methods of information-gathering and usage to devise investigative tactics and prevention strategies. Those who commit horrific acts, such as terrorists, may be quickly labeled as "crazy." More likely, the person believes they are working toward a logical path to change some condition they object to; they see themselves as rational. And while rank-and-file terrorists within groups are not generally found as suffering from mental illness, the lone wolf is more likely (though this is not a majority) to be suffering from a mental illness (Corner & Gill, 2015). Regardless, violence of the type perpetrated or attempted by single terrorists is frightening and difficult, if at all possible, to predict.

An example from October 2014 perhaps illustrates the difficulty in properly classifying some attacks and attackers. A 32-year-old New Yorker used a hatchet to assault four New York police officers. The man injured two of the officers before he was shot dead. While the man was described as a "Muslim extremist," there were also indications that he struggled with mental illness. His conversion to Islam and self-radicalization had occurred in only the two years prior to his attack on the police officers. In such instances, was the motivation to attack the result of a perverted sense of religious zeal, or one caused by an interplay of mental illness and the conceptualization (by the individual) of a group's mission? For additional examples of lone wolf attacks see Table 11.1.

Hate Crimes

Other crimes motivated by ideology rather than profit or emotion based on an immediate conflict may be categorized as hate crimes. Some extremist groups victimize others based on race, ethnicity, or religion. Hate crimes are cloaked in the mantle of religion, at times, or reflect a worldview that projects blame onto people based on their demographic characteristics. Targeting people in this way is not the same thing as terror-motivated crime, yet the topic deserves discussion. It is an appropriate point to compare the similarity of hate crime to terrorist crime. In both cases the offenders want to harm innocents to "make a point," and they strike not just at one individual but at a class or group of people.

How did you hear about the most recent terrorism attack?

©iStockphoto.com/Cineberg

Role of the Media

The media play a role in how citizens learn about acts of terror, and also what we understand about the motivation for such acts. Terror groups launch attacks with a clear goal of garnering media attention. Terrorists also use the Internet to coordinate their activities, including recruitment, fund-raising, meetings, and training. Media coverage in the United States is seen to benefit from coverage of dramatic events reported in dramatic ways. The variation in terminology used by the media can also confuse the public at times. Poland (2011) notes "groups and individuals are often referred to as rebels, freedom fighters, commandos, guerrillas, protestors, dissidents, rioters,

TABLE 11.1 United States Examples of Lone Wolf Attacks

1995	Oklahoma City	Timothy McVeigh	Killed 168 and injured hundreds more.
1978–1995	Various	Theodore Kaczynski	Known as the "Unabomber"; mailed bombs that killed three and wounded 23.
1996–1998	Various in Southeast	Eric Rudolph	Bombings killed three and injured at least 150.
1999	Los Angeles, CA	Buford Furrow	White supremacist; injured five at Jewish day care and then shot and killed a mail carrier.
2006	UNC, Chapel Hill	Mohammed Reza Taheri-azar	Drove a vehicle into a crowd of students, injuring nine.
2009	Little Rock, AR	Abdul Mujahid Muhammad	Shot two soldiers at recruiting office, killing one.
2009	Fort Hood, TX	Nidal Malik Hassan	Shot and killed 13 and wounded 30 others.
2010	Austin, TX	Joseph Stack	Flew small airplane into IRS building, killing one and injuring 13.
2013	Boston, MA	Dzhokhar and Tamerian Tsarnaev	Set off two homemade bombs at finish line of Boston Marathon, killing three and wounding 260.
2014	New York City	Zale Thompson	Attacked four police officers with a hatchet before being shot dead.
2015	Chattanooga, TN	Mohammad Abdulazeez	Fired on military members outside of a recruiting center, killing four U.S. Marines and one Navy sailor.
2015	San Bernardino, CA	Syed Rizwan Farook and Tashfeen Malik	Apparently self-radicalized couple opened fire on party-goers, killing 14.
2016	Orlando, FL	Omar Mateen	In what until then was the worst attack since 9/11, killed 49 people and wounded many others at a nightclub. Killer claimed allegiance to ISIS.
2017	Charlottesville, VA	James Fields Jr.	Drove into a crowd of protesters, killing one and injuring 19 others at a rally of far-right white nationalists and others.
2017	New York City	Sayfullo Habibullaevic Saipov	Claiming the attack for ISIS, drove truck onto a bicycle path near the World Trade Center, killing eight and injuring many more.
2017	Las Vegas, NV	Stephen Paddock	Fired on a concert crowd from hotel window above. Fifty-nine people were killed and more than 500 injured as a result of the attack.
2018	Pittsburgh, PA	Robert D. Bowers	Killed 11 people and injured six others in a synagogue.
2019	El Paso, TX	Patrick Wood Crusius	Killed 22 shoppers and injured 24 others at a Walmart. White nationalist killer was targeting Latinos.
2019	Dayton, OH	Connor Stephen Betts	Killed nine in historic Dayton, including sister, 13 hours after El Paso shooting.

extremists, or jihadists" (p. 52). If the average citizen is not up to date on different groups across the country or around the world, and he or she hears a group referred to as "dissidents" or "freedom fighters," it may not be at all clear to the person the relative legitimacy of the group. Whether discussing violent groups or acts of hate crime, competent news reporting can raise public awareness and forward the aims of a community or society as a whole.

Homeland Security

An important task, perhaps the most important task, of homeland security is coordinating efforts. Agencies at the federal, state, and local level as well as private sector organizations, the public, business groups, and more all have a role to play in preventing terrorist acts or investigating ones that have occurred. The federal Department of Homeland Security is the primary agency with overall responsibility for this coordination.

While much of the effort at preventing terrorism is often focused on international cooperative efforts, a great deal of effort is also required within the United States by state and local law enforcement agencies. Because the majority of the nation's **critical infrastructure** is controlled by private companies, there is an ongoing need for partnerships and close coordination of efforts between these organizations and government agencies. Critical infrastructure consists of things such as roads and bridges and other aspects of public transit along with water, power, communications systems, and more. Contemplating ways to protect these many components of the nation's functioning is daunting and underscores the need for each sector of society to take part. Weapons of mass destruction are frightening to consider, but no less attention should be paid to the potential for people with bad intentions trying to disrupt things such as the supply of food, water, or the provision of public health.

Critical infrastructure Assets necessary for society to function.

Airplanes and other transportation remain at risk of attack. What strategies are used to reduce this risk?

steviepics/500px/Getty Images

State and Local Level

At the local level, for example, agencies of law enforcement and government must keep up to date with a community inventory of critical infrastructure, and other properties or systems that may make for a meaningful target of a domestic or international terror group. At the state level, investigative efforts will include following up on intelligence leads from local agencies as well as coordinating with federal departments to obtain resources for hardening targets in the state and to find prevention and investigation efforts. And while critical infrastructure covers many services and items, some, considered **key assets**, rise to the top of the priority list in each region and community.

Key assets Vital elements of infrastructure whose damage or loss would create significant effects on society.

Federal Level

Remaining aware of potential dangers and threats can be emotionally fatiguing. This is true for those who work within the law enforcement and national security communities as well as by average citizens who have concerns about safety within their communities. Familiar to most Americans after the attacks of 9/11 is the color code Homeland Security advisory system. This National Terrorism Advisory System (NTAS) provides an ongoing assessment of the threat of an imminent terror attack and is based on the intelligence efforts of the Department of Homeland Security as they work to evaluate gathered information from across the country and around the world.

The investigation of terrorist crimes and efforts to prevent violent acts relies, in part, on an understanding of what motivates the offender. Harming or killing others based upon a bias against someone not like you (e.g., hate crime), and those whose worldview is different (terrorism), is difficult to grasp emotionally, but investigators have to intellectually understand that some few people can be motivated in these ways. Hough and McCorkle (2017) made this observation:

> Some individuals come to a belief system or political movement at a point in their life course that makes them susceptible to the fervor of rhetoric. Some cling to the message or to the messenger. Some terrorists come from poor backgrounds and choose to join a group perhaps based on limited options. Other terrorists come from middle or even upper-class socioeconomic groups in their respective society. (p. 147)

There is no shortage of historical examples of how groups or those who are aligned with certain groups have used violence and destruction to further their goals. Such instrumental use of violence and chaos can be intended to draw attention to the group's ideology or unsettle the public more than an aim to actually succeed in causing large-scale damage. The investigation of ongoing terrorist efforts, as well as examination of such events after they have occurred, presents many challenges to the law enforcement community. Agencies work to deter terror attacks through intelligence-gathering, educational efforts, and physical security measures.

Much as federal agencies came to realize the need to integrate and encourage local law enforcement efforts in the investigation of terrorism, the community is a resource in this as well. In the case of Muslim citizens and communities, a useful perspective mirrors the one contemporary police face during community policing efforts generally—that of perceptions of legitimacy of those who police the community (Madon, Murphy, & Cherney, 2017). If Muslim citizens perceive that law enforcement agencies follow concepts of procedural justice, greater cooperation between citizens and police is likely. Being treated with respect and equality engenders the view of police as legitimate (Madon et al., 2017).

In addition to information that emanates from the community, agencies must share information and intelligence gathered and developed that can impact investigations that are ongoing. Long known to local agencies and later validated by the report of the 9/11 Commission, intelligence-sharing among agencies must stay at the forefront of each agency's priorities. The 2018 report of the DOJ's Office of the Inspector General emphasized the need for agencies to cooperate in sharing counterterrorism information. The benefits of mindful sharing of information and collaborative development of intelligence and strategies are clearly important between countries as well to more effectively counter terrorist crimes (Jensen, 2016; Parkin, Gruenewald, & Jandro, 2017).

Fusion Centers

Federal, state, and local entities have come a long way in the analysis and sharing of information related to potential or ongoing terrorist activities. Following the September 11, 2001 attacks, a National Network of Fusion Centers was established in an effort to investigate terrorist activity and coordinate response by multiple agencies. On the frontlines of prevention, fusion centers provide a space and method for multiple agencies to combine efforts at identifying and stopping threats. In a 2017 review of the national network, it was noted that "Many fusion centers have expanded capabilities to address all crimes and threats, recognizing that early indicators of terrorism often include criminal activity" (p. 4).

> **EXPLORE THIS**
>
> Visit the website of the U.S. Department of Homeland Security (DHS), https://www.dhs.gov/taxonomy/term/3588/all/feed, and search for resources on Homegrown Violent Extremists. Find and read a number of the documents addressing HVE, terrorists, terrorism prevention, and more.

SUMMARY

Homeland security efforts and the investigation of threats to the country's infrastructure and other important locations or properties is a job shared by each level of law enforcement and agencies around the country. While the crimes committed are at times similar, the motivations may differ and those that are committed for profit may see the proceeds go to a political extremist group rather than in the pockets of a normal criminal. The crimes done by group members will likely be directed or coordinated by a leader or as part of a hierarchy for the terrorist organization. Since many crimes committed by members of an organization involve communicating between and among various individuals, law enforcement may find opportunities to intercept communications or leverage individuals to work with them to catch other group members. The lone wolf may act or carry out some type of crime or attack with little to alert authorities to his presence before he strikes.

The terrorist group or individual who attacks a location or people may use a wide range of weapons, including explosives or vehicles. These types of attacks have led government groups to harden their facilities and buildings with physical features to block or slow down such attackers. Similarly, airplane cockpit doors were strengthened following the 9/11 attacks, and it is now commonplace for schools to design in or add physical security features in their buildings. The tracking of compounds or devices used to construct explosives has become more comprehensive and again illustrates the need for each level of government to share information among enforcement and regulatory agencies to discern patterns or red flag purchases of certain items. The Department of Homeland Security (DHS) and the Transportation Safety Administration (TSA) work constantly to gather information and intelligence to assist in preventing acts of terrorism.

The media will report significant acts of violence within our local communities and for larger attacks that may signal concerted terrorist activities. Terrorists, however, often seek the publicity and notoriety that come from media coverage. At the same time, disseminating information through the media can assist law enforcement agencies as well as encourage the public to provide information that may be helpful in ongoing investigations. Legislation is used to deter terrorist violence by regulating acts and access to materials and methods used to carry out violence. Law enforcement agencies including the Department of Homeland Security use a number of intelligence-gathering methods as they work to prevent attacks; they now recognize the importance of involving local police departments, sheriffs' offices, the public, and private sector organizations to combat terrorists.

Investigative strategies and methods continue to evolve. What seems clear is that everyone in society has a role to play in homeland protection and the prevention and investigation of terrorist crimes. Police legitimacy as perceived by citizens can encourage citizens to come forward with information. "See something, say something" may only work if you trust law enforcement agencies. Technology improvements bring about more effective investigations through enhanced abilities to monitor the movement and activities of people. Intelligence-gathering and information-sharing among agencies at all levels of government as well as between countries remain a paramount concern.

KEY TERMS

Critical infrastructure 239
Domestic terrorism 232
Genocide 230
Hate crime 230
International terrorism 230
Key assets 239
Left-wing group 233
Lone wolf terrorist 234
Right-wing group 233
Single-issue terrorism 233

DISCUSSION QUESTIONS

1. How have the rates of terrorist incidents changed in the last 25 years and why?
2. List and discuss several legal factors that impact terrorism investigations.
3. How do local law enforcement agencies investigate terrorist activity?
4. In what ways do terrorists use social media?
5. How can agencies improve information-sharing?
6. What agencies and groups coordinate in homeland security efforts?

SAGE edge™

- Get the tools you need to sharpen your study skills. SAGE edge offers a robust online environment featuring an impressive array of free tools and resources.
- Access practice quizzes, eFlashcards, video, and multimedia at **edge.sagepub.com/houghci**

PRACTICE AND APPLY WHAT YOU'VE LEARNED

▶ edge.sagepub.com/houghci

SAGE edge™

WANT A BETTER GRADE ON YOUR NEXT TEST?

Head to the study site where you'll find:

- **eFlashcards** to strengthen your understanding of key terms.
- **Practice quizzes** to test your comprehension of key concepts.
- **Videos and multimedia content** to enhance your exploration of key topics.

©iStockphoto.com/Image Source

12 Criminal Investigation in Court

RUNNING CASE STUDY: POSSIBLE HOMICIDE AT THE FLORIDAN—COURTROOM TRIAL

Introduction

The Process: Screening and Filing
Stages
The Pretrial Stage
The Arraignment Stage
Juror Selection
The Trial
Building the Case and Charging Decisions

How It's Done: Overarching Principles for Prosecutors

Evidentiary Issues
Defense Attorney Functions and Tactics

Defenses
Failure of Proof Defenses
Alibi
Justification Defenses
Excuse Defenses

Ethical Issues

Plea Bargaining

Sentencing

Explore This

Summary

Key Terms

Discussion Questions

LEARNING OBJECTIVES

After reading this chapter, students will be able to

12.1 Discuss factors in the screening of cases.

12.2 Explain the types of evidentiary issues.

12.3 Discuss the types of criminal defenses.

12.4 Describe the ethical issues related to criminal trials.

12.5 Explain the purpose and process of plea bargaining.

12.6 Identify the sentencing options in a criminal trial.

Running Case Study: Possible Homicide at the Floridan—Courtroom Trial

At the pretrial meeting, Assistant State Attorney (ASA) Corbin Kelsea took notes as Detectives Bradley Macon and Richard Ashley narrated the actions taken in the homicide investigation from nearly 1 year before.

After Frank Denney's arrest, Corbin Kelsea had conducted an assessment to gauge whether the case had the potential to succeed in court.

Thirty-two hours after his arrest, Frank Denney appeared in court and was told he was being charged with second-degree murder. As Denney had no one to represent him, the court appointed public defender Hank Bell.

A grand jury met behind closed doors and heard from only the prosecution. Based on the gun matching Denney's registration, the witnesses to his argument with Bill Johnson, and Denney's confession, the grand jury members determined there was enough evidence and probable cause that he had committed the crime.

Six months later, an arraignment had followed where Frank Denney entered a plea of not guilty.

"Not guilty?" Bradley Macon had asked. "He confessed to shooting the man with his own gun."

"He probably wants to give his lawyer more time to investigate the case and seek a plea agreement," Richard Ashley had said.

As a plea could not be agreed upon, a trial had been scheduled.

The defense had offered pretrial motions requesting the court to suppress the gun or the statements of Frank Denney, but the prosecution maintained this evidence should be allowed.

Corbin Kelsea thanked the detectives for the summary, then headed into the courtroom for jury selection.

After the jury was selected, the trial began with opening statements. The prosecutor and defense attorney summarized their positions. Corbin Kelsea presented direct evidence. He called witness Miley Denis to corroborate that Frank Denney had argued with Bill Johnson. Bubba Paul made a statement as to the positions of the men in the alley. Officer Carl Jayden relayed how Frank Denney had remained nearby after the crime. Corbin Kelsea showed Denney's gun and registration. He played the CCTV film of the men arguing on the dance floor. He played Frank Denney's confession to Richard Ashley.

In rebuttal, Hank Bell claimed that the defendant was not fully aware of his right not to talk to the detectives. He maintained that Frank Denny's argument with Bill Johnson was provoked, and that Denney was only trying to scare the victim.

Corbin Kelsea presented circumstantial evidence that Miley Denis saw the two men argue, and that Frank Denney was found near the crime scene.

Corbin Kelsea conducted direct examination of each officer and police employee taking part in the case, as well as Miley Denis and Bubba Paul.

Hank Bell followed with his own examination. In questioning Miley Denis, he tried to establish doubt. "Are you sure Bill wasn't the one to start the argument? Surely, your judgment was colored by your feelings for him."

"I only met him that night," Miley Denis said. "My judgment was objective."

The jury seemed to believe her.

Both sides offered closing arguments. Corbin Kelsea summarized the case and how witnesses saw the two men argue. He noted that evidence was collected that connected Denney to the crime, including his confession.

Hank Bell attempted again to categorize Miley Denis as misunderstanding the argument between Denney and Johnson.

Following the closing arguments, the judge instructed the jury on the available verdicts—first-degree murder would involve establishing that Denney had lain in wait to kill Johnson while second-degree murder would indicate the killing resulted without premeditation, but with a "depraved mind."

The jury retired to deliberate and discuss the case. In a homicide trial, the verdict needed to be unanimous. After almost two days of deliberation, the jury found Frank Denney guilty of second-degree murder.

The judge set the sentencing hearing for the following month to give both sides sufficient time to prepare. Corbin Kelsea met with Bill Johnson's parents and his sister Linda, and explained that their victim impact statements could influence the jury's recommendations in sentencing. On the day of sentencing, Linda testified through tears, "You took away my brother. The person who always protected me. Since his murder, I have been alone in the world."

The same jury that had heard the evidence reconvened behind closed doors and agreed upon a penalty.

The judge stood and read the sentence, looking at Frank Denney. "For the crime of second-degree murder, you will be sentenced to 20 to 25 years imprisonment."

Why are the rights that are afforded to each defendant so important in the American system of criminal justice? Think about the different responsibilities of the prosecutor, the defense attorney, and the judge. Why is each role essential?

Introduction

Criminal investigation in the modern era involves the determination of a violation of a criminal statute, identifying who is responsible for that violation, compiling evidence that is relevant and legally admissible, and then working with prosecutors to effectively present the case to the court. This definition seems straightforward, but accomplishing this list is often quite challenging. In the United States and other democratic nations the process of the criminal investigation is very much concerned with ensuring that the rights of accused persons are protected at critical stages of the case. Searching for physical evidence must follow established rules, and taking testimonial evidence likewise has guidelines to protect against law enforcement from inappropriately compelling to speak against their own interest. An investigation can amass quite a bit of information, but not all of it may be allowed to be considered in determining the guilt or innocence of an accused person.

The Process: Screening and Filing

Everyone growing up or living in America has more than a passing familiarity with the steps of the criminal justice system, especially when it comes to a criminal court case and trial. An unending stream of television shows and movies present to us the process of a murder suspect arrested and then taken to court, where a hard-nosed prosecutor faces a determined criminal defense lawyer. Okay, this seems to leave quite a bit out, doesn't it?

How do attorneys work with investigators to determine which cases to try?

Hero Images/Hero Images/Getty Images

As discussed earlier in the book, the interest we all have in not just cases of criminal homicide, but, even more broadly, the criminal justice system has been shaped and is fed by the steady diet of Hollywood creations and embellishments. The reality is sometimes far less gripping and unquestionably tragic. Notorious cases of murder or homicides involving people known to the general public can drive a significant amount of media coverage. The vast majority of homicide cases, however, are rarely covered past an initial mention on the local news or in a local newspaper. By the time the average homicide case goes to trial there is little interest from the public save for the immediate friends and family of the people affected by the event.

The decision about what specific charge to file in response to the actions of someone who has killed another person is the province of the prosecutor. Some jurisdictions call this individual the district attorney, and some refer to the position as the state attorney. In any event, the prosecutor will look at the known facts of the case and decide which category of murder or manslaughter and what level of degree of seriousness will be formally charged. Each stage of a homicide case is important in our adversarial system and falls in a specific sequence of stages that ultimately disposes of a case via one of the options available by law. Multiple crimes committed within one incident can bring multiple charges. Often, the charges do not carry as heavy a penalty as murder or manslaughter and may be used to negotiate a plea arrangement with the defendant; plea to the most serious and the prosecutor will drop the less serious.

Stages

THE PRETRIAL STAGE

Following the arrest of a suspect, the pretrial stage begins. The prosecutor conducts an assessment to gauge whether the case has potential to succeed in court. If the prosecutor does not believe the evidence is sufficient to prove the case **beyond a reasonable doubt**, she may use discretion to dismiss charges. In some cases, as mentioned, the prosecutor may file additional charges to those initially listed by law enforcement. This may serve as leverage for an eventual plea arrangement. An initial court appearance follows within 48 hours of the arrest, during which the accused individual is made aware of the charges brought against him; if the person cannot afford to hire an attorney the court will appoint one to represent the defendant. The court will also determine based on legislative guidelines if a defendant is eligible to be released prior to trial. This pretrial release may be after the defendant puts up one of several types of bond to guarantee his appearance for future court dates, or he may be released on his own recognizance, though this is rare in a homicide case. A Bureau of Justice Statistics (BJS) special report in 2007 noted that just 19% of murder defendants in state court nationwide were granted pretrial release. The report identifies a decreasing number of pretrial releases as the bail amount rose above $50,000 and then above $100,000. This level of pretrial release denial is not surprising and reflects the court's assessment of risk to the public or flight by the defendant. Though uncommon in a murder case, the judge may assign bail or not. If ordered by the court, the amount of bail for a murder suspect is usually significant given the seriousness of the charge.

> **Beyond a reasonable doubt** Legal standard in a criminal prosecution for the jury to find someone guilty.

About half of the states and the federal criminal courts utilize a **grand jury** system. The grand jury meets behind closed doors and hears from only the prosecution. The grand jury's role is to determine if there is sufficient evidence to move forward with a trial for the defendant. If the grand jury members determine there is enough evidence and probable cause that an individual committed a specific crime, an indictment or "true bill" is rendered. Evidence is only presented by the prosecutor during a grand jury. Fewer rules apply to the conduct of the jury, and it is held behind closed doors. Given the manner by which the grand jury operates, it is nearly a foregone conclusion in American jurisprudence that a prosecutor seeking a grand jury indictment will get one.

> **Grand jury** Group of citizens impaneled within a jurisdiction to examine information and determine whether the prosecutor should bring charges against an individual or entity.

THE ARRAIGNMENT STAGE

If charges go forward against an individual an **arraignment** will follow where the defendant can enter a plea. The plea will either be guilty, not guilty, or *nolo contendere* or no contest. If a guilty plea is entered and accepted by the court, a sentencing hearing will come next. Likewise with a plea of no contest, the judge will move the process to sentencing. The defendant may enter a plea of not guilty as a result of believing themselves innocent of the crime for which they are charged or to force the prosecutor to fulfill her role and prove the case against the defendant. If this last option is taken a trial will take place. Usually a plea arrangement or bargain will be sought during the arraignment stage, though a plea can be agreed upon at any time during the process. The defense counsel must provide his client with information regarding all offers of a plea deal by the state. If the defendant wishes to accept a plea arrangement it is the responsibility of the court to ensure he or she has accepted the arrangement willingly and knowingly, with no promises or threats to do so. Increased sentence severity in cases showing a greater likelihood of conviction reduces the chances that

> **Arraignment** A hearing where a formal charge against an individual is entered.

JUROR SELECTION

Following the pretrial stages, prosecution and defense will offer various pretrial motions requesting the court suppress or allow evidence in the trial. And then, before the trial can begin, a jury must be seated. A group of citizens is selected to weigh the evidence presented and render a verdict. This panel of prospective jurors is known as the **venire**. Choosing jurors is important to the integrity of the process as the court searches for citizens who can be impartial. Lawyers from both sides directly question prospective jurors in what is called **voir dire**. Ideally, a juror will not favor either the prosecution or the defense in their views. Lawyers on both sides are allowed a set number of peremptory challenges or strikes, which can be used for almost any reason other than a juror's race, religion, disability, or other protected class. Lawyers can also use **challenges for cause**, which are strikes that can be used when it is apparent that juror cannot be fair and impartial. Judges must rule on all challenges to make sure they are lawful. In some cases, mainly significant civil trials, so-called scientific jury selection has been used. The potential high-dollar amounts of settlements in such cases and the resources available to defendants make the use of the jury consultant more likely. Scientific jury selection (SJS) employs a consultant who is often a social scientist to conduct survey research of the profiles of eligible jurors and assess the most suitable ones for one side or the other. Research is mixed regarding scientific jury selection, as jurors still rely predominantly on the evidence presented.

Venire Gathering a pool of potential jurors.

Voir dire Usually refers to the initial procedure of selecting jurors for a trial.

Challenge for cause Disqualification of a potential juror with a specific reason given.

THE TRIAL

The trial begins with opening statements in which the prosecutor and defense attorney briefly summarize their positions to the jury. The attorneys will generally outline for the jury what they will show during the trial as well as the evidence or testimony they will present. A preview of how the case will proceed is helpful to the jury in understanding the issues of the case and possibly to begin to orient them to have the attorneys want the jurors to see the case. Presentation of evidence by the prosecution follows the opening statements. The rules of evidence guide its use during a trial. The prosecution seeks to establish that the defendant is accountable through the presentation of the evidence. The defense works to show the court or jury that there is reasonable doubt of the defendant's guilt. The defense strategy may be making the case that the defendant is guilty of a lesser crime. Attorneys on both sides may present what are called direct evidence and circumstantial evidence. While direct evidence tends to prove a fact, circumstantial evidence is considered indirect and therefore calls for inferences of what happened.

If the defendant enters a plea of guilty, he is admitting the charges and a trial is no longer necessary. The court will schedule a sentencing hearing to announce the sanction to be imposed. The defendant may plead *nolo contendere*, or no contest. While this is not an admission of guilt, the plea acknowledges responsibility for the charge but the plea may not be used in any subsequent civil suit relating to the death. Finally, the defendant may plead not guilty and both sides prepare for trial. Plea bargaining can take place at any point in the process.

As the trial nears its completion, both sides offer a closing summary. The prosecution confidently tells the jury that it has been proven beyond a reasonable doubt that the defendant committed the crime. The defense points out weaknesses of the case and assures the jury that the evidence is not convincing. Following the closing arguments the judge instructs the jury members on what verdicts are available to them and what laws are relevant for them to consider. The jury retires to deliberate and discuss the case. Note that in a homicide trial, the jury verdict must be unanimous whether that is guilty, not guilty, or a conviction on a lesser offense than what was charged.

Factors about jurors, as well as about the defendant, the victim, or criminal justice players in the case, have all been found to hold the potential to influence the verdict delivered by the jury (Stauffer, Cochran, Fogel, & Bjerregaard, 2006). A verdict harsher than otherwise expected has been noted when the socioeconomic status of the jurors vis-à-vis the defendant is observed (Foley & Powell, 1982). The harsh verdict may reflect a perceived threat of economic encroachment by the defendant (Litwin, 2004). And while religious references or imagery are often frowned upon in a trial, attorneys have been known to use various biblical or religious references in an attempt to sway jurors (Chavez & Miller, 2009).

If a defendant is found guilty, sentencing will follow. Victim impact statements would be presented at this stage of the proceedings. In serious violent crimes such as murder, it is common for family members or friends to testify about how their loved one's murder has affected them (Newman, 1995). The victim impact statement has been shown to influence jurors' recommendations in sentencing, and this calls for the mitigating or controlling practice of the court providing victim impact instructions to jurors (Berman, Narby, & Cutler, 1995; Newman, 1995; Platania & Berman, 2006). It should also be pointed out that victims of crime and their family members in the case of homicide may feel insecure submitting statements (Mesmaecker, 2012). Murder cases are hard on family members of the deceased as well as the family of the defendant. The victim's family members may have to testify and endure media attention, which is a burden, no matter how sympathetic it may be.

Building the Case and Charging Decisions

After the investigative phase, prosecutors make charging decisions in cases. Most students of criminal law know that there are varying levels or categories of many crimes, ranging from the most serious to the less serious. Every state has criminal statutes that define these levels, and explain the minimum and maximum penalties for the crime. Law enforcement officers and prosecutors often work together to review the details and circumstances of the crimes so that prosecutors can determine the appropriate charge.

Something known to those who study criminal justice is the concept of the crime funnel. The funnel is wide at the top where many crimes are discovered, and it narrows as fewer crimes are reported, fewer still are investigated, a yet smaller number develop a suspect and evidence in a case, which then leads to an even smaller group of cases where charges are brought, followed through to prosecution (let alone trial), a possible conviction secured, and then and only then is a tiny portion compared to the original number convicted let alone sentenced to a sanction that statistically improbably would be incarceration!

O'Neill (2004) examined declines to prosecute in federal prosecution offices. He notes that the reason given by prosecutors for some declinations is a lack of investigative resources yet also comments on the hesitancy by some prosecutors to devote the effort to low-level cases. On the other hand, news coverage and an increased profile that comes from involvement in a homicide case also requires a great deal of effort and resources, including the writing of search warrants, the subpoenaing and managing of witnesses, and ongoing coordination with law enforcement and other criminal justice system actors. While caseload can certainly be a factor for prosecutors as well as any position constrained by the numbers of cases one is assigned, the message O'Neill derives also has to do with the value or lack of value placed on crimes considered low level, such as drug cases. As an investigation proceeds, the prosecutor has a dual role of advising and assisting law enforcement and conducting the specific duties of the people's representative. Pope (2011) addressed the prosecutorial investigation standards to which attorneys should adhere.

How It's Done
OVERARCHING PRINCIPLES FOR PROSECUTORS

The American Bar Association (ABA) developed a set of five overarching principles to guide prosecutors and followed with detailed factors to consider along the way in a criminal case. The principles that to Pope (2011) seemed most important to the task force of attorneys and judges are:

1. The prosecutor has a duty to view the facts with completely disinterested eyes during an investigation. Much injustice is done when a prosecutor falls in love with a case or a solution, and ignores the truth in favor of a pet theory.

2. A prosecutor must understand the potential harmful collateral consequences of an investigation, consider them in choosing investigative steps, and attempt to minimize or mitigate them.

3. A prosecutor should use a sense of proportionality in choosing investigative steps, much in the same way discretion informs prosecutive decisions. Particularly intrusive or damaging choices may not be warranted in less serious cases. Not everything that can be done, should be done.

4. A prosecutor is not a free-lancer, but a member of an office and a tradition that existed before the line assistant and will exist after the line assistant leaves. Any individual prosecutor is no more than a temporary custodian of the powerful and intrusive tools placed in his or her hands. It is, consequently, the prosecutor's duty to continually consult with supervisors and peers, and learn and follow the rules and mores of the office.

5. A prosecutor's craft and ethics are inseparable. Good craft leads to better truth finding and the prevention of unwarranted intrusions on privacy and liberty. (p. 5)

The prosecutor has the power to file or drop a case in part based on the evidence assembled by law enforcement officers during their investigation. This power is often observed to be almost unchecked and presents the classic risk of abuse that wide discretion holds (Van Patten, 2010). The prosecutor evaluates whether he can secure a guilty verdict by the fruits of the investigation, including witness accounts and any physical evidence. The prosecutor should never pursue a case based solely on the belief of being able to gain a conviction, but on being convinced of the defendant's guilt. Rasmusen, Raghav, and Ramseyer (2009) commented that conviction rates at trial (convictions/prosecutions) do not tell us about the strength of the prosecutors' cases, "... only about the strength of those few cases the prosecutor chooses to pursue and the defendant chooses to contest" (p. 50).

Given the prosecutor's decision to pursue a case based in large part on what the police assemble, it is clear that a high level of cooperation is desired. Vecchi (2009) describes it this way:

> No matter how well conceived and successful the investigation, the investigator must be able to articulate the facts of the case in a manner that enables prosecution. Prosecutors are the gatekeepers of the courts who ultimately determine whether a case is accepted and prosecuted. As such, it is essential that the investigator coordinate and collaborate with the prosecutor early on in the investigation in order to provide the prosecutor with a sense of ownership and participation throughout the investigation. In this way, the prosecutor becomes a stakeholder in the success of the investigation and will be more willing to see it through even if weaknesses exist. (p. 10)

The role of the prosecutor has not always been agreed upon. Some of the inconsistency surrounding this role comes from the lack of legislative direction provided by states as to the specifics guiding the prosecutor's quest for justice. The gaps in the law that allow a prosecutor latitude in charging decisions are rarely a subject of intense public scrutiny. The lack of legislative direction is a key component of the fairly unrestricted decision-making power of prosecutors (Green & Zacharias, 2008; Levenson, 1998). Discussed but rarely pursued is placing meaningful limits on a prosecutor's discretion in plea bargains as well as in charging. However much prosecutorial behavior and guidelines are addressed through legislation, communities frequently grade their local prosecutor on the statistics or "batting average" they compile.

Many jurisdictions have experimented with or committed to the use of specialized prosecution units to focus on one single crime type or grouping. Many of these are well known, such as those dealing with drugs, domestic violence, gun violence, and gang crime. Pyrooz, Wolfe, and Spohn (2011) noted in their examination of a gang prosecution unit in Los Angeles that the use of specialized units and the seriousness of the crimes involved do affect charging decisions.

Both sides in a criminal case must prepare their witnesses for deposition and potential courtroom testimony. And each side prepares differently. Prosecution witnesses often have trial testimony experience allowing for potentially more impactful performance (Campbell, 2007). The defense witnesses are less experienced or familiar with court procedures. The lack of trial experience on the part of defense witnesses can provide the opportunity for the attorneys to construe the testimony as lacking reliability or appearing inconsistent with previous testimony (Campbell, 2007; Wheatcroft & Ellison, 2012). The weight given to any witness's testimony is a decision for the juror, which underscores the need for witness preparation. Preparation must be within ethical guidelines and not merely coaching a witness what to say (Campbell, 2007). Proper witness preparation is not likely to cause the witness to alter his testimony. And as long as the defense attorney does not "rehearse" the defendant's testimony there is less likelihood that the defendant would at some point accuse the defense attorney of telling him what to say on the stand (Campbell, 2007).

Juries assess a case through the actions of lawyers: who used the evidence most effectively and who made the most compelling argument. The jury makes the final decision based on the presentations of both legal teams doing their best to effectively put forth their side of a case.

Evidentiary Issues

The lay public often thinks of a weapon with the suspect's fingerprints on it when evidence is mentioned. In fact, there are a number of forms of evidence. Cases are routinely decided on the basis of circumstantial evidence that does not directly link a defendant to a victim but does so indirectly. The value or weight of circumstantial evidence is a matter for the jury to decide; or, if the case is a judge-trial, the judge determines if the evidence is believable. There is generally not an abundance of direct evidence in cases that go to trial. Direct evidence includes eyewitness testimony, confessions, or videotape evidence. In criminal cases the burden of proof rests on the prosecution, and so they must present evidence showing proof beyond a reasonable doubt.

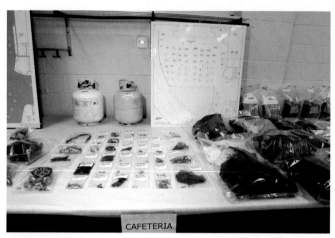

Evidence collected throughout the investigation is brought out for the trial. What would make this evidence inadmissible, and what impact can that have on a trial?

Mark Leffingwell/Stringer/Getty Images

Throughout the course of a trial various challenges to or complications with evidence can arise. Anyone who has watched a crime drama in the last 50 years is aware that law enforcement officers must read to suspects their warning of constitutional rights, colloquially referred to as the "Miranda warning." If the police fail to do this, or it is later shown the suspect did not comprehend his rights, any information or statements gained from the suspect will be held inadmissible in court (Weis, 2005). Any leads or further evidence discovered as the result of evidence initially obtained illegally will also be found inadmissible under the doctrine known as fruit of the poisonous tree. Such constitutional violations of a defendant's right can result in the case being dismissed (Weis, 2005).

Evidence presented in court may include witness testimony that must be based on personal knowledge, various documents including victim and witness statements, and the familiar physical evidence. Each state and the federal court system have rules of evidence that govern admissibility issues. In addition to the rules, the admissibility of different forms of evidence is guided by case precedents. Scientific standards for evidence address tests or observations offered by experts and that conform to the scientific method. DNA evidence is one example.

Previously, we discussed the "CSI Effect" on laypeople (and criminal justice actors) often believing they understand forensic and technical detection and analysis equipment and methods (such as DNA) more than they actually do. Curtis (2014) points out that since "there is relatively limited insight into the public perceptions and expectations of forensic DNA use" we have reason to be concerned about the impact on case outcomes. The presence of DNA evidence was previously shown to increase the chances a case would reach court and on the decision by juries to convict (Briody, 2004). Because DNA evidence is viewed as so significant by jurors, the process of collection, storage, and analysis is critical to the integrity of DNA evidence. DNA evidence may be seen as compromised if strict procedural steps are not followed.

Defense Attorney Functions and Tactics

Federal or state prosecutors represent the community, and sometimes making the case is not just during preparation or in trial but also through statements to the media and to the public directly. Attorneys from both sides of a murder case may work to create some level of public support outside of the court house. Prosecutors at their best are also trying to get justice for the victim and the victim's family. The prosecution must make charging decisions, including when to pursue a case through to trial. If a case has relatively less direct evidence or lacks evidence overall, it may be practical to offer a plea to a lesser charge. Defense attorneys are essential to provide a vigorous defense for anyone accused of such a serious crime as causing the death of another person. Defense attorneys are required by the codes of legal ethics to provide clients with zealous representation. Our adversarial system of justice is said to be effective as both sides advocate diligently to represent their clients.

Whether private attorneys or public defenders, the defendant's legal representatives are there to provide fair and competent representation. Discussion continues over the relative benefit to a defendant of a public versus a private attorney to represent him (Hartley, Miller, & Spohn, 2010). History is replete with examples of insufficient or nonexistent representation of persons charged with a crime by government authorities. The percentage of clients who actually are innocent of the charges brought against them is not the central feature of the role of the defense attorney. The check and balance is key to our adversarial system of justice. The effectiveness of the public versus the private defense attorney is often a subject of commentary. It is important to recognize that public defenders have as their sole responsibility the representation of criminal defendants. On the other hand, private practice attorneys may assist clients in a wide range of legal matters. The depth and familiarity with contemporary issues in criminal practice may favor the public defender. What a private attorney may provide that is frequently unavailable to the public defender is the resource of time. The private attorney will typically have a lighter caseload, allowing them to focus more attention on their selected cases. In addition, they may have more investigators and staff to assist them in preparing the defense case.

The investigators and support staff for prosecutors or defense attorneys are important in assembling information, checking leads, and confirming facts. The team members locate and coordinate with witnesses for the attorneys to interview. The preparation of witnesses for depositions and trial is crucial as a case moves forward. Wheatcroft and Ellison say that this "familiarization of witnesses to cross-examination processes increased accurate responses and reduced errors" (Wheatcroft & Ellison, 2012, p. 821). In the effort to arrive at the truth, witness familiarity with courtroom procedures and attorney tactics can reduce the discomfort of the witness and attendant lack of focus and accurate statements that come from feeling intimidated or unsure.

Direct examination can be thought of as storytelling wherein attorneys explain a version of events using the testimony of witnesses. Attorneys for both the prosecution and the defense in murder cases must be well organized and have their presentation reflect that organization (Twiss, 2007). During a direct examination, the attorney generally asks specific questions to establish the foundation of the case. This function of direct examination is critical in a jury trial. Once a witness has been directly examined the opposing counsel will cross-examine to clarify and validate or attempt to discredit and impeach all or parts of the witness testimony. The cross-examination allows opposing counsel to probe issues raised during the direct examination that may benefit their side of a case. This may come as the result of showing weaknesses in the opposing attorney's case (Wheatcroft & Ellison, 2012).

At the beginning of a case each attorney will make a statement to the jury or judge about what they intend to prove or show during the trial. The prosecutor will explain to the jury why the defendant is guilty and describe the prosecution's view of how the defendant committed the crime. This may involve showing physical evidence to the jury or eliciting testimony from witnesses. It is this role as essentially an interviewer that often determines how effective the attorney is perceived to be. The outcome of the trial may very well hinge on each attorney's ability to present their side of the case. At the end of the trial each attorney will make a closing argument or statement summarizing the key points he or she wants the judge or jury to consider.

The American Bar Association and many legal writers over the years have provided guidelines and advice on controlling the testimony of witnesses. During direct examination of witnesses the attorney for the prosecution is attempting to present evidence that meets the burden of proof of the charge and convince jurors that the evidence and the witness are reliable. As direct and cross-examination of witnesses proceeds, each side is laying the groundwork for their final summation of the case as they bring together the points they have laid out.

A challenge faced by the defense attorney is providing some type of explanation for his client's actions that might make sense to jurors. If the defense attorney can provide a scenario that makes some type of sense, the jury will be left to consider whether the defendant acted reasonably (Duck, 2009). The defense may argue that the defendant's mental state was such that he did not know right from wrong when he acted. This would often fall under a temporary insanity defense. An argument might be made in a murder case, for example, that the killing was spur of the moment as the result of a distraught or emotional individual. Perhaps a claim of self-defense where the defendant asserts that he was in fear for his life and that he saw no other course of action but to take the life of someone else. The laws in many states may provide an opportunity to assert self-defense, as so many have incorporated so-called stand your ground laws. The defense attorney still must make the case that the defendant's actions were reasonable and that most people (including the jurors) would have done the same in his place (Duck, 2009). While the prosecutor may depict the defendant as a remorseless killer, the defense explains the actions of her client in the context of the defendant's life history (Costanzo & Peterson, 1994).

Defenses

As in all criminal cases, there are numerous legal defenses available to the criminal defendants charged with various crimes. While there are many texts devoted to the detailed study of criminal law and procedure, this chapter will provide a brief overview of the most common defenses used specifically in homicide cases. There are many ways to categorize criminal defenses, so we have used the following categories for the sake of simplicity and consistency with other research: failure of proof defenses, alibi, **justification defenses**, and **excuse defenses**.

Failure of Proof Defenses

This defense holds that the prosecution has failed to prove its case beyond a reasonable doubt and is used frequently by defendants in criminal cases. It is often used effectively in homicide cases where prosecutors have many elements to prove beyond a reasonable doubt, which is a very high burden of proof. Skilled and experienced defense attorneys will continue to remind the jurors throughout the trial about how high this burden is. In first-degree murder cases, for instance, prosecutors must prove the element of **premeditation**, most often using circumstantial evidence to show that a defendant's actions before the crime

Justification defenses The defendant admits committing the illegal act but argues that the result was positive or that he does not deserve the blame.

Excuse defenses A claim by the accused of acting without criminal intent.

Premeditation Evidence of planning a crime ahead of time.

prove deliberation. Defendants can argue that the evidence does not prove premeditation or planning, and that the defendant should not be convicted of the crime charged but perhaps of a lesser included offense instead, like second-degree murder or manslaughter.

Alibi

The word **alibi** comes from Latin and literally means "elsewhere." Defendants assert this defense when they claim they could not have committed the crime because they were somewhere else when the crime was committed. The defendant attempts to prove he was not present at the crime scene, at least during the commission of the crime. The defendant does not carry the burden of proof for an alibi defense. The government continues to bear the burden of proving that the crime was committed by the defendant beyond a reasonable doubt. Often, there will be conflicting testimony at trial about a defendant's possible alibi, with a defense witness testifying that she was with the defendant when the crime occurred and state's witnesses testifying that they saw the defendant near the crime scene. Many defendants assert alibi defenses at trial, but these defenses are not always effective, especially when the state offers other compelling evidence of the defendant's guilt.

Alibi Information or evidence intended to show that the accused was not in a place where he could have committed the alleged crime.

Justification Defenses

The justification defenses include self-defense or the defense of others, defense of home and property (commonly known as "stand your ground" laws), consent, and necessity. A justification defense means that the defendant admits committing the act, but he asserts that he had just cause in committing the act and is therefore not criminally liable.

The self-defense and defense of others is considered an affirmative defense, which requires a defendant to prove that he used reasonable actions to protect himself or a third person and that his actions were necessary to protect himself or the third person. The burden of proof is on the defendant for this defense. It can be difficult to prove self-defense because a person has to prove that he did not cause or start the conflict that led to the event, and he must prove that his actions were reasonable given the circumstances he faced. The category of "stand your ground" laws like the one in the Florida statute is relatively recent, as Florida's law was enacted in 2005. These laws do not require people to retreat if they are in their homes or their property, and allow them to use deadly force. There has been wide concern about these laws, with some commentators suggesting we are returning to the Wild West and legalizing vigilante justice. Since Florida's enactment of the first stand your ground law, at least 23 states have passed some form of this law.

The necessity defense is defined as "[a] justification defense for a person who acts in an emergency that he or she did not create and who commits a harm that is less severe than the harm that would have occurred but for the person's actions" (*Black's Law Dictionary*, 1999). Generally when a defendant asserts the necessity defense, the courts require him to demonstrate that the pressure arose from a physical force of nature and not from other human beings.

Excuse Defenses

There are also several defenses that are known as excuse defenses. These defenses may include age or infancy, involuntary intoxication, provocation, and insanity. Excuse defenses mean that the person admits to committing the criminal act at issue but claims he cannot be held legally responsible because he lacked the criminal intent necessary to be guilty of

Mens rea Knowledge or intent to commit the crime.

the crime. The age or infancy defense hails from the common law where there was a strong belief that young children were incapable of forming the required **mens rea**, or guilty mind, to be guilty of a crime. Under the common law, there is a presumption that children under the age of 7 are incapable of forming the required mental state to form criminal intent, and there was rebuttable presumption that children between the ages of 7 and 14 could not do so.

Presenting adequate evidence to convince jurors that a person was involuntarily drugged or induced into a state of intoxication is challenging. If this threshold can be overcome, the defense hopes to show that the defendant was incapable of forming the requisite intent to commit the charged crime. Similarly, the defense of provocation seeks excusing the defendant's behavior with the claim that any reasonable person would have lost control temporarily in the same circumstances and act in a similar way to the defendant.

M'Naghten rule Testing the accused to determine if they were sane at the time of a criminal act.

The most common excuse defense in homicide crimes is the insanity defense, though it is infrequently used overall. Each state establishes its own definition of insanity. Currently, 26 states use the **M'Naghten rule** in their statutes, and this standard is often referred to as the "right/wrong test." The M'Naghten rule is based on a British case from 1843 where a criminal defendant was found not guilty by reason of insanity and holds that offenders who are unable to know right from wrong at the time of the offense should not be held criminally responsible. States that do not use the M'Naghten standard instead use the **substantial capacity** test from the American Law Institute's Model Penal Code. Currently, 23 states use some version of the substantial capacity test. This test is considered a less stringent standard than M'Naghten and holds "a person is not responsible for criminal conduct if at the time of such conduct as a result of mental disease or defect he lacks substantial capacity either to appreciate the criminality of his conduct or to conform his conduct to the requirements of the law" (art. 4, sect. 4.01).

Substantial capacity Supposes the inability to control one's behavior.

It is important to note that in an insanity defense, the burden to prove the defense rests with the defendant. Defendants must prove insanity by clear and convincing evidence. Research suggests that juries are reluctant to find defendants not guilty by reason of insanity (Torry & Billick, 2010).

Ethical Issues

The legal systems of all countries have known bias, perjury, use of false evidence, and misconduct of various sorts by different actors within the system. Ongoing improvements in the law, in procedures, and in the professionalism and training of personnel throughout the criminal justice system have led to improved access to justice and equity in the American criminal justice system. None of the components of the criminal justice system are perfect, nor do they run flawlessly at all times. With that said, the American system of jurisprudence works fairly well and is examined, envied, and emulated by many other countries.

Planting evidence to incriminate a suspect is unlawful and weakens the faith people have in their legal system. Stories in the media about such occurrences, while not frequent, are known to us. Some officers have used what Delattre (1996) calls "noble cause corruption" to attempt to achieve a conviction of a known criminal who for various reasons has not been held sufficiently accountable in the eyes of many people. An officer's rationale for trying to remove an offender from society is a flawed one because if citizens are expected to obey the law they must feel confident that the officers they entrust with enforcing the law also respect and follow the law.

The prosecution and defense teams must both conduct themselves in their case in accordance with ethical principles governing the conduct of lawyers and case law. When a prosecutor uncovers evidence that tends to exonerate a defendant they are obliged to share this information with the defense. This is known as exculpatory evidence, and the ethical codes are very clear that prosecutors must share this information with defense counsel. For their part, as the defense team prepares their case they may not claim ignorance of information made available to them or that they discovered during their own preparations. This also goes to the matter of counsel effectively representing their clients. If the defendant in a case is found guilty, it is not uncommon to attempt to appeal their conviction based on a claim of ineffective counsel even if the defense attorney thoroughly and professionally carried out their duties and advised their client in the best way they could. There is an enormous amount of case law relating to ineffective assistance of counsel, especially in criminal felony cases like homicide cases.

Plea Bargaining

Both prosecutors and defense attorneys face a difficult and resource-intensive task in many cases. One of the challenges facing the prosecutor obviously involves proof that the suspect was the one who committed the crime. At the same time, the defense is working to disprove this accusation in any legal way possible. Of course, some cases are resolved through **plea bargaining**.

Plea bargaining The defendant agrees to plead guilty or no contest and in return will generally receive conviction on a crime less serious than that originally charged.

It is common that members of the public misunderstand the full dimensions of the use of plea bargains. Plea bargains not only provide a lesser sanction for a defendant but also complete a case without expending unnecessary public resources and allowing the prosecutor to move on to other cases. While different people have varying views on what a case outcome should be, quicker justice and imposition of sentence may be helpful to a victim as compared to the ordeal of a lengthy trial. The plea bargain may ensure a lesser sanction, but it also guarantees some sanction. A clear incentive for the defendant is that fewer charges might be brought and he would therefore receive a reduced sentence. Through no fault of either the prosecutor or the law enforcement officers who investigated a case, there simply may not be sufficient evidence to reasonably go to court and expect to win a conviction. In these cases a plea bargain may be an appropriate compromise.

Plea bargaining is considered by some to be coercive and may result in false convictions and misuse of this mechanism. Plea bargaining under certain circumstances may induce self-incrimination and a plea of guilty in hopes of a lighter sentence. Even the death penalty is used in cases to leverage a plea bargain. It is understood that in order to avoid the potential imposition of a death penalty sentence, a defendant will often plea bargain to a lesser offense to achieve a lesser sentence.

As you may recall from Chapter 1, the conviction of a suspect is considered one way of measuring the success of an investigation. Considering all that you now know about investigations, which of the four measures mentioned on page 8 do you think is the most important?

©iStockphoto.com/RichLegg

As described, plea bargains can work to benefit the state and the defense. However, there is also controversy regarding their use. While the victim's family may feel the relief of avoiding an involved criminal trial, they may also be upset at the killer receiving less than the maximum sentence. Even though some semblance of closure may come to the victim's family more quickly and the defendant will receive some sanction, the family may still feel the killer should receive greater punishment. Controversy about plea bargains includes the defendant waiving his right to a jury trial, against self-incrimination, and to confront witnesses. Still, the Supreme Court has stated that plea bargaining passes constitutional muster. The courts insist that the defendant must knowingly and willingly plead guilty with full knowledge of the consequences. On balance, though plea bargaining has critics, its usage keeps the system moving, reducing cost to taxpayers; time and suffering for victims' families; and, arguably, achieving little different outcome in sanctions imposed, given that many cases if proceeding to trial would not have won convictions or received more of a punishment.

Roberts (2013) comments on the view of further regulations on plea bargaining and efforts by counsel. Roberts notes some believe "that regulating bargaining will open floodgates to future litigation" (p. 2650), but that "[w]hile real, these are manageable challenges that do not outweigh the need to give meaning to the constitutional right to effective counsel. After all, in a criminal justice system that is largely composed of plea bargains, what is effective assistance of counsel if it does not encompass effectiveness within the plea negotiation process?" (p. 2650). A plea bargain must be approved by the defendant but also by the court to guard against undue influence by the attorneys in the case.

A consideration in deciding on a plea arrangement is the penalty faced by the defendant. Sometimes a defendant will be able to offer evidence in a larger case, which can influence what level of sanction the prosecution is willing to offer. Absent a plea, the case will move forward toward a trial.

Hoffman, Rubin, and Shepherd examined case outcomes for all felony cases in Denver, Colorado, for the year 2002. Their findings indicate "that public defenders achieved poorer outcomes than their privately retained counterparts, measured by the actual sentences defendants received" (p. 223). The differences were significant. The study pointed out that the Supreme Court more than 50 years ago identified representation by counsel as a constitutional guarantee. The public defender systems in most states evolved from this view. Implicit is that the counsel must be effective.

Research has been varied about whether public defenders or private counsel is more effective. As we mentioned, private attorneys often have greater resources (including time) available to mount a defense in a homicide case. The study supported the proposition that private attorneys achieve significantly more positive outcomes, measured in actual length of sentences, than public defenders. The authors raise the question of whether a partial explanation for the difference in outcomes has to do with clients "self-selecting." The idea goes that if a "marginally indigent" defendant has money available and is faced with a relatively serious charge, he will hire a private attorney and gain the benefit of the attorney's lighter caseload and (hopefully) expanded resources. If that same defendant is faced with the reality that the evidence of his guilt is substantial, he may hesitate to seek out money to hire an attorney that may be essentially of no great help. Public defenders may be representing, on the whole, people with worse cases, which in turn leads to worse case outcomes. And so, the authors assert, "Thus, in a system that tries only 5% of all criminal cases, the most important skill for a lawyer on either side is the ability to evaluate a case before entering into plea negotiations, not the ability to shine at trial" (p. 245).

Sentencing

If the case proceeds to a trial, the jury must find the defendant guilty by a unanimous verdict. Jurors may be impacted in their decision not by the facts alone but the tactics of the attorneys in the case. Perhaps a victim fails to show "enough" emotion; jurors may see the victim as less sympathetic or deserving of justice (Rose, Nadler, & Clark, 2006). The defense attorney being friendly toward his client has been found to increase juror receptivity to mitigation in a case (Brewer, 2005). The defense may raise the issue of the victim's socioeconomic status in subtle ways—or perhaps that the victim identified as LGBTQ—to influence jurors to think of the victim as less deserving of justice (Gallagher, 2002).

Not only are the arresting and prosecution of individuals accused of murder important, criminal sentencing is part of the sober process of making formal social control decisions (Ulmer & Johnson, 2004). A conviction in a murder case can result in the harshest of sentences. Sentences, as we know, vary from state to state and even within jurisdictions, based in part on state guidelines enacted by legislatures but also with some discretion through judges acting within those state guidelines. A guilty verdict in a first-degree murder case in some states may bring the death penalty or life in prison without the possibility of parole. Seeking the death penalty in a particular case is a decision made by the prosecutor's office based on a variety of factors.

Sentencing options vary from state to state in homicide cases. In Florida, for example, first-degree murder provides the option to assign the death penalty. Classification of the crime is the main consideration driving the charge. Three potential options are the death penalty (depending on if the state uses the death penalty), life in prison without the possibility of parole (LWOP), and typically a sentence of 20 to 25 years extending to life imprisonment. Disagreement on appropriate sentencing in homicide cases remains animated.

A conviction for murder may bring a sentence of a number of years in prison up to and including life. And for states not utilizing the death penalty a sentence of life in prison may come as the result of a conviction of first-degree murder and based on any of a number of aggravating factors. Penalties assigned after a conviction of homicide will vary based on the degree charged. For example, a manslaughter conviction may bring a prison term that is less than a sentence of life behind bars.

> **EXPLORE THIS**
>
> The American Bar Association (ABA) maintains a public education page on its website. Go to http://www.americanbar.org/portals/public_education.html and examine the section "About the Courts" to learn more about the steps in a trial.
>
> PBS and KQED Public Television created the documentary *Presumed Guilty* covering 3 years of work by public defenders. Go here to learn more: http://www.pbs.org/kqed/presumedguilty/index.html

SUMMARY

Most readers began this chapter with somewhat of a grasp of the process of how a criminal case makes its way through the court system from arrest through arraignment and pretrial motions to an actual trial and, if a conviction occurs, to sentencing. We did not intend a state-by-state review of court rules and requirements and homicide cases, but rather to familiarize you with the actual sequence of steps in homicide cases and point out the importance of various parts of the process.

We have again discussed the interest people have in the criminal justice process and certainly in cases of murder. This extends to the portion of the process involving the courts and trial preparation, process, and outcomes. Hollywood and the news media certainly fanned the flames of sometimes misinformed interest in the criminal trial, but unless the case is notorious it often receives relatively little attention. The verdict in notorious cases is often the single thing of interest to most members of the public. Even though homicide case clearance rates are higher than other serious crimes, the rate at which such cases are cleared has dropped significantly in the past 40 years (Litwin, 2004). Advances in science and technology, as well as the continuing protections of the American justice system, result in significant numbers of defendants being found not guilty subsequent to the court process.

Preparation and conduct of a homicide case for trial is not a simple or quick process. While most people are never aware of the tremendous amount of work that goes into such cases, the coordination required between the prosecution and law enforcement is substantial. Law enforcement enters a homicide investigation with the goal of finding out who is accountable for the killing and, along with other actors in the criminal justice system, determining whether the death was the result of a criminal act. However, in addition or as part of this goal is preparing each and every step of the way for potentially a trial to determine whether or not a particular suspect is guilty of a crime. To that end, law enforcement must always be thinking about the prosecutor's office and how best to help the prosecutors assemble the most complete case possible. The television viewing public will typically see police dramas where law enforcement catches the killer. Some shows also provide an image of the high drama of a witness on the stand in a murder trial. We do well to remember, however, that the prosecution is also considering the public's resources in deciding whether or not to offer a plea arrangement with the defendant and considering whether the strength of available evidence would support going all the way through to a trial regardless of available resources.

Whether or not a case actually goes all the way to the trial setting is not always a reflection of the work done by law enforcement or the prosecutor. There simply may be insufficient evidence to prove a case.

Homicide prosecutions are clearly of great importance in society. Reflecting the seriousness of the crime and the impact on victims and community, adequate evidence must be discovered and assembled to prove the case in court before a jury. The preparation of the case by both prosecution and defense involves a great deal of work with members of the criminal justice system, preparing lay and expert witnesses, and presenting a coherent case to the jury. It is vital to the success of a case that the attorneys scrupulously pursue the work of representing either the people or the defendant and avoid any ethical lapse.

KEY TERMS

Alibi 257
Arraignment 249
Beyond a reasonable doubt 249
Challenge for cause 250
Excuse defenses 256

Grand jury 249
Justification defenses 256
M'Naghten rule 258
Mens rea 258
Plea bargaining 259

Premeditation 256
Substantial capacity 258
Venire 250
Voir dire 250

DISCUSSION QUESTIONS

1. How should a prosecutor evaluate a charging decision in a case of homicide?

2. What is the difference between direct and indirect evidence? Give an example of each.

3. Do the principles and rules guiding attorney behavior benefit the trial process in homicide cases? If so, how?

4. Describe and discuss the factors considered in plea bargaining.

5. What are aggravating or mitigating factors in deciding the sentence in a murder case?

6. Briefly discuss the activities at each stage of the homicide case process following an arrest.

- Get the tools you need to sharpen your study skills. SAGE edge offers a robust online environment featuring an impressive array of free tools and resources.
- Access practice quizzes, eFlashcards, video, and multimedia at **edge.sagepub.com/houghci**

Glossary

Accelerant: Any substance used to increase the speed that a fire burns or spreads. The substance or mixture may be liquid or solid.

Actus reus: Element that is the criminal act.

Admission: Acknowledgment, such as to the guilt of committing an act.

Aggravated assault or battery: If the person committing an assault or a battery does so with the intention to inflict severe injury or by using any item as a weapon to cause injury.

Algor mortis: Body temperature change after death.

Alibi: Information or evidence intended to show that the accused was not in a place where he could have committed the alleged crime.

Alternate light source (ALS): Used to discover various types of evidence; ultraviolet light, for example, illuminates natural substances such as semen or saliva.

Anger-excitation rapist or sexual sadist: Wants to inflict pain and suffering on his victim to achieve his sexual satisfaction.

Anger-retaliatory rapist: Typically a man who is clearly hostile to women and commits his crime to intentionally demean and punish.

Arraignment: A hearing where a formal charge against an individual is entered.

Asphyxiation: Suffocation resulting in loss of oxygen.

Assault: An attempted physical attack.

Asset forfeiture: Seizure of property or funds used in furtherance of a crime.

Automated Fingerprint Identification System (AFIS): Database that allows computer matching of unknown to known fingerprints.

Battery: One person's unlawful touching of another person.

Beyond a reasonable doubt: Legal standard in a criminal prosecution for the jury to find someone guilty.

Biological theory: Perspective that examines physical aspects of humans to explain psychological reasons for the behavior.

Biometrics: Measurements of the human characteristics that serve to identify individuals.

Blood feud: Ongoing series of attacks typically between two families.

Blunt force trauma: Nonpenetrating injury to the body.

Body-worn camera (BWC): Camera devices typically attached to the uniform of a police officer to gather evidence, enhance officer safety, and improve public relations through accountability of officer actions.

Burglary: Entry into a building or conveyance or remaining within it with the intent to commit a crime.

Carjacking: Involves robbing an individual of the vehicle they are driving.

Case management software (CMS): Allows the coordination and management of digitally collected and stored information.

Case review: Set time periods to evaluate all information gathered and actions taken during an investigation.

Challenge for cause: Disqualification of a potential juror with a specific reason given.

Circumstantial evidence: Evidence that alone may infer a variety of facts. The trier of fact will reach his or her own conclusion about what the presence of a piece of evidence ultimately means in a case.

Clandestine lab: A laboratory set up secretly to illegally produce controlled substances.

Class characteristics: Those attributes of an item that place it within a broader group of objects.

Cognitive bias, to include confirmation bias: Each individual's perception of social reality. Includes the information processing shortcuts that allow faster response to previously encountered circumstances.

Cold case: Generally, a crime that is no longer being actively investigated, usually due to a lack of leads or evidence.

Combined DNA Index System (CODIS): Searchable database of stored DNA profiles.

Community-oriented policing (COP): An approach that relies on partnering with public, private, and volunteer organizations as well as individual outreach to neighborhoods. Officers are often assigned to work in the same area of a community for extended periods of time so they can become thoroughly familiar with the residents and businesses, and so those businesses and residents may come to know and trust the officers.

Computer-aided drawing (CAD), or computer-assisted drafting and design (CADD): Software used to facilitate design.

Computer voice stress analyser (CVSA): Pseudoscientific device that purports to measure stress from a person's voice.

Confession: Typically refers to admitting to all elements of a crime.

Confidential informant: An individual who provides verifiable information to law enforcement.

Confirmation bias: Interpreting evidence as confirming an established belief.

Confrontational homicide: One generally male individual who kills another in a conflict that likely escalated from an argument over what is perceived to belong to each person or a slight to honor; the most common form of criminal homicide.

Control specimen: Allows comparison between a known sample and an unknown item of evidence.

Controlled buy: The use of a police officer or informant, under strict conditions, to purchase illegal drugs or other contraband from an offender.

Controlled substances: Drugs or chemicals produced in a manner regulated by government.

Conveyance: A vehicle used to transport people.

Coroner: Elected or appointed officials who inquire into the cause and manner of death.

Crime Victims' Rights Act (CVRA): Enacted in 2004, the act lists the rights of victims in federal crimes.

Criminal mischief/vandalism: Unlawful damaging of another's property.

Criminalist: An individual (usually degreed) who applies scientific knowledge to analyze evidence.

Critical infrastructure: Assets necessary for society to function.

Cybercrime: Benefit monetarily by victimizing others through the use of computers and the Internet.

Danger Assessment: An instrument that helps to quantify and illustrate the level of danger posed by an intimate partner of killing the abused person.

Deoxyribonucleic acid (DNA): Two coiled chains of molecules that carry genetic functions of an organism.

Detection of deception device: One of several pieces of technology that measure physiological response as a way to detect stress, which may, in turn, be caused in part by speaking or acting deceptively.

Differential association theory: In criminal justice, perspective that interacting with others provides a vehicle to learn and adopt their values, attitudes, or motivations to commit deviant acts.

Direct evidence: Evidence such as personal knowledge or information that may prove or disprove a fact. Direct evidence does not require inference.

Domestic terrorism: U.S.-based group of extremists who perpetrate crimes based on, generally, a single issue or a belief system.

Domestic violence (DV): Assault and violent crimes committed by family members against other family members.

Drug laboratory: Government or private laboratory that conducts analysis of substances suspected to be illegal drugs.

Embezzlement: Theft or conversion of funds under a person's care that was entrusted to them by another.

Equivocal death: Manner of death open to interpretation.

Evidence: Facts supporting the proof of a thing.

Excited utterance: A spontaneous comment or statement by someone as a reaction to a startling or shocking circumstance. Usually an exception to the hearsay rule.

Exclusionary rule: Prohibits the use at trial of evidence that was not obtained legally.

Excuse defenses: A claim by the accused of acting without criminal intent.

Exhibitionism: Obtaining sexual gratification from the exposure of genitals to strangers.

Exigent circumstances: May justify law enforcement entering a building without permission if people are in danger, evidence is at risk of imminent destruction, or if a suspect may escape.

Expert witness: A person whose experience, training, and education provide the basis for acceptance by a court to testify to his or her opinion on matters before the court.

Fence: Person who makes a living or routinely receives stolen goods for resale, or an individual trying to benefit from the low-cost item.

Fetishism: A sexual attraction to various objects, very often articles of clothing.

Field identification: Viewing a suspect by a victim or witness within minutes of a crime. Also known as a showup identification.

Fifth Amendment: To the U.S. Constitution: enumerates certain rights such as the right against self-incrimination, the right to due process, and the protection against being tried twice for the same crime.

Follow-up investigation: Additional resources or an extended investigation may occur in serious cases or those where sufficient evidence exists.

Forensic science: The application of science to criminal and civil matters of law.

Fourth Amendment: To the U.S. Constitution: prohibits unreasonable searches and seizures.

Framing: The filters individuals apply when viewing situations or that organizations or individuals may use to present an issue in a particular way.

Fraud: Involves some element of deception or trickery to gain access to the money or resources of the victim.

Frotteurism: Involves rubbing against another person who is not aware of the action, usually in a crowded public place.

Genocide: Killing a large group of nationally or ethnically similar people.

Good faith exception: When faced with the potential blocking of evidence due to the Exclusionary Rule, if officers relied on a search warrant they believed to be valid, evidence seized illegally may be admitted by the court.

Graffiti: Words or symbols drawn or painted onto buildings and objects with the intention to mark the territory of a gang or group or to convey a message to others.

Grand jury: Group of citizens impaneled within a jurisdiction to examine information and determine whether the prosecutor should bring charges against an individual or entity.

Grow house: Generally, the use of a residential house to cultivate marijuana indoors.

Gunshot residue (GSR): Particulates left on the skin or clothing of someone discharging a firearm or in close proximity to one that has been fired.

Hard surveillance: Traditional methods of investigation.

Hate crime: Nature of such crimes goes beyond the single offense to symbolically target whole demographics of people.

Individual characteristics: Narrow down the class characteristic that put an item into a group of objects to one emanating from a single source.

Informants: Persons who gives information to authorities.

International terrorism: Variously defined, but generally including violent or dangerous acts intended to frighten a civilian population and influence government action; primarily occurs outside of the United States.

Interrogation: Questioning an individual who is suspected of committing a crime.

Interview: Questioning or talking to an individual who is believed to have information regarding some incident or crime.

Intimate partner homicide (IPH): Fatal assault often the end result of an ongoing current or previous abusive intimate relationship.

Intimate partner violence (IPV): Various forms of abuse or violence committed by an individual with whom the victim is currently or had previously been in an intimate relationship.

Justification defenses: The defendant admits committing the illegal act but argues that the result was positive or that he does not deserve the blame.

Key assets: Vital elements of infrastructure whose damage or loss would create significant effects on society.

"Knock and talk": Police investigative method of asking a resident if the police may come in and search the premises.

Larceny/theft: Unlawful taking of property.

Latent fingerprint: Visible print after powder is applied to the residue from oils of the hand.

Launder: Filtering illegally obtained money through a legitimate business to obscure its source.

Law Enforcement Assistance Administration (LEAA): Was a federal agency that administered funding to state and local government entities for research, training, and programs.

Learning Management System (LMS): Software to assist in the delivery and documentation of training and education.

Left-wing group: Communist- or socialist-influenced group that seeks to overthrow capitalist societies or systems.

Ligature: Something used to tie or bind.

Lineup: A presentation of the suspect and filler individuals all at once or one after the other for a witness or victim to identify the offender.

Lividity: Pooling of body and the blood that can help determine body position after death.

Lone wolf terrorist: May act consistent with the ideology of a terror group but without interacting with or being directed by that group.

M'Naghten rule: Testing the accused to determine if they were sane at the time of a criminal act.

Major case squad: Unit tasked with investigating crimes that require significant resources. Such squads in larger agencies may handle, for example, homicides.

Medical examiner: Trained pathologist and medical doctor appointed at the county, regional, or state level in many U.S. states.

Mens rea: Knowledge or intent to commit the crime.

Miranda warning: Advisement of rights given by the police to a criminal suspect prior to questioning.

Modus operandi (MO): Method of operation.

Motor vehicle theft: stealing or attempting to steal a vehicle.

Munchausen Syndrome by Proxy (MSP): Also referred to as factitious disorder, MSP is a mental illness in which a caregiver claims or induces illness of someone whose well-being he or she is in charge of so that the caregiver may receive attention.

National Integrated Ballistics Information Network (NIBIN): Computerized network available to law enforcement for the comparison of ballistic evidence.

Necrophilia: An instance of someone being sexually aroused by a dead body.

Neighborhood canvass: Door-to-door contact with businesses or residents to gather information.

Organizational groupthink: Group members maintaining the status quo at the possible cost of effective decision-making or positive outcomes.

Paraphilia: Deviant acts that are sexually arousing to various individuals.

Patent print: transfer of material from a print to the surface leaving an impression.

Perimortem: At, or near, the time of death.

Personal protective equipment (PPE): Items such as goggles, helmets, and protective clothing that shield an individual from various hazards or injuries.

Personal retribution: Punishment for a crime or offense carried out by the victim against the one who committed the crime.

Persons crimes: A common way of referring to crimes committed against the body of a victim, such as an assault, battery, robbery, sexual battery, or homicide.

Plain feel exception: If, during a pat-down, an officer feels something that he can immediately identify such as a weapon, the officer may confiscate or seize the object.

Plain view: If an officer is in a location he may legally be in, and sees something he recognizes as evidence, he may seize it.

Plastic print: impression of a print left in a soft surface.

Plea bargaining: The defendant agrees to plead guilty or no contest and in return will generally receive conviction on a crime less serious than that originally charged.

Polygraph: Device that measures physiological responses during questioning to detect stress levels.

Postmortem: After death; also refers to the examination of a dead body.

Power-assertive rapist: Uses his assault to prove his masculinity as his primary motivation.

Power-reassurance rapist: Has the primary motivation of exhibiting or holding power over his victim.

Preliminary investigation: The initial inquiry into a reported crime, usually conducted by a patrol officer.

Premeditation: Evidence of planning a crime ahead of time.

Premortem: Before death.

Presumptive test: Analyzes a sample and determines that it is either not a particular substance or that it probably is a particular substance.

Primary aggressor: An individual determined through investigation to pose the most threat or who has shown a history of abuse to the victim.

Probability errors: The probability of making a wrong decision, which can result from rejecting a true hypothesis or failing to reject a false hypothesis.

Probable cause: The level of information needed for a court to issue a warrant, for an officer to conduct certain searches or arrests, and the standard for a grand jury indictment.

Problem-oriented policing (POP): A strategy built on the idea of identifying the underlying causes of a crime or other police problem in devising a strategy to combat that problem rather than the symptoms.

Property crimes: Generally refer to those offenses that affect the property of a person or business rather than the actual person. Some examples include theft, vandalism, burglary, or arson.

Psychological theory: Describes and predicts behavior based on human thought and emotion.

Rape or sexual battery: Sexual assault involving penetration.

Reasonable suspicion: Lower standard of proof than probable cause based on specific and articulable facts.

Records Management System (RMS): Generally a relational database system for record storage.

Reference: An item of evidence from a known source that allows comparison from a suspect source.

Reliable confidential informant (RCI): Sometimes distinguished from a confidential informant by virtue of having been utilized a set number of times and found to have provided accurate information.

Representativeness: Decision-making based on how alike one event is to similar events. This can lead to decision-making errors.

Restraining order: Court order intended to protect a person or organization from assault or harassment.

Right-wing group: Far-right groups such as neo-Nazi and white nationalists that seek to overthrow government and replace it with a system of their own ideologies.

Rigor mortis: Temporary stiffening of muscles and joints following death.

Road rage: A version of confrontational violence erupting between drivers or manifested by one driver acting aggressively toward someone else.

Robbery: Taking the property of another by force or the threat of force.

Routine activities theory (RAT): The appearance at one time and in one place of a likely offender, a suitable target for that offender, and the absence of a capable guardian of the target.

Satisficing: In decision-making, accepting the first adequate alternative.

Scatologia: When a caller is sexually aroused by the fearful reaction of the person receiving the call.

Schedules: Categories of various substances, chemicals, and drugs that are used to make other drugs. Many classified drugs have a medical use but may also have the potential to be abused.

Search warrant: Court order authorizing law enforcement to search a particular person, place, or vehicle for evidence of a specific crime.

Security threat group (STG): Group or gang within a jail or prison setting that has a pattern of violent or disruptive behavior.

Showup: Viewing a suspect by a victim or witness within minutes of a crime. Also known as a field identification.

Shrinkage: Theft of inventory by employees of a business.

Single-issue terrorism: Violent acts carried out focused on one issue, such as animal or environmental rights.

Sixth Amendment: To the U.S. Constitution: guarantees the right of a speedy trial, an attorney to represent the accused, the ability to face one's accusers, and the selection of a jury of one's peers, among other protections.

Social control perspective: The use of sanctions by individuals or institutions to guide or control behavior through the informal means of socialization or the formal means of the external sanctions by government agencies.

Social control theories: Various and include: containment, neutralization, and self-control.

Social disorganization theory: Examines the circumstances of the environment where a person lives to postulate criminal involvement.

Social learning theory: Learning through observing and imitating others, especially admired individuals.

Soft surveillance: New collection technologies.

Solvability factors: Those pieces of information or evidence that provide adequate information to move an investigation from the preliminary to the follow-up stage. Examples include pictures of a suspect or the tag number from a car involved in a crime, a viable way to identify stolen property.

Stalking: Following or monitoring someone's activities who interprets the stalking as harassment or intimidation.

Stand-your-ground laws: Such laws often establish legal protection for a person to use force or not retreat from a threat or perceived threat.

Stockholm Syndrome: A captive becoming psychologically dependent on or aligned with their captor.

Stop and frisk: Temporarily detain a person and pat-down their outer clothing for the presence of weapons.

Strain theory: The concept that various societal pressures influence behavior.

Strangulation: Compression of the neck.

Strong arm robbery: Using violence or the threat of violence when taking something from a person.

Subcultural perspective: Posits the existence of alternative values and attitudes by groups or subcultures in a larger society.

Substantial capacity: Supposes the inability to control one's behavior.

Surveillance; hard and soft: Respectively, use of traditional criminal investigative techniques and the use of new technologies to collect and use information.

Team policing: Patrol officers and detectives permanently assigned to neighborhoods to improve police community relations.

Venire: Gathering a pool of potential jurors.

Vice crimes: Generally, those crimes made unlawful as the result of a community's moral stance on issues. Examples include gambling, prostitution, and pornography.

Victim advocate: Professionals often employed in law enforcement or prosecutorial agencies to support victims throughout their interactions with the criminal justice process through direct services and providing referrals.

Victim blaming: Suggests that the eventual victim is wholly responsible for whatever occurred.

Victimology: The study of victims and their experiences, including interacting with offenders.

Victim-precipitated: During the exchange the victim contributed in some significant way to the escalation of a situation.

Voir dire: Usually refers to the initial procedure of selecting jurors for a trial.

Voluntary consent: Permission by the person with proper authority to control property to allow the search of that property.

Voyeurism: Known to many as Peeping Toms, voyeurs are sexually aroused by secretly observing people in various stages of undress or having sex.

White-collar crime: Money is taken by someone in a position of trust at a business or organization without the use of violence.

Witness: A person who either sees something take place or has information about the event or people involved.

References

Chapter 1

Bloch, P., & Anderson, D. (1974). *Policewomen on patrol*. Washington, DC: Police Foundation.

Boydston, J., Sherry, M., & Moelter, N. (1977). *Patrol staffing in San Diego: One- or two-officer units*. Washington, DC: Police Foundation.

Brookman, F., & Innes, M. (2013). The problem of success: What is a "good" homicide investigation? *Policing and Society, 23*(3), 292–310.

Eck, J. E. (1983). *Solving crimes: The investigation of burglary and robbery*. Washington, DC: Police Executive Research Forum.

Federal Bureau of Investigation. (2017). *Crime in the United States, 2015*. Table 25: Percent of offenses cleared by arrest or exceptional means by population group, 2015. Clarksburg, WV: Author.

Greenwood, P. (1979). *The RAND criminal investigation study: Its findings and impacts to date*. Santa Monica, CA: RAND Corporation. Retrieved from https://www.rand.org/pubs/papers/P6352.html

Greenwood, P., & Petersilia, J. (1975). *The criminal investigation process*. Santa Monica, CA: RAND Corporation.

International Association of Chiefs of Police. (2003). Training key #558. Criminal investigations.

Kelling, G. L. (1974). *Kansas City preventive patrol experiment*. Police Foundation. Retrieved from https://www.policefoundation.org/publication/the-kansas-city-preventive-patrol-experiment/

Kelling, G. L., & Moore, M. H. (1988). The evolving strategy of policing. *Perspective on Policing*, No. 4. Washington, DC: National Institute of Justice and Harvard University.

Murphy, G. R., & Wexler, C. (2004). *Managing a multijurisdictional case*. Washington, DC: Police Executive Research Forum.

Pate, T., Ferrara, A., Bowers, R., & Lorence, J. (1976). *Police response time: Its determinants and effects*. Washington, DC: Police Foundation.

Pawson, R. (2006). *Evidence-based policy: A realist perspective*. London, England: Sage.

Piliavin, I., & Blair, S. (1964). Police encounters with juveniles. *American Journal of Sociology, 70*, 206–214.

Reaves, B. A. (2015). *Local police departments, 2013: Personnel, policies, and practices*. Washington, DC: U.S. Department of Justice, NCJ 248677.

Rhodes, W., Dyous, C., Chapman, M., Shively, M., Hunt, D., & Wheeler, K. (2009). *Evaluation of the Multijurisdictional Task Forces (MJTFs), Phase II: MJTF performance monitoring guide*. Cambridge, MA: Abt Associates Inc.

Saferstein, R. (2007). *Criminalistics: An introduction to forensic science*. Upper Saddle River, NJ: Pearson.

Sherman, L., & Berk, R. (1984). The specific deterrent effects of arrest for domestic assault. *American Sociological Review, 49*, 261–272.

Skolnik, J. (1966). *Justice without trial: Law enforcement and democratic society*. New York, NY: Wiley.

Telep, C. W., & Lum, C. (2014). The receptivity of officers to empirical research and evidence-based policing: An examination of survey data from three agencies. *Police Quarterly, 17*(4), 359.

Thistlewaite, A. B., & Wooldredge, J. D. (2010). *Forty studies that changed criminal justice*. Upper Saddle River, NJ: Prentice Hall.

U.S. Department of Justice, Bureau of Justice Statistics. (2015). *Local police departments, 2013: Personnel, policies, and practices*, NCJ 248677.

Van Maanen, J. (1973). Observations on the making of policemen. *Human Organization, 33*, 407–418.

Wilson, J. (1968). *Varieties of police behavior: The management of law and order in eight communities*. Cambridge, MA: Harvard UP.

Wilson, J. Q., & Kelling, G. L. (1982, March). Broken windows: The police and neighborhood safety. *Atlantic Monthly, 127*, 29–38.

Chapter 2

Adcock, J. M., & Chancellor, A. S. (2013). *Death investigations*. Burlington, MA: Jones and Bartlett Learning.

Aronson, J., & Cole, S. A. (2009). Science and the death penalty: DNA, innocence, and the debate over capital punishment in the United States. *Law and Social Inquiry*.

Cole, S. A. (2007). How much justice can technology afford? The impact of DNA technology on equal criminal justice. *Science and Public Policy, 34*. Retrieved from http://www.ingentaconnect.com

Daubert v. Merrell Dow Pharmaceuticals, 509 U.S. 579 (1993).

Fisher, B. A. J. (2003). *Techniques of crime scene investigation* (7th ed.). Boca Raton, FL: CRC Press Inc.

Gabriel, M., Boland, C., & Holt, C. (2010). Beyond the cold hit: Measuring the impact of the national DNA data bank on public safety at the city and county level. *The Journal of Law, Medicine & Ethics, 38*, 396–411.

Hayes-Smith, R. M., & Levett, L. M. (2011). Jury's still out: How television and crime show viewing influences jurors' evaluations of evidence. *Applied Psychology in Criminal Justice, 7*, 29–46.

Hough, R., McCorkle, K. D., & Harper, S. (2019). An examination of investigative practices of homicide units in Florida. *Homicide Studies, 232*, 175–194.

Kirsch, L. (2006, Summer). Heating up cold cases. *The Forensic Examiner*. Retrieved April 13, 2014, from http://findarticles.com

Kumho Tire Co. v. Carmichael, 526 U.S. 137 (1999).

Lovgren, S. (2004, September 23). "CSI Effect" is mixed blessing for real crime labs. *National Geographic News*. Retrieved from http://www.wcsap.org/events/workshop2008/Effective%20Advocacy%20in%20a%20Hospital%20Setting%203.pdf

Maclin, T. (2006). Special needs search under the Fourth Amendment? What should (and will) the Supreme Court do? *Symposium*. American Society of Law, Medicine and Ethics. Retrieved April 13, 2014, from http://www.ncbi.nlm.nih.gov/pubmed

National Institute of Justice. (2000). *Crime scene investigation: A guide for law enforcement*. NCJ 178280. Washington, DC: U.S. Department of Justice, Office of Justice Programs, National Institute of Justice, Technical Working Group on Crime Scene Investigation, 58.

Nichols, R. G. (2007). Defending the scientific foundations of the firearms and tool mark identification discipline: Responding to recent challenges. *Journal of Forensic Sciences, 52*(3), 586–594.

Roane, K. R. (2005, April 25). The CSI Effect. *U.S. News & World Report*.

Shelton, D. E. (2008). The "CSI Effect": Does it really exist? *National Institute of Justice Journal, 259*. Retrieved from http://www.nij.gov/journals/259/csi-effect.htm

Singer, J. A., Miller, M. K., & Adya, M. (2007). Impact of DNA and other technology on the criminal justice system: Improvements and complications. *Technology and Criminal Justice, 17*.

Smith, S. M., Patry, M. W., & Stinson, V. (2007). The CSI Effect: Reflections from police and forensic investigators. *The Canadian Journal of Police & Security Services, 5*(3), 125–133.

Widyanto, M. R., Soedarsono, N., Katayama, N., & Nakao, M. (2010). Various defuzzification methods on DNA similarity matching using fuzzy inference system. *Journal of Advanced Computational Intelligence and Intelligent Information, 14*(3).

Young, T. J., & Ortmeier, P. J. (2011). *Crime scene investigation: The forensic technician's field manual*. Boston: Prentice Hall.

Chapter 3

Bornstein, B. H., & Hamm, J. H. (2012). Jury instructions on witness identification. *Court Review, 48*, 48–53.

Brickner, D. R., Mahoney, L. S., & Moore, S. J. (2010). Providing an applied-learning exercise in teaching fraud detection: A case of academic partnering with IRS criminal investigation. *Issues in Accounting Education, 25*(4), 695–708.

Brodeur, J. P. (2010). *The policing web*. New York, NY: Oxford University Press.

Brookman, F., & Innes, M. (2013). The problem of success: What is a "good" homicide investigation? *Policing and Society, 23*(3), 292–310.

Camber, R. (2017, June). Be a detective . . . with NO police training! *Daily Mail*, 5.

Carson, D. (2009). The abduction of Sherlock Holmes. *International Journal of Police Science & Management, 11*(2), 193–202. doi:10.1350/ijps.2009.11.2.123

Cassell, P. G., Mitchell, N. J., & Edwards, B. J. (2014). Crime victims' rights during criminal investigations? *Journal of Criminal Law & Criminology, 104*(1), 59–103.

Cook, T., & Tattersall, A. (2016). *Blackstone's senior investigating officers' handbook* (4th ed.). Oxford, England: Oxford University Press.

Cutler, B. L., & Kovera, M. B. (2010). *Evaluating eyewitness identification*. Oxford, England: Oxford University Press.

Dean, G., Fahsing, I. A., Glomseth, R., & Gottschalk, P. (2008). Capturing knowledge of police investigations: Towards a research agenda. *Police Practice & Research, 9*(4), 341–355.

Fisher, R. P., & Geiselman, R. E. (1992). *Memory-enhancing techniques for investigative interviewing: The cognitive interview*. Springfield, IL: Chas. C. Thomas.

Florida Criminal Justice Standards and Training Commission. Retrieved from http://www.fdle.state.fl.us/cms/CJSTC/Curriculum/Active-Courses.aspx

Geiselman, R. E., & Nielsen, M. (1986). Cognitive memory retrieval techniques. *Police Chief, 53*(3), 191–206.

Gottschalk, P. (2007). Organizational culture as determinant of enterprise information systems use in police investigations. *Enterprise Information Systems, 1*(4), 443–455.

Jackall, R. (2000). Investigating criminal violence. *Social Research, 67*(3), 849–875.

Kneller, W., & Harvey, A. J. (2016). Lineup identification accuracy: The effects of alcohol, target presence, confidence ratings, and response time. *The European Journal of Psychology Applied to Legal Context*, 811–118.

Mason, M. (2016). The "preparatory" and "argumentation" stages of police interrogation: A linguistic analysis of a criminal investigation. *Language and Communication, 48*, 79–87.

Miller, S. (2014). Police detectives, criminal investigations and collective moral responsibility. *Criminal Justice Ethics, 33*(1), 21–39.

Miranda, D. (2015). Criminal investigation through the eye of the detective: Technological innovation and tradition. *Surveillance & Society, 13*(3/4), 422–436.

Poos, J. M., van den Bosch, K., & Janssen, C. P. (2017). Battling bias: Effects of training and training context. *Computers & Education, 111*, 101–113.

Tong, S., & Bowling, B. (2006). Art, craft and science of detective work. *Police Journal, 79*(4), 323–330.

Westera, N. J., Kebbell, M. R., Milne, B., & Green, T. (2016). Towards a more effective detective. *Policing & Society, 26*(1), 1–17.

Chapter 4

Bloch, P. B., & Bell, J. (1976). *Managing investigations: The Rochester system*. Washington, DC: The Police Foundation.

Braga, A. A., Flynn, E. A., Kelling, G. L., & Cole, C. M (2011). *Moving the work of criminal investigators towards crime control*. Harvard's Kennedy School of Government and the National Institute of Justice.

Cawley, D. F., Miron, H. J., Araujo, W. J., Wasserman, R., Mannello, T. A., & Huffman, Y. (1977). *Managing criminal investigations manual*. Washington, DC: National Institute of Law Enforcement and Criminal Justice.

Chaiken, J. M., Greenwood, P. W., & Petersilia, J. (1977). The criminal investigation process: A summary report. *Policy Analysis, 3*(2), 187–217.

Commonwealth of Pensylvania, Pennsylvania Justice Network. (2018). Retrieved from https://www.pajnet.pa.gov/Pages/About.aspx

Cook, T., & Tattersall, A. (2016). *Blackstone's senior investigating officers' handbook* (4th ed.). Oxford, England: Oxford University Press.

Davis, R. C., Jensen, C. J., Burgette, L., & Burnett, K. (2014). Working smarter on cold cases: Identifying factors associated with successful cold case investigations. *Journal of Forensic Sciences, 59*(2), 375–382.

Eck, J. E. (1983). *Solving crimes: The investigation of burglary and robbery*. Washington, DC: Police Executive Research Forum.

Gottschalk, P., Holgersson, S., & Karlsen, J. T. (2009). How knowledge organizations work: The case of detectives. *The Learning Organization, 16*(2), 88–102.

Greenberg, B., et al. (1975). *Felony investigation decision model: An analysis of investigative elements of information*. Menlo Park, CA: Stanford Research Institute.

Greenwood, P., Chaiken, J., & Petersilia, J. (1977). *The investigation process*. Lexington, MA: Lexington Books.

Groenendaal, J., & Helsloot, I. (2014). Investigating decision-making mechanisms and biases in Dutch criminal investigation teams by using a serious game. *Journal of Police Studies/Cahiers Politiestudies, 2*(2), 133–157.

Horvath, F., Meesig, R. T., & Lee, Y. H. (2001). *A national survey of police policies and practices regarding the criminal investigation process: Twenty-five years after RAND*. Washington, DC: National Institute of Justice, Office of Justice Programs.

Hough, R. M., McCorkle, K. D., & Harper, S. (2019). An examination of investigative practices of homicide units in Florida. *Homicide Studies, 23*(2), 175–194.

Hough, R. M., & Tatum, K. M. (2014). Murder investigation and media: Mutual goals. *Law Enforcement Executive Forum, 14*(3).

Kahneman, D. (2011). *Thinking, fast and slow*. New York, NY: Farrar, Straus and Giroux.

Kingshott, B. F., Walsh, J. P., & Meesig, R. T. (2015). Are we training our detectives? A survey of large law enforcement agencies regarding investigation training and training needs. *Journal of Applied Security Research, 10*(4), 481–509.

Liederbach, J., Fritsch, E. J., & Womack, C. L. (2011). Detective workload and opportunities for increased productivity in criminal investigations. *Police Practice & Research, 12*(1), 50–65.

McClellan, J. (2008). Evidence-based practice in criminal investigations. *Journal of Applied Security Research, 3*(1), 37–56.

McDevitt, D. S. (2005). *Managing the investigative unit*. Springfield, IL: Charles C Thomas, Publisher, Ltd.

Phillips, R. G. (1988, August). Training priorities in state and local law enforcement. *FBI Law Enforcement Bulletin*, 10–16.

Rassin, E., Eerland, A., & Kuijpers, I. (2010). Let's find the evidence: An analogue study of confirmation bias in criminal investigations. *Journal of Investigative Psychology & Offender Profiling, 7*(3), 231–246.

Reaves, B. A. (2015). Local police departments, 2013: Personnel, policies, and practices (NCJ 248677). U.S. Department of Justice, Office of Justice Programs, Bureau of Justice Statistics.

Rossmo, D. K. (2009). *Criminal investigative failures*. Boca Raton, FL: CRC Press.

Salet, R., & Terpstra, J. (2014). Critical review in criminal investigation: Evaluation of a measure to prevent tunnel vision. *Policing: A Journal of Policy & Practice, 8*(1), 43–50.

Simon, H. A. (1956). Rational choice and the structure of the environment. *Psychological Review, 63*(2), 129–138.

Sparrow, M. K., Moore, M. H., & Kennedy, D. M. (1990). *Beyond 911: A new era for policing*. New York, NY: Basic Books.

Stelfox, P. (2011). Criminal investigation: Filling the skills gap in leadership, management, and supervision. *Policing: A Journal of Policy & Practice, 5*(1), 15–22.

Willman, M. T., & Snortum, J. R. (1984). Detective work: The criminal investigation process in a medium-size police department. *Criminal Justice Review (Georgia State University), 9*(1), 33–39.

Womack, C. (2007). Criminal investigations: The impact of patrol officers on solving crime. (M.S. dissertation). University of North Texas.

Chapter 5

Bill of Rights. United States Constitution. (1789).

Brown v. Mississippi, 297 U.S. 278 (1936).

California v. Prysock, 453 U.S. 355 (1981).

Carlson, C. A., Carlson, M. A., Weatherford, D. R., Tucker, A., & Bednarz, J. (2016). The effect of backloading instructions on eyewitness identification from simultaneous and sequential lineups. *Applied Cognitive Psychology, 30*(6), 1005–1013.

Carroll v. United States, 267 U.S. 132 (1925).

Chimel v. California, 395 U.S. 752 (1969).

Doornbos, C. (2017). New Florida law requires photo-lineup conductors to be unaware of suspect's identity. *Orlando Sentinel.* Retrieved from http://www.orlandosentinel.com/news/breaking-news/os-eyewitness-identification-law-20170615-story.html

Florida v. Powell, 559 U.S. 50 (2010).

Forrest, K. D., & Woody, W. D. (2010, November). Police deception during interrogation and its surprising influence on jurors' perceptions of confession evidence. *The Jury Expert*, 9–20.

Frazier v. Cupp, 394 U.S. 731 (1969).

Illinois v. Caballes, 543 U.S. 405 (2005).

Innocence Project (2016). DNA exonerations worldwide. Retrieved from http://www.innocenceproject.org/Content/DNA_Exonerations_Nationwide.php

Mapp v. Ohio, 367 U.S. 643 (1961).

Maryland v. Buie, 494 U.S. 325 (1990).

Michigan Department of State Police v. Sitz, 496 U.S. 444 (1990).

Minnesota v. Dickerson, 508 U.S. 366 (1993).

Miranda v. Arizona, 384 U.S. 436 (1966).

Morris, M. (2012). The decision zone: The new stage of interrogation created by Berghuis v. Thompkins. *American Journal of Criminal Law, 39*(2), 271–299.

Mu, E., Chung, T. R., & Reed, L. I. (2017). Paradigm shift in criminal police lineups: Eyewitness identification as multicriteria decision making. *International Journal of Production Economics, 184*, 95–106.

New York v. Belton, 453 U.S. 454 (1981).

Nix v. Williams, 467 U.S. 431 (1984).

Oliver, W. M. (2016). Prohibition, stare decisis, and the lagging ability of science to influence criminal procedure. *Journal of Criminal Law & Criminology, 105*(4), 993–1030.

Oliver v. United States, 466 U.S. 170.

Saucier v. Katz, 533 U.S. 194 (2001).

Slobogin, C. (2017). Manipulation of suspects and unrecorded questioning: After fifty years of Miranda jurisprudence, still two (or maybe three) burning issues. *Boston University Law Review, 97*(3), 1157–1196.

Strömwall, L. A., & Willén, R. M. (2011). Inside criminal minds: Offenders' strategies when lying. *Journal of Investigative Psychology & Offender Profiling, 8*(3), 271–281.

Terry v. Ohio, 392 U.S. 1 (1968).

Thornton v. United States, 541 U.S. 615 (2004).

United States v. Ash, 413 U.S. 300 (1973).

United States v. Leon, 468 U.S. 897 (1984).

United States v. Martinez-Fuerte, 482 U.S. 543 (1976).

United States v. Wade, 388 U.S. 218 (1967).

Weeks v. United States, 232 U.S. 383 (1914).

Wells, G. L., Steblay, N. K., & Dysart, J. E. (2012). Eyewitness identification reforms: Are suggestiveness-induced hits and guesses true hits? *Perspectives on Psychological Science, 7*(3), 264–271.

Wells, G. L., Steblay, N. K., & Dysart, J. E. (2015). Double-blind photo lineups using actual eyewitnesses: An experimental test of a sequential versus simultaneous lineup procedure. *Law and Human Behavior, 39*, 1–14.

Wex Free Legal Dictionary. Legal Information Institute (LII) of the Cornell Law School, https://www.law.cornell.edu/search/site/fifth%2520amendment%2520rights

Wilson v. Arkansas, 514 U.S. 927 (1995).

Wixted, J. T., Mickes, L., Dunn, J. C., Clark, S. E., & Wells, W. (2016). Estimating the reliability of eyewitness identifications from police lineups. *Proceedings of The National Academy of Sciences of The United States of America, 113*(2), 304–309.

Chapter 6

Braga, A. A., & Dusseault, D. (2016). Can homicide detectives improve homicide clearance rates? *Crime and Delinquency, 64*(3), 283–315.

Brookman, F., Maguire, E. R., & Maguire, M. (2019). What factors influence whether homicide cases are solved? Insights from qualitative research with detectives in Great Britain and the United States. *Homicide Studies, 23*(1), 145–174.

Farrell, M. (2017). *Criminology of homicidal poisoning.* New York, NY: Springer.

Federal Bureau of Investigation. (2018). *Crime in the United States, 2017.*

Hickey, E. (2015). *Serial murderers and their victims* (7th ed.). San Francisco, CA: Wadsworth.

Hough, R. M., McCorkle, K. D., & Harper, S. (2019). An examination of investigative practices of homicide units in Florida. *Homicide Studies, 23*(1), 175–194.

Hough, R. M., & Tatum, K. M. (2014). Murder investigation and media: Mutual goals. *Law Enforcement Executive Forum, 14*(3).

Lum, C., & Nagin, D. S. (2017). Reinventing American policing. *Reinventing American Criminal Justice, 46*, 339–393.

National Institute of Mental Health. (2015). *Suicide in America.* NIH Publication No. QF 18-6389.

Paraschakis, A., Michopoulos, I., Douzenis, A., Christodoulou, C., Koutsaftis, F., & Lykouras, L. (2012). Differences between suicide victims who leave notes and those who do not: A 2-year study in Greece. *Crisis, 33*(6), 344–349.

Wellford, C., & Cronin, J. (1999). *An analysis of variables affecting the clearance of homicides: A multisite study.* Washington, DC: Justice Research and Statistics Association.

Chapter 7

Alvarez, A., & Bachman, R. (2017). *Violence: The enduring problem* (3rd ed.). Thousand Oaks, CA: Sage.

Baskin, D., & Sommers, I. (2012). The influence of forensic evidence on the case outcomes of assault and robbery incidents. *Criminal Justice Policy Review, 23*(2), 186.

Federal Bureau of Investigation. (2016). *Crime in the United States, 2015.*

Federal Bureau of Investigation. (2017). *Crime in the United States, 2016.*

Groth, A. N. (1979). Sexual trauma in the life histories of rapists and child molesters. *Victimology: An International Journal, 4*, 10–16.

Horvath, F., Meesig, R. T., & Lee, Y. H. (2001). *A national survey of police policies and practices regarding the criminal investigation process: Twenty-five years after RAND.* Washington, DC: National Institute of Justice, Office of Justice Programs.

Hough, R. M., & McCorkle, K. D. (2017). *American homicide.* Thousand Oaks, CA: Sage.

Jenkins, L. (1996). Violence in the workplace. *Current Intelligence Bulletin, 57*, 96–100. National Institute for Occupational Safety and Health.

National Forensic Science Technology Center. (2013). *Crime scene investigation: A guide for law enforcement.*

National Institute for Occupational Safety and Health. (1996). Violence in the workplace: Risk factors and prevention strategies. *Current Intelligence Bulletin, 57*, DHHS (NIOSH) Publication No. 96-100.

Polk, K. E. (1994). *When men kill: Scenarios of masculine violence.* Cambridge, England: Cambridge University Press.

U.S. Department of Justice. Office on Violence Against Women. (2013). *National protocol for sexual assault medical forensic examinations* (2nd ed.). Available from http://www.ncjrs.gov/pdffilesI/avw/24l903.pdf

U.S. Department of Justice. Office of Justice Programs. Bureau of Justice Statistics. (2017, December 14). *National crime victimization survey, 2016.* Ann Arbor, MI: Inter-university Consortium for Political and Social Research [distributor]. Retrieved from https://doi.org/10.3886/ICPSR36828.v1

Wolfgang, M. E. (1958). *Patterns of criminal homicide.* Philadelphia: University of Pennsylvania Press.

Chapter 8

Baum, K., Catalano, S., Rand, M., & Rose, K. (2009). *Stalking victimization in the United States.* (NCJ 224527). Retrieved from the CDC website at: https://www.cdc.gov/violenceprevention/pdf/NISVS_Report2010-a.pdf

Breiding, M. J., Smith, S. G., Basile, K. C., Walters, M. L., Chen, J., & Merrick, M. T. (2014). Prevalence and characteristics of sexual violence, stalking, and intimate partner violence victimization—national intimate partner and sexual violence survey, United States, 2011. *Morbidity and Mortality Weekly Report. Surveillance Summaries, 63*(8), 1–18.

Breitman, N., Shackelford, T. K., & Block, C. R. (2004). Couple age discrepancy and risk of intimate partner homicide. *Violence and Victims, 19*(3), 321–342.

Campbell, J. C. (2004). *Danger assessment.* Retrieved from http://www.dangerassessment.org

Daly, J. M. (2010). Violence and the elderly. In C. J. Ferguson (Ed.), *Violent crime: Clinical and social implications.* Los Angeles, CA: Sage.

Florida State Statute 784.049. Sexual cyberharassment.

International Association of Chiefs of Police. (2018). *Domestic violence: Model policy concepts and issues paper.* IACP Law Enforcement Policy Center.

McMahon, S., & Armstrong, D. Y. (2012). Intimate partner violence during pregnancy: Best practices for social workers. *Health & Social Work, 37*(1), 9–17.

National Center for Victims of Crime. (2002). *Creating an effective stalking protocol.* Washington, DC: U.S. Department of Justice, Office of Community Oriented Policing Services.

National Center for Victims of Crime. (n.d.). *Help for victims.* Stalking Resource Center. Available from http://victimsofcrime.org/our-programs/past-programs/stalking-resource-center/help-for-victims

United States Department of Justice. Office of Justice Programs. Bureau of Justice Statistics. (2018). *National Crime Victimization Survey, 2017.* Ann Arbor, MI: Inter-university Consortium for Political and Social Research

O'Leary, K. D., Smith-Slep, A. M., & O'Leary, S. G. (2007). Multivariate models of men's and women's partner aggression. *Journal of Consulting and Clinical Psychology, 75*(5), 752–764.

Smith, S. G., Chen, J., Basile, K. C., Gilbert, L. K., Merrick, M. T., Patel, N., Walling, M., & Jain, A. (2017). *The National Intimate Partner and Sexual Violence Survey (NISVS): 2010-2012 State Report.* Atlanta, GA: National Center for Injury Prevention and Control, Centers for Disease Control and Prevention.

U.S. Department of Commerce, U.S. Census Bureau. (2014). *An aging nation: The older population in the United States* (PDF).

U.S. Department of Health and Human Services, Administration for Children and Families, Administration on Children, Youth and Families, Children's Bureau. (2018). *Child maltreatment 2016.* Available from https://www.acf.hhs.gov/cb/research-data-technology/statistics-research/child-maltreatment

U.S. Department of Justice, Bureau of Justice Statistics. (2017). *Criminal victimization, 2016.* Revised (NCJ 252121). Retrieved from BJS website: https://www.bjs.gov/content/pub/pdf/cv16re.pdf

Chapter 9

Dimarino, F. (2015). White-collar and financial fraud investigations. In J. A. Eterno and C. Roberson (Eds.), *The detective handbook*. Boca Raton, FL: CRC Press/Taylor & Francis Group, pp. 117–132.

FBI.gov. *What we investigate, cyber crime*. Available from https://www.fbi.gov/investigate/cyber

FBI.gov. *What we investigate, white-collar crime*. Available from https://www.fbi.gov/investigate/white-collar-crime

Federal Bureau of Investigation. (2017). *Crime in the United States, 2016*. Retrieved from https://ucr.fbi.gov/crime-in-the.u.s/2017/crime-in-the.u.s.-2017

Strategic Plan, 2014/2018. [electronic resource]. (n.d.). U.S. Department of the Treasury, Financial Crimes Enforcement Network.

U.S. Department of Justice, Office of Public Affairs. *Justice Department coordinates nationwide elder fraud sweep of more than 250 defendants*. Press Release Number: 18–225.

U.S. Department of the Treasury, Financial Crimes Enforcement Network. *Strategic plan 2014–2018*.

USA.gov. *Identity theft*. Available from https://www.usa.gov/identity-theft

Chapter 10

Bonney, L. (2015). Undercover policing: Practices and challenges. *Journal of the Utah Academy of Sciences, Arts & Letters*, 92, 273–284.

Caulkins, J. P., Kilmer, B., Reuter, P. H., & Midgette, G. (2015). Cocaine's fall and marijuana's rise: Questions and insights based on new estimates of consumption and expenditures in US drug markets. *Addiction*, 110(5), 728–736.

Chase, C. A. (2015). Cops, canines, and curtilage: What *Jardines* teaches and what it leaves unanswered. *Houston Law Review*, 52(5), 1289–1312.

Chistova, L. E. (2017). The concept and tasks of the basic method of investigation of crimes related to illegal trafficking narcotic drugs. *Interaktivnaâ Nauka, Iss* 5(15), 141–144.

Darst, A. (2013). Covert investigations: Training to go un-noticed. *Kentucky Law Enforcement Magazine*, 12(4), 62.

Doran, G. S., Deans, R., De Filippis, C., Kostakis, C., & Howitt, J. A. (2017). Work place drug testing of police officers after THC exposure during large volume cannabis seizures. *Forensic Science International*, 275, 224–233.

Florida State Statute 874.03. Criminal gang enforcement and prevention, definitions.

Green, M. K., Kuk, R. J., & Wagner, J. R. (2017). Collection and analysis of fire debris evidence to detect methamphetamine, pseudoephedrine, and ignitable liquids in fire scenes at suspected clandestine laboratories. *Forensic Chemistry*, 4, 82–88.

Gustin, B. (2010, June). The hazards of grow houses. *Fire Engineering*, 69–72.

Hughes, C. E., Barratt, M. J., Ferris, J. A., Maier, L. J., & Winstock, A. R. (2018). Drug-related police encounters across the globe: How do they compare? *International Journal of Drug Policy*, 56, 197–207.

Kelly, B. D., & Kole, M. (2016). The effects of asset forfeiture on policing: A panel approach. *Economic Inquiry*, 54(1), 558–575.

Krause, M. (2009). History and evolution of the FBI's undercover safeguard program. *Consulting Psychology Journal: Practice & Research*, 61(1), 5–13.

Logue, D. (2008). The hidden badge: The undercover narcotics operation. *Law Enforcement Technology*, 35(2), 94.

Martyny, J. W., Serrano, K. A., Schaeffer, J. W., & Van Dyke, M. V. (2013). Potential exposures associated with indoor marijuana growing operations. *Journal of Occupational & Environmental Hygiene*, 10(11), 622–639.

Michael, J. D. (2008). Firefighting in clandestine drug labs. *Fire Engineering*, 6, 119–122.

Mirel, D. (2011). Meth: Is it the new mold? Clandestine drug labs continue to pose risks to residential rental units. *Journal of Property Management*, 3, 40.

NIDA. (2014, April 18). Principles of drug abuse treatment for criminal justice populations—a research-based guide. Retrieved from https://www.drugabuse.gov/publications/principles-drug-abuse-treatment-criminal-justice-populations-research-based-guide

Puddister, K., & Riddell, T. (2012). The RCMP's "Mr. Big" sting operation: A case study in police independence, accountability and oversight. *Canadian Public Administration*, 55(3), 385.

Reynolds, T. (2007). The Fourth Amendment: The appropriate use of drug dogs to search vehicles on school grounds. *Journal of Law & Education*, 36(4), 589–594.

Reynolds, V., & Carlson, M. S. (2018). Atlanta area DA: We need to do something about gangs. *Prosecutor: Journal of the National District Attorneys Association*, 1, 10.

Salas-Wright, C. P., Vaughn, M. G., Cummings-Vaughn, L. A., Holzer, K. J., Nelson, E. J., AbiNader, M., & Oh, S. (2017). Trends and correlates of marijuana use among late middle-aged and older adults in the United States, 2002–2014. *Drug and Alcohol Dependence*, 171, 97–106.

Schreiber, S. (2013). Reflections of an undercover officer. *Law Enforcement Technology*, 40(8), 20–24.

United Nations Office on Drugs and Crime. Retrieved from https://www.unodc.org/e4j/en/organized-crime/module-8/key-issues/special-investigative-techniques/undercover-operations.html

Chapter 11

Alvarez, A., & Bachman, R. (2016). *Violence: The enduring problem* (3rd ed.). Thousand Oaks, CA: Sage.

Bjelopera, J. P. (2017). *Domestic terrorism: An overview*. Congressional Reference Service, 7-5700, R44921.

Cahyani, N. D. W., Rahman, N. H. A., Glisson, W. B., & Choo, K. R. (2017). The role of mobile forensics in terrorism investigations involving the use of cloud storage service and communication apps. *Mobile Network Applications, 22*, 240–254.

Corner, E., & Gill, P. (2015). A false dichotomy? Mental illness and lone-actor terrorism. *Law and Human Behavior, 39*(1), 23–34.

Hough, R. M., & McCorkle, K. D. (2017). *American homicide*. Thousand Oaks, CA: Sage.

International Association of Chiefs of Police. (2008). *Homegrown violent extremism*. Awareness brief. Washington, DC: Office of Community Oriented Policing Services.

Jensen, T. (2016). National responses to transnational terrorism: Intelligence and counterterrorism provision. *Journal of Conflict Resolution, 60*(3), 530–554.

Laqueur, W. (1999). *The new terrorism: Fanaticism and the arms of mass destruction*. New York, NY: Oxford University Press.

Laqueur, W. (2006). Terrorism: A brief history. *Foreign Policy Agenda, 12*(5), 20–23.

Madon, N. S., Murphy, K., & Cherney, A. (2017). Promoting community collaboration in counterterrorism: Do social identities and perceptions of legitimacy mediate reactions to procedural justice policing? *British Journal of Criminology, 57*, 1144–1164.

Martin, G. (2003). *Understanding terrorism: Challenges, perspectives, and issues*. Thousand Oaks, CA: Sage.

Neumayer, E., & Plümper, T. (2011). Foreign terror on Americans. *Journal of Peace Research, 48*(1), 3–17.

Norris, J. J., & Grol-Prokopczyk, H. (2018). Entrapment allegations in right-wing terrorism cases: A mixed methods analysis. *International Journal of Law, Crime, and Justice, 53*, 77–88.

Parkin, W. S., Gruenewald, J., & Jandro, E. (2017). Extremist violence from the fatherland to the homeland: A comparison of far-right homicide in Germany and the United States. *International Criminal Justice Review, 27*(2), 85–107.

Poland, J. M. (2011). *Understanding terrorism: Groups, strategies, and responses* (3rd ed.). Upper Saddle River, NJ: Prentice Hall.

Spitzmuller, M., & Park, G. (2018). Terrorist teams as loosely coupled systems. *American Psychologist, 73*(4), 491–503.

U.S. Department of Justice. (2018). *Top management and performance challenges facing the Department of Justice-2018*. FY 2018 Agency Financial Report.

Chapter 12

Baumer, E. P., Messner, S. F., & Felson, R. B. (2000). Role of victim characteristics in the disposition of murder cases. *Justice Quarterly, 17*(2), 281–307.

Berman, G., Narby, D. J., & Cutler, B. L. (1995). Effects of inconsistent eyewitness statements on mock-jurors' evaluations of the eyewitness, perceptions of defendant culpability and verdicts. *Law and Human Behavior, 19*(1), 79–88.

Black's Law Dictionary. (1999). (7th ed.). St. Paul, MN: West Academic Publishing, p. 1053.

Brewer, T. (2005). The attorney–client relationship in capital cases and its impact on juror receptivity to mitigation evidence. *Justice Quarterly, 20*(2), 74–88.

Briody, M. (2004). The effects of DNA evidence on homicide cases in court. *Australian & New Zealand Journal of Criminology, 37*(2), 231–252.

Bureau of Justice Statistics. (2007). Special report: State court processing statistics, 1990–2004, pretrial release of felony defendants in state courts and CJ 214994. Washington, DC: U.S. Department of Justice.

Campbell, J. F. (2007). Ethical concerns in grooming the criminal defendant for the witness stand. *Hofstra Law Review, 36*(2), 265–274.

Chavez, L., & Miller, M. (2009). Religious references in death sentence phases of trial: Two psychological theories that suggest judicial rulings and assumptions may affect jurors. *Criminal Justice Periodicals, 13*(4), 1037–1083.

Costanzo, M., & Peterson, J. (1994). Attorney persuasion in the capital penalty phase. *Journal of Social Issues, 3*, 305–316.

Curtis, C. (2014). Public understandings of the forensic use of DNA: Positivity, misunderstandings, and cultural concerns. *Bulletin of Science, Technology & Society, 34*(1–2), 21–32.

Delattre, E. J. (1996). *Character and cops*. Washington, DC: AEI Press.

Duck, W. (2009). "Senseless violence": Making sense of murder. *Ethnography, 10*(4), 417–434.

Foley, L. A., & Powell, R. S. (1982). The discretion of prosecutors, judges, and juries in capital cases. *Criminal Justice Review, 7*(2), 16–22.

Gallagher, J. (2002). Homophobia for the defense. *Law Review, 39*(5), 34–37.

Green, B., & Zacharias, F. (2008). "The U.S. attorneys scandal" and the allocation of prosecutorial power. *Ohio State Law Journal, 69*.

Hartley, R., Miller, H., & Spohn, C. (2010). Do you get what you pay for? Type of counsel and its effects on criminal court outcomes. *Journal of Criminal Justice, 38*(5), 1063–1070.

Hoffman, M. B., Rubin, P. H., & Shepherd, J. M. (2005). An empirical study of public defender effectiveness: Self-selection by the "marginally indigent." *Ohio State Journal of Criminal Law, 3*, 223–255.

Levenson, L. (1998). Working outside the rules: The undefined responsibilities of federal prosecutors. *Fordham Urban Law Journal, 26*(3), 551–572.

Litwin, K. J. (2004). A multilevel multivariate analysis of factors affecting homicide clearances. *Journal of Research in Crime and Delinquency, 41*(4), 327–351.

Mesmaecker, V. (2012). Antidotes to injustice? Victim statements' impact on victims' sense of security. *International Review of Victimology, 18*(2).

Newman, D. W. (1995). Jury decision making and the effect of victim impact statements in the penalty phase. *Criminal Justice Policy Review, 7*(3–4), 291–300.

O'Neill, M. (2004). Understanding federal prosecutorial declinations. *The American Criminal Law Review, 41*(4), 1439–1498.

Platania, J., & Berman, G. L. (2006). The moderating effect of judge's instructions on victim impact testimony in capital cases. *Applied Psychology in Criminal Justice, 2*(2), 84–101.

Pope, P. B. (2011). Prosecutorial investigation standards. *Criminal Justice, 26*(1), 4–11.

Pyrooz, D. C., Wolfe, S. E., & Spohn, C. (2011). Gang-related homicide charging decisions: The implementation of a specialized prosecution unit in Los Angeles. *Criminal Justice Policy Review, 22*(1), 3–26.

Rasmusen, E., Raghav, M., & Ramseyer, M. (2009). Convictions versus conviction rates: The prosecutor's choice. *American Law and Economics Review, 11*(1), 47–78.

Roberts, J. (2013). Effective plea bargaining counsel. *The Yale Law Journal, 122*(8), 2650–2673.

Rose, M., Nadler, J., & Clark, J. (2006). Appropriately upset? Emotion norms and perception of crime victims. *Criminal Law Review, 30*(2), 203–219.

Stauffer, A. R., Smith, M. D., Cochran, J. K., Fogel, S. J., & Bjerregaard, B. (2006). The interaction between victim race and gender on sentencing outcomes in capital murder trials. *Homicide Studies, 10*(2), 98–117.

Torry, Z. D., & Billick, S. B. (2010). Overlapping universe: Understanding legal insanity and psychosis. *Psychiatric Quarterly, 81*(3), 253–262.

Twiss, R. (2007). A view from the bench: Keys to a successful direct. *The Army Lawyer, 3*(2), 28–41.

Ulmer, J. T., & Johnson, B. (2004). Sentencing in context: A multilevel analysis. *Criminology, 42*(1), 137–177.

Van Patten, J. (2010). Suing the prosecutor. *South Dakota Law Review, 55*(2), 214–252.

Vecchi, G. (2009). Principles and approaches to criminal investigation, part 1. *Forensic Examiner, 18*(2), 8–13.

Weiss, S. J. (2005). *Missouri v. Seibert*: Two-stepping towards the apocalypse. *Journal of Criminal Law & Criminology, 95*(3), 945–984.

Wheatcroft, J. M., & Ellison, L. E. (2012). Evidence in court: Witness preparation and cross examination style effects on adult witness accuracy. *Behavioral Sciences & the Law, 30*(6), 821–840.

Index

Abductive inferential reasoning, 60
Accelerant, 35
Accused, rights of, 57, 62, 64–65, 99, 254. *See also* Counsel, right to; Suspects; *Individual amendments*
Actus reus, 56
Adcock. J. M., 26
Administration on Aging (AoA), 181
Admission, 66
Affinity fraud, 197
Affirmative defense, 257
African Americans, disproportionate punishment of, 211
Agencies
    contemporary, 16
    drug crimes and, 215–216
    early, 4
    interaction with Muslim citizens, 240
    responsibilities of, 9–10
    sharing of information and intelligence, 240–241
    size of, 14, 77, 209
    training and, 93
Agencies, federal
    development of, 5
    homeland security and, 239–240
    responsibilities of, 9–10
    See also Federal Bureau of Investigation (FBI)
Agencies, local
    homeland security and, 239
    size of, 14
    terrorism and, 229
Agencies, state
    development of, 4
    drug crimes and, 210
    gang activity and, 210
    homeland security and, 239
    responsibilities of, 9
Aggravated assault, 146–148, 147 (table). *See also* Intimate partner violence (IPV)
Alcohol, 210
Algor mortis, 129
Alibi, 257
Al-Qaeda, 234, 235
Alternate light source (ALS), 36
Alvarez, A., 155, 234
American Automobile Association (AAA), 151
American Bar Association (ABA), 5, 252, 256, 261
American Society of Industrial Security (ASIS International), 196
American Society of Lab Directors, 38
Anger-excitation rapist, 157
Anger-retaliatory rapist, 156
Animal Liberation Front (ALF), 233
Anthropometry, 27
Application for search warrant, 100
Army of God (AOG), 233
Arrest
    in IPV, 172
    for possession of drugs, 214
    resisting, 107
Arson, 15, 35
Aryan Nations (AN), 233
ASIS International (American Society of Industrial Security), 196
Asphyxiation, 131–132
Assault, 145, 146–151. *See also* Intimate partner violence (IPV)
Asset forfeiture, 218–219
Asset protection, 195
Assumptions, 86, 88
Autoerotic asphyxiation, 133
Automated Fingerprint Identification System (AFIS), 32
Autopsy, 128
Avulsion, 131
Awareness Brief, Homegrown Violent Extremism (HVE), 233

Bachman, R., 155, 234
Bail, 249
Ballistics, 28, 34
Bank Secrecy Act (BSA), 199–200
Battery, 145, 146–151
Behavioral Analysis Units (BAU), 39
Behavioral science unit (BSU), 138
Be on the lookout (BOLO) broadcast, 125, 148
Bertillon, Alphonse, 27
Best practices, 18
Beyond 911: A New Era for Policing (Sparrow, Moore, and Kennedy), 92
Bias, 258
Biker gangs, 220. *See also* Gangs
Bill of Rights, 99

Biological theories of crime, 54
Biometrics, 28
Blood feud, 52
Blunt force trauma, 131
Body language, 66, 67, 86
Body-worn cameras (BWC), 41, 42
Boko Haram, 235
BOLO (be on the lookout) broadcast, 125, 148
Bomb Threat Checklist, 236
Boston Marathon bombing, 235
Bow Street Runners, 3
Brickner, D. R., 60
Brodeur, J. P., 58
Broken windows concept, 17
Brookman, F., 58
Brown v. Mississippi, 110
Burden of proof, 254
Bureau of Alcohol, Tobacco, Firearms and Explosives (ATF), 9, 15
Bureau of Immigration, 5
Bureau of Justice Assistance, 79, 91
Bureau of Justice Statistics (BJS), 18, 169
    crime statistics, 7
    Law Enforcement Management and Administrative Statistics (LEMAS), 18, 77
    on pretrial release, 249
    on size of agencies, 14
Burglars, types of, 191
Burglary, 187, 188, 189–192, 193

Cahyani, N. D. W., 236
California v. Prysock, 113
Cameras, 28
Campbell, Jacqueline, 182
Cannabis grow houses (CGH), 217–218
Cannabis/marijuana, 210, 211. *See also* Drugs, illegal
Canvass, 126, 152
Capture technology, 41
Cardiograph, 68
Carjacking, 151, 195
Carroll Doctrine, 106
Carroll v. United States, 106
Caseload management, 82, 134
Case management, 14. *See also* Investigation, managing
Case management software (CMS), 40, 84–85, 98, 127
Cases, court. *See* Courtroom trials
Census Bureau, 181
Centers for Disease Control and Prevention (CDC), 121, 174, 211
Chaiken, J., 82
Challenges for cause, 250
Chancellor, A. S., 26
Charges, filing, 248. *See also* Courtroom trials
Charging decisions, 251–253
Charleston church shooting, 138, 139

Checklists, 127, 168
    Bomb Threat Checklist, 236
    Domestic Violence Report Checklist, 168, 169
Checkpoints, 105
Child abuse, 177–181
    incidence of, 177
    investigating, 179–180
    legislation, 178–179
    Munchausen Syndrome by Proxy (MSBP), 135, 180–181
    relationship with IPV, 171
    risk factors for, 178
    sexual abuse of minors, 180
    See also Domestic violence (DV)
Child Maltreatment Report, 177
Child pornography, 15
Chimel v. California, 103, 107
Choking, 132, 172
Choo, K. R., 236
Citizens, role of, 17
Civilian employees, 90. *See also* Crime scene technicians
Civil rights violations
    investigation of, 10
    See also Individual amendments
Clandestine labs, 217
Class characteristics, 32
Clearance, 7, 7 (figure), 76, 90, 218
    of aggravated assault cases, 147–148
    of burglary cases, 189
    of homicide cases, 135
    investigation as process and, 77–79
    of robbery cases, 154
Closed-circuit television (CCTV), 193
Closed-mindedness, 87
Closing summary, 250
CMS (case management software), 40, 84–85, 98, 127
Cocaine, 211. *See also* Drugs, illegal
Cognitive biases, 86
Cognitive interview technique, 61
Cold cases, 14, 139
    defined, 7
    DNA evidence and, 32, 137–138
    homicide cases, 137
    technology and, 137
    witness interaction and, 79
College rape, 159
Combined DNA Index System (CODIS), 29, 33, 137
Community Oriented Policing Services (COPS), 16–17, 18, 222
Comprehensive Drug Abuse Prevention and Control Act, 210, 215
Computer-aided dispatch (CAD) system, 40, 84
Computer-aided drawing (CAD), 42
Computer-assisted drafting and design (CADD), 42
Computers, 40
Computer voice stress analyzer (CVSA), 68

Conan Doyle, Arthur, 27
Confession, 66
　voluntariness of, 109
Confidential informant (CI), 62, 82
Confirmation bias, 86, 87–88
Confrontational violence, 150–151
Congressional Reference Service (CRS), 232
Consent, 102
Constitution, U.S., 99, 254. *See also Individual amendments*
Continuing Criminal Enterprise Statute, 215
Controlled buys, 216
Controlled substances, 211. *See also* Drug crimes; Drugs, illegal
Controlled Substances Act (CSA), 211
Control question technique (CQT), 68
Control specimen, 32
Contusion, 131
Conveyance, 189
Cook, T., 58, 85
Cornell Law School, 108
Coroner, 120, 127–130. *See also* Death investigations
Corporate fraud, 197
Counsel, right to, 112, 255. *See also* Defense attorneys
Court cases. *See* Courtroom trials
Courtroom trials
　arraignment stage, 249–250
　defenses, 256–258
　ethical issues, 258–259
　evidentiary issues, 254–256
　plea bargains, 249–250, 259–261
　preparing for, 10
　pretrial stage, 249
　steps in, 261
Crime
　changing nature of, 92
　clearance of cases. *See* Clearance
　discovering, 10
　local nature of, 76
　theories of, 54
　unreported, 7
Crime, reported, 7, 7 (figure), 10, 52, 121–122, 200
　Uniform Crime Report (UCR), 121, 135, 147, 150, 154, 155, 187, 189, 192, 195, 196, 213
Crime analysis, 78–80
Crime fiction, 27
Crime funnel, 251
Crime in the United States, 2016 (FBI), 155
Crime mapping, 40
Crime prevention through environmental design (CPTED) movement, 17
Crime rate, 17, 76, 121–122, 188 (figure)
Crime scene
　access to, 31
　contamination of, 11–12

　documenting, 31, 40–44, 149. *See also* Photography; Sketches
　establishing perimeter of, 29
　multiple, 30–31
　outdoor, 29
　personnel safety and, 37, 213–214, 217
　processing, 36–37, 38–39
　searches of, 29–31, 30 (figure)
　securing, 29, 37, 79, 125–126
　vehicle as, 195
Crime scene tape, 37
Crime scene technicians, 13, 29, 31, 38–39, 90
Crime scene unit (CSU), 13
Crime Victims' Rights Act (CVRA), 53
Criminal investigation
　in court. *See* Courtroom trials
　defined, 5, 247–248
　goals of, 6–8
　history of, 3–6
　stages of, 10
　successful, 7–8. *See also* Clearance
　types of, 8–10
　See also Investigation
Criminal Investigation (Gross), 28
Criminal Investigative Failures (Rossmo), 86
Criminalists, 39
Criminal Justice Standards and Training Commission, 110
Criminal justice system
　goal of, 53
　perceptions of, 248
Criminal mischief, 190
Criminal street gang, 221. *See also* Gangs
Criminal street gang associate, 221
Criminal street gang member, 221
Critical infrastructure, 239
Cronin, J., 127
Cross-examination, 255
CSI (television series), 38
CSI Effect, 41, 44–45, 254
Curtilage, 31, 105
Curtis, C., 254
Cybercrime, 187, 200–203
Cyberstalking, 172

Danger Assessment, 171, 182
Daubert v. Merrell Dow Pharmaceutical, 36
D.C. Sniper shootings, 16
Death
　manner of, 120, 127–128
　medical classifications of, 123–124
　time of, 128–130
　See also Homicide; Suicide
Death investigations, 120, 122
　challenges in, 134–136
　clearing homicide cases, 135

cold cases, 137
death notifications, 136
equivocal death, 134–135
establishing time of death, 128–130
identifying victim, 126
intimate partner homicides (IPH), 138
investigators/detectives in, 127, 134
managing suspects in, 125
media and, 135–136
medical examiner/coroner (ME/C), 127–130
methods for, 130–133
preliminary investigation in, 124–126
serial murders, 138
specialists in, 130
Sudden Infant Death Syndrome (SIDS), 134, 180
Sudden Unexplained Infant Death (SUID), 134, 180
suicide, 133–136
survivors, 136
See also Homicide; Medical examiner/coroner (ME/C); Murder
Death notifications, 136
Death penalty, 259, 261
Deception, detection of, 67–68
Decision-making, 87, 88
Deductive reasoning, 60
Defense attorneys, 249–250
ethics and, 259
functions and tactics of, 254–256
interviews and, 63
lineup procedures and, 61
private counsel, 260
public defenders, 255, 260–261
See also Courtroom trials; Plea bargains
Defense of others, 257
Defenses, 256–258
Delattre, E. J., 258
Democracies, 5, 16
Department of Health and Human Services, 177
Department of Homeland Security (DHS), 233, 239, 240, 241
Department of Justice (DOJ), 16, 197
Bureau of Justice Assistance, 79, 91
creation of, 5
National Institute of Justice (NIJ), 19, 33, 213
Office of Community Oriented Policing Services (COPS), 17, 18, 175, 176
Office of Sex Offender Sentencing, Monitoring, Apprehending, Registering and Tracking (SMART), 161
Office of the Inspector General (OIG), 233, 240
Office on Violence Against Women, 160, 173
terrorism and, 232
violations investigated by, 10
Department of the Treasury, 5, 198
Department of Transportation (USDOT), 151

Detection of deception devices, 67–68
Detectives. *See* Investigators/detectives
Diagramming, 42–43
Differential association theory, 54
Direct examination, 255, 256
Discovery, inevitable, 106–107
Discovery process, 63
Discussion Questions
Assault, Battery, Robbery, and Sex Crimes, 163
Burglary, Theft, White-Collar Crime, and Cybercrime, 204
Criminal Investigation in Court, 263
Criminal Investigation Then and Now, 21
Drug Crimes, Organized Crime, and Gangs, 224
Familial Crimes, 183
Forensic Evidence, 46
Homicide, Death, and Cold Case Investigations, 140
Managing the Criminal Investigation, 94
People in the Process, 70
Searches, Seizures, and Statements, 114
Terrorism and Homeland Security, 242
District attorney. *See* Prosecutors
DNA evidence, 254
cold cases and, 32, 137–138
collecting and preserving, 33
Combined DNA Index System (CODIS), 29, 33, 137
ethics and, 32
use of, 28–29
Documentation, 31
of aggravated assault, 147–148
case management software, 40, 84–85, 98, 127
of crime scene, 40–44, 149
informants and, 62
photography/video, 41–42
of robbery, 152
of sexual battery/rape, 158
of witness interviews, 63
Dogs, drug, 106, 214, 217
Domestic violence (DV)
arrest in, 172
described, 167–168
elder abuse, 181–182
specialized prosecution and, 253
See also Child abuse; Intimate partner violence (IPV)
Domestic Violence Report Checklist, 168
Drowning, 132
Drug cartels, 234
Drug crimes, 210
asset forfeiture, 218–219
controlled buys, 216
drug laboratories, 217–218
evidence, 215
grow houses, 217–218
investigating, 214–219

knock and talk, 218
risk to personnel, 217
specialized prosecution and, 253
undercover operations, 215–216
Drug dogs, 106, 214, 217
Drug Enforcement Administration (DEA), 9, 210, 211, 217, 223
Fentanyl Briefing Guide, 213–214
Drug laboratories, 217–218
Drugs, illegal
danger to personnel, 213–214
dealers, 214
defining and classifying controlled substances, 211
exposure to, 217
history of, 210–211
schedules, 211, 213
scope of problem, 213
Drug trafficking, 15
DV (domestic violence). *See* Child abuse; Domestic violence (DV); Intimate partner violence (IPV)
Dyer Act, 196

Earth Liberation Front (ELF), 233
Eck, J. E., 13
Edgar Hoover, J., 5
Edged or penetrating weapons, 131
Eighteenth Amendment, 210
Elder abuse, 181–182
Elderly, fraud targeting, 197–198
Election interference, 202–203
Electronic article surveillance (EAS), 193, 194
Ellison, L. E., 255
Embezzlement, 195, 196, 198–199. *See also* White-collar crime
Emotional abuse, 180
Employee theft, 194–195
Environmental Protection Agency (EPA), 217
Equivocal death, 134–135
Evidence
admissible, 6, 248
collection, methods of, 29–31
compiling, 5
defined, 25
inadmissible, 106–107, 254
knock and talk, 218
narcotics, 215
planting, 258
processing, 32–36, 40–41
scientific standards for, 254
searching for, 6
types of, 5–6, 254
See also Eyewitnesses; Witnesses
Evidence, circumstantial/indirect, 6, 40–41, 254
Evidence, direct, 6, 254
Evidence, false, 258

Evidence, forensic. *See* Forensic evidence
Evidence, physical
admissible, 248
arson/explosives and, 35
firearms, 33–35
identifying suspect and, 56–57
lack of, 80
in preliminary investigation, 79–80
processing, 32–36
of rape, 157, 158
types of, 32
See also DNA evidence
Evidence, testimonial, 40, 248
Evidence-based policing, 18
Evidence-based research, 89
Excited utterance, 109
Exclusionary rule, 106–107
Excuse defenses, 256, 257
Exhibitionism, 159
Exigent circumstances, 102, 103–104, 110
Expert witnesses, 35–36
Explore This
American Bar Association public education page, 261
cold case/unsolved murder investigations, 139
Danger Assessment, 182
detective training, 69
Drug Enforcement Administration, 223
FBI's Handbook of Forensic Services, 45
Federal Law Enforcement Training Centers, 93
homegrown violent extremists, 241
identity theft, 203
intimate partner violence (IPV), 182
lineup procedures, 113
National Institute of Justice (NIJ), 19
Office of Sex Offender Sentencing, Monitoring, Apprehending, Registering and Tracking (SMART), 161
Explosives, 35, 235, 236
Extortion, 219–220
Extremism, 231
hate crimes and, 237
See also Terrorism
Eyewitnesses, 60–62
lineup procedures and, 61
reliability of, 110–112
See also Lineups; Showups

Facial expressions, 86
Failure of proof defenses, 256–257
Fair and Accurate Credit Transactions Act (FACT), 200
Familial crimes. *See* Child abuse; Domestic violence (DV); Intimate partner violence (IPV)
Farrell, M., 133
FBI (Federal Bureau of Investigation). *See* Federal Bureau of Investigation (FBI)

Federal Bureau of Investigation (FBI)
   behavioral science unit (BSU), 138
   Crime in the United States report, 121
   crime statistics, 7
   development of, 5
   Handbook of Forensic Services, 45
   investigative priorities of, 5
   Joint Terrorism Task Force (JTTF), 16
   National Crime Information Center (NCIC), 28
   protocols for working undercover, 216
   rape classifications, 156
   responsibilities of, 9
   terrorism and, 232, 233
   Uniform Crime Report (UCR), 121, 135, 147, 150, 154, 155, 187, 189, 192, 195, 196, 213
   Violent Criminal Apprehension Program (ViCAP), 15–16, 39
Federal Law Enforcement Training Centers (FLETC), 57
Federal Trade Commission (FTC), 197, 199, 203
Fence, 193
Fentanyl, 213
Fentanyl Briefing Guide, 213–214
Fetishism, 160
Field identification, 57
Fielding, James, 3
Field/presumptive tests, 26
Field training officer (FTO), 2
Fifth Amendment, 64, 98, 99, 106, 108–112. *See also* Miranda warning; Statements
Fighting words, 150
Financial Crimes Enforcement Network (FinCEN), 198
Fingerprints, 27–28
   Automated Fingerprint Identification System (AFIS), 32
   equipment for gathering, 36
   Integrated Automatic Fingerprint Identification System (IAFIS), 28
Firearms, 33–35, 130–131. *See also* Ballistics
Fires. *See* Arson
First 48 (television series), 25
First responders, 11–13. *See also* Patrol officers
Fisher, B. A. J., 34
Florida v. Powell, 113
Follow-up investigation, 13–14, 76, 80, 81–83
Food and Drug Administration (FDA), 211
Forensic evidence
   CSI Effect and, 44–45
   expectations of vs. reality, 41, 44–45
   overreliance on, 26
   See also DNA evidence; Evidence
Forensic sciences, 28
Forgery, 196. *See also* White-collar crime
Formal charges, 10
Fourteenth Amendment, 99

Fourth Amendment, 31, 32, 99, 104, 105, 106, 107. *See also* Probable cause; Searches; Search warrants; Seizures; Unreasonable searches and seizures
Framing, 87
Fraud, 196, 197–198. *See also* White-collar crime
Frazier v. Cupp, 110
Free speech, 150
Frisk, 104
Fritsch, E. J., 93
Frotteurism, 160
Fruit of the poisonous tree, 106, 254
Fusion centers, 240–241

Galton, Francis, 27–28
Galvanic skin response, 68
Gambling, 219
Gangs, 220–223, 253
Genocide, 230
Glisson, W. B., 236
Goddard, Calvin, 28
Good cop–bad cop, 67
Good faith exception, 106
Graffiti, 221
Grand jury system, 249
Grand theft auto, 195–196
Greenberg, B., 80
Greenpeace, 233
Greenwood, P., 11, 80, 82
Grid pattern, 30
Grol-Prokopczyk, H., 232
Gross, Hans, 28
Groth, A. N., 156
Groupthink, 86, 87
Grow houses, 217–218
Guilty mind, 258
Guilty plea, 249, 250, 259
Gunshot residue (GSR), 34, 131
Gun violence, 253

Hanging, 132
Haptics, 66
Harrison Narcotics Act, 5
Hate crimes, 230, 233, 237
   compared to terrorism, 232
   motivations for, 240
Heroin, 211. *See also* Drugs, illegal
Heuristics, 86
Hickey, E., 138
Hoffman, M. B., 260
HOLMES (Home Office Large Major Enquiry System), 85
Holmes, Sherlock, 27, 58
Homegrown violent extremists (HVE), 233, 241
Homeland security, 239–241. *See also* Hate crimes; Terrorism

Homicide, 120, 124
    clearance rate, 135
    cold cases, 137
    establishing time of death, 128–130
    intimate partner homicides (IPH), 138
    media and, 89, 248, 251
    methods of investigation, 130–133
    vs. murder, 122
    research into, 120–122
    sentencing options, 261
    trials, 251
    victim-precipitated, 150
    See also Death investigations
Homicide, confrontational, 150
Homicide, excusable, 123
Homicide, justifiable, 123
Homicide, negligent, 122
Homicide rates, 119–120
Hostage-taking, 152
Hot pursuit, 102, 103
Hough, R., 136, 150, 240
How It's Done
    bomb threat checklist, 236
    burglary investigation, 192
    crime scene documentation, 149
    crime scene photography, 42
    determining level of risk for intimate partner violence, 170
    first responding officers, 12
    handling narcotics evidence, 215
    interrogation, 67
    investigating suspected stalking crimes, 175
    live lineup, 112
    National Protocol for Sexual Assault Medical Forensic Examinations, 156
    organized retail crime (ORC), 194
    overarching principles for prosecutors, 252
    presumptive/field tests, 26
    processing crime scenes, 31
    search warrant exceptions, 102
    sexual assault examination, 160–161
    solvability factors, 81
    stalking protocol, 176
    undercover investigations, 216
    victims' rights, 55
Human trafficking, 15
Hypnosis, 61
Hypoxia, 132

ICE (Immigration and Customs Enforcement), 199
Identity theft, 199–200
Illinois v. Caballe, 106
Immigrants, 4
Immigration, 5, 199
Immigration and Customs Enforcement (ICE), 199

Incident to arrest, 102, 107
Incision wounds, 131
Independent source exception, 107
Individual characteristics, 32
Indoor marijuana growing operations (IMGO), 217
Inductive reasoning, 60
Industrial Revolution, 4
Inevitable discovery, 106–107
Informants, 62, 82
Information communication technology (ICT), 236
Information managers, 58
Initial court appearance, 249
Innes, M., 58
Innocence Project of Florida, 110
Insanity, 258
Integrated Automatic Fingerprint Identification System (IAFIS), 28
Intelligence-led policing, 18, 222
Internal Revenue Service (IRS), 16, 60
Internal security personnel, 195
International Association of Chiefs of Police (IACP), 12, 110
    Awareness Brief, Homegrown Violent Extremism (HVE), 233
    determining level of risk for intimate partner violence, 170
    Domestic Violence Report Checklist, 168, 169
    Model Policy on Suicide Bombings, 236
    See also Checklists
Internet
    cybercrime and, 200–203
    undercover investigations conducted on, 216
Internet Research Agency, 202–203
Interrogation, 67
    accused's rights in, 109. *See also* Fifth Amendment; Sixth Amendment
    deception by officers during, 110
    detecting deception in, 67–68
    good cop–bad cop, 67
    vs. interview, 62
    methods, 65–67
    Miranda warning, 57, 62, 64–65, 254
    planning, 65
    See also Confession; Statements
Interviews, 61, 62–64
    cognitive interview technique, 61
    detecting deception in, 67–68
    rape victims and, 158
Intimate partner homicides (IPH), 138
Intimate partner violence (IPV), 147, 148, 167
    arrest in, 172
    assessing danger of, 170, 171, 182
    changing views on, 170–171
    LGBTQ victims, 176
    primary aggressor in, 172
    rates of, 169–170

relationship with child abuse, 171
reporting of, 170
responding officers' safety, 173
restraining orders, 171–172
stalking, 173–175
victims in, 55, 171, 172, 176
See also Assault; Domestic violence (DV); Sexual assault
Investigation
of burglary, 190–191, 192
of child abuse, 179–180
of cybercrimes, 201–202
of drug crimes, 214–219
follow-up, 13–14, 76, 80, 81–83
management of. *See* Investigation, managing
of motor vehicle theft, 196
of organized crime, 220
preliminary, 10, 11–13, 79–80, 124–126
of stalking crimes, 175
of terrorist crimes, 235, 236, 240–241
of theft, 192–193
of white-collar crime, 196–197
See also Criminal investigation
Investigation, managing
avoiding failures, 88–89
case assignment, 80
coordination with prosecutor, 83
follow-up investigation, 81–83
improving productivity, 89–93
investigation as process and, 77–79
managing available resources, 78, 153
preliminary investigation, 79–80
relationship with media and, 89. *See also* Media
selection of detectives, 83–84, 134
skill sets for, 84
sources of information, 82–83
supervising investigative unit, 83–84
weaknesses/failures in, 86–89
working with external agencies, 85
See also Case management software (CMS)
Investigative detention, 104
Investigative unit, organizing, 14–15
Investigators/detectives, 13
caseload management, 82, 134
characteristics of, 58
in death investigations, 127
general vs. specialized, 14
at initial crime scene, 13
perception of, 52, 77
responsibilities of, 77
role of, 29, 57–58
selection of, 83–84, 134
shifting functions away from, 90
soft skills of, 59
specialized, 14, 81–82
training of, 57, 59–60, 69, 83, 88, 91, 93

Jacob Wetterling Crimes Against Children and Sexually Violent Offender Registration Act., 155
Jeanne Clery Disclosure of Campus Security Policy and Campus Crime Statistics Act, 159
Jeffries, Alec, 28
Joint Terrorism Task Force (JTTF), 16, 235
Jurisdiction, 15
drug crimes and, 210
gang activity and, 210
task force investigations and, 15
Juror selection, 44, 250
Jury, 251, 261
CSI Effect and, 44–45
DNA evidence and, 254
evidence and, 44–45
eyewitness identification and, 61
witness's testimony and, 253
Justice Department. *See* Department of Justice (DOJ)
Justice Network (JNET), 84
Justification defenses, 256, 257
Juveniles
burglary and, 191
drug crimes and, 214

Kahneman, D., 86
Kanka, Megan, 155
Kansas City Preventive Patrol Experiment, 18
Kennedy, D. M., 92
Key Terms
Assault, Battery, Robbery, and Sex Crimes, 162
Burglary, Theft, White-Collar Crime, and Cybercrime, 203
Criminal Investigation in Court, 262
Criminal Investigation Then and Now, 21
Drug Crimes, Organized Crime, and Gangs, 223
Familial Crimes, 183
Forensic Evidence, 46
Homicide, Death, and Cold Case Investigations, 139
Managing the Criminal Investigation, 94
People in the Process, 69–70
Searches, Seizures, and Statements, 114
Terrorism and Homeland Security, 242
Kinesics, 66
Kingpin Statute, 215
Kingshott, B. F., 91
Knives, 131
Knock and announce, 100
Knock and talk, 218
KQED Public Television, 261

Laboratories, crime, 26, 31
availability of, 39
contemporary, 37–38
development of, 28
establishment of, 12
personnel, 39

Laboratory technicians, 39
Lane search, 30, 30 (figure)
Laqueur, W., 231
Larceny/theft. *See* Theft
Las Vegas shooting, 138, 139
Latent fingerprint, 36
Laundering, money, 220
Law enforcement
    CSI Effect and, 45
    need to adapt, 10
    training in, 91
Law enforcement agencies. *See* Agencies
Law Enforcement Assistance Administration (LEAA), 78, 80
Law Enforcement Management and Administrative Statistics, 2013 (LEMAS), 18, 77
LEAA (Law Enforcement Assistance Administration), 78, 80
Learning management system (LMS), 84
Left-wing groups, 233
Legal Information Institute (LII), 108
LGBTQ victims, 176, 261. *See also* Orlando nightclub shooting
Liederbach, J., 93
Lifestyle, 54
Life without possibility of parole (LWOP), 261
Ligature, 132
Lincoln, Abraham, 4
Lineups, 110, 112, 113
    accuracy of identification in, 62
    defense attorneys and, 61
    procedures, 61
Lividity, 129
Livor mortis, 129
Loan sharking, 219
Locard, Edmond, 12, 28
Locard's Exchange Principle, 12
London, 3
London Metropolitan Police Department (The Met), 3, 59
Lone wolf terrorists, 234, 236–237, 238 (table)
Lovgren, S., 44

Mahoney, L. S., 60
Mail theft, 193
Major cases, 145
Major case squad, 9
Malice aforethought, 122
Managing criminal investigations. *See* Investigation, managing
Managing Criminal Investigations Manual, The (Cawley et al.), 81
Manslaughter, 122, 124, 132. *See also* Death investigations
Mapp v. Ohio, 106
Marijuana/cannabis, 210, 211, 217. *See also* Drugs, illegal
Marijuana Tax Act, 210

Martin, G., 234
Maryland v. Buie, 106
Mass murder, 138–139
McCorkle, K. D., 150, 240
McVeigh, Timothy, 35
Media
    death investigations and, 135–136
    defense attorneys and, 254–255
    homicide cases and, 248, 251
    relationship with, 89
    terrorism and, 235, 237, 239
Medical examiner/coroner (ME/C), 120, 127–130. *See also* Death investigations
Meesig, R. T., 91
Megan's Law, 155
Memory, 61
Mens rea, 56, 258
Methamphetamine, 217. *See also* Drug crimes; Drugs, illegal
Michigan Department of State Police v. Sitz, 105
Minnesota v. Dickerson, 104
Miranda, D., 58
Miranda v. Arizona, 109, 113
Miranda warning, 57, 62, 64–65, 254
M'Naghten rule, 258
Mobile computer terminal (MCT), 40
Model Policy on Suicide Bombings, 236
Modus operandi (MO), 152–154
Money laundering, 220
Moore, M. H., 92
Moore, S. J., 60
Motive, 56
Motor vehicle theft, 195–196
    carjacking, 151, 195
    investigating, 196
    rates of, 188
MS-13, 220. *See also* Gangs
Multijurisdictional task forces, 15, 16
Munchausen Syndrome by Proxy (MSBP), 135, 180–181
Murder, 122, 124
    conviction for, 261
    vs. homicide, 122
    See also Death investigations; Homicide
Murder, first-degree, 122
Murder, mass, 138–139
Murder, second-degree, 122
Murder, serial, 138
Murder, third-degree, 122
Muslim citizens, 240

Narcotic Division of the Internal Revenue Bureau, 5
Nassar, Larry, 159
National Center for Health Statistics (NCHS), 121
National Center for the Analysis of Violent Crime (NCAVC), 39
National Center for Victims of Crime, 55, 173, 175, 176

National Center on Elder Abuse (NCEA), 181
National Crime Information Center (NCIC), 28, 201
National Crime Victimization Survey (NCVS), 154, 155, 168, 169, 174
National Forensic Science Technology Center, 149
National Incident-Based Reporting System (NIBRS), 121, 122
National Institute for Occupational Safety and Health (NIOSH), 151
National Institute of Justice (NIJ), 19, 33, 213
National Institute of Mental Health (NIMH), 133
National Institute on Drug Abuse (NIDA), 213
National Insurance Crime Bureau, 196
National Integrated Ballistics Information Network (NIBIN), 34
National Intimate Partner and Sexual Violence Survey, 174
National Motor Vehicle Theft Act, 196
National Protocol for Sexual Assault Medical Forensic Examinations, A, 156, 160
National Resource and Technical Assistance Center for Improving Law Enforcement Investigations (NRTAC), 91
National Retail Federation, 194
National Terrorism Advisory System (NTAS), 240
National Violent Death Reporting System (NVDRS), 121
Necessity defense, 257
Necrophilia, 160
Neighborhood canvass, 126, 152
Neighborhood watch program, 17
Neumayer, E., 234
News coverage. *See* Media
New York v. Belton, 107
Nichols, Terry, 35
9/11, 234, 235, 240
Nixon, Richard, 210, 211
Nix v. Williams, 107
Nolo contendere (no contest), 249, 250
Norris, J. J., 232
Notetaking, 43–44, 172
Not guilty, 249, 250

Occupational Safety and Health Administration (OSHA), 132
Office for Victims of Crime Training and Technical Assistance Center, 16
Office of Community Oriented Policing Services (COPS), 175, 176
Office of Justice Programs, 16
Office of Sex Offender Sentencing, Monitoring, Apprehending, Registering and Tracking (SMART), 161
Office of the Inspector General (OIG), 233, 240
Office on Violence Against Women, 160, 173
Oklahoma City bombing, 35
Oliver v. United States, 105
O'Neill, M., 251–252
Open fields exception, 105

Operation Intercept, 210
Opioids, 211, 212 (figure), 213. *See also* Drugs, illegal
Orfila, Mathieu, 28
Organizational groupthink, 86, 87
Organized crime (OC), 219–220
Organized retail crime (ORC), 194
Orlando nightclub shooting, 138, 139
Ortmeier, P. J., 34
Other-awareness, 64
Outlaw motorcycle gangs (OMG), 220. *See also* Gangs
Overdose, 211, 212 (figure). *See also* Drugs, illegal

Painkillers, 211. *See also* Drugs, illegal
Paraphilias, 159
Parents of Murdered Children (POMC), 136
Parkland school shooting, 138–139
Pat-down searches, 104
Patent print collection, 36
Patriot Act, 232
Patrol officers, 11
    arrests for possession of drugs by, 214
    death investigations and, 124–125
    importance of, 80
    increased investigative duties, 90
    as investigators, 58
    responsibilities of, 77
    role of, 29, 79–80
    training of, 91
    See also First responders; Preliminary investigation
Pawn shops, 193
Pawson, R., 18
PBS, 261
Pedophiles, 180
Peel, Robert, 3, 27
People for the Ethical Treatment of Animals (PETA), 233
Perimeter, 29
Perimortem, 129
Perjury, 258
Personal protective equipment (PPE), 37
Personal retribution, 52
Personnel, safety of, 37, 173, 213–214, 217
Persons crimes, 8, 145
    follow-up investigation and, 81
    See also Assault; Battery; Domestic violence (DV); Rape; Robbery; Sex crimes; Sexual battery/rape
Petersilia, J., 11, 82
Photography, 31
Photography/video, 41–42
    CSI Effect and, 41
Pinkerton, Allan, 4
Pinkerton's National Detective Agency, 4
Plain feel exception, 104
Plain smell exception, 104
Plain view exception, 102, 104–105
Plastic print collection, 36

Plea, 249, 250, 259
Plea bargains, 249–250, 259–261
Plümper, T., 234
Poisoning, 132–133
Poland, J. M., 232, 237
Police departments. *See* Agencies
Police Executive Research Forum (PERF), 16, 90, 110
Policing research, 18–19
Polk, Kenneth, 150
Polygraphs, 68
Pope, P. B., 252
Postal Service (USPS), 5, 193, 197
Postmortem, 130
Power-assertive rapist, 156
Power-reassurance rapist, 156
Prediction, prevention and preemption, 8
PRELIMINARY acronym, 12
Preliminary investigation, 10, 11–13, 79–80, 124–126
Premeditation, 122, 256
Premortem, 129
Prescription medications, 211. *See also* Drugs, illegal
Presumed Guilty (documentary), 261
Presumptive/field tests, 26
Pretrial release, 249
Primary witness, 60
Private eyes, 4
Probability errors, 86, 87–88
Probable cause, 62, 98, 100, 107
Problem-oriented policing (POP), 17–18
Productivity, improving, 89–93
Prohibition era, 210
Property, abandoned, 105
Property crimes, 8
 clearance rate, 7, 53
 follow-up investigation and, 81
 rate of, 188, 188 (figure)
 types of, 188
 See also Burglary; Motor vehicle theft; Theft
Prosecutors, 255
 burden of proof and, 254
 caseload of, 251–252
 charging decisions, 251–253
 closing summary, 250
 coordination with, 83, 253
 CSI Effect and, 45
 ethics and, 259
 filing of charges and, 248
 interaction with, 13–14
 jury selection and, 44
 overarching principles for, 252
 specialized, 253
 working with, 251
 See also Courtroom trials
Prostitution, 219
Protective sweeps, 105

Psychological theories of crime, 54
Public
 death investigations and, 135–136
 defense attorneys and, 254–255
Public defenders, 255, 260–261. *See also* Defense attorneys
Puncture wounds, 131
Pure Food and Drug Act, 210
Pyrooz, D. C., 253

Racketeer Influenced and Corrupt Organizations Act (RICO), 220
Radio frequency identification (RFID), 194
Raghav, M., 252
Rahman, N. H. A., 236
Ramseyer, M., 252
RAND Corporation study, 11, 18, 80, 82
Rape, 145, 154–160
 college rape, 159
 investigating, 157–159
 physical evidence of, 157, 158
 reporting, 154
 sexual assault examination, 160
 typologies, 155–157
 victims of, 155, 157–159
 See also Sex crimes; Sexual assault; Sexual battery/rape
Rape classifications, 156
Rasmusen, E., 252
Reagan, Nancy, 211
Reagan, Ronald, 211
Reasonable doubt, 249
Reasonable person, 100, 107
Reasonable suspicion, 100
Reasoning, abductive inferential, 60
Reasoning, deductive, 60
Reasoning, inductive, 60
Reasoning, rational, 86
Recording devices, 67. *See also* Cameras; Technology
Records management systems (RMS), 40, 84–85
Rectangular coordinate method, 43
Reference, in photography, 36
Reliable confidential informant (RCI), 62, 82
Report Review Checklists, 168. *See also* Checklists
Report writing/notetaking, 43–44
 in IPV cases, 172
Representation, competent, 255. *See also* Defense attorneys
Representativeness, 87
Restraining orders, 171–172
Reviews, external, 88–89
Rights of accused, 57, 62, 64–65, 99, 254. *See also* Counsel, right to; Suspects; *Individual amendments*
Right-wing groups, 233
Rigor mortis, 129
Roadblocks, 105

Road rage, 151
Robbery, 145, 151–154
   clearance rate for, 154
   documentation of crime scene, 152
   modus operandi (MO), 152–154
   statistics, 154
   victims in, 152
   witnesses in, 152, 153
Roberts, J., 260
Rossmo, D. K., 86, 87–88
Routine activities theory (RAT), 54, 191
Rubin, P. H., 260
Running Case Study
   Assault, Battery, Robbery, and Sex Crimes, 143–145
   Burglary, Theft, White-Collar Crime, and Cybercrime, 185–187
   Criminal Investigation in Court, 245–247
   Criminal Investigation Then and Now, 1–3
   Drug Crimes, Organized Crime, and Gangs, 207–209
   Familial Crimes, 165–167
   Forensic Evidence, 24–25
   Homicide, Death, and Cold Case Investigations, 118–119
   Managing the Criminal Investigation, 73–76
   People in the Process, 49–51
   Searches, Seizures, and Statements, 97–99
   Terrorism and Homeland Security, 227–229
Russian election interference, 202–203

Saferstein, R., 19
Safety, personnel, 37, 173, 213–214, 217
Sanchez, Raphael, 199
Sandusky, Jerry, 159
Satisficing, 87
Saucier v. Katz, 107–108
Scanning electron microscope (SEM), 35
Scatologia, 160
Scene contamination, 11–12
Science, 19, 25, 27–29. *See also* Forensic evidence; Laboratories, crime; Technology
Scientific jury selection (SJS), 250
Scotland Yard, 3
Searches
   consent to, 218
   probable cause and, 62
Search warrant exceptions, 102–106
Search warrant return, 100
Search warrants, 100–102, 107, 176
Secondary witness, 60
Secret Service, 5
Security threat groups (STGs), 220. *See also* Gangs
Seizures, 99–100, 101, 107. *See also* Arrest
Self-awareness, 64
Self-defense, 123, 256, 257
Self-incrimination, 108, 259
Sentences, mandatory, 211

Sentencing, 251, 261
September 11, 2001, 234, 235, 240
Sex crimes, 145, 154–161
   college rape, 159
   criminal deviant sex acts, 159–160
   underreporting of, 155
   See also Rape; Sexual assault; Sexual battery/rape
Sex offenders, 155, 161
Sex trafficking, 219
Sexual abuse, 159
   of minors, 180
Sexual assault, 154–160
   investigation of, 156
   reporting, 154
   sexual assault examination, 160
   typologies, 155–157
   victims of, 156
   See also Intimate partner violence (IPV); Rape; Sex crimes; Sexual battery/rape
Sexual assault nurse examiners (SANE), 158
Sexual assault response teams (SART), 158
Sexual battery/rape, 145, 154–161
   investigating, 157–159
   sexual assault examination, 160
   victims of, 157–159
   See also Rape; Sex crimes; Sexual assault
Sexual sadist, 157
Sex workers, 219
Shepherd, J. M., 260
Sheriff, origins of, 4
Shoplifting, 192–193
Showups, 110
Shrinkage, 194–195
Sicarii Zealots, 231
Simon, Herbert, 87
Single issue, 233
Sixth Amendment, 99, 109, 112–113. *See also* Counsel, right to; Trial, right to
Sketches, 31, 42–43
Sobriety checkpoints, 105
Social control perspective, 54
Social control theories, 54
Social disorganization, 54
Social learning theory, 54, 171
Social media, 193
   election interference and, 202–203
   terrorism investigation and, 235
Socioeconomic status, verdicts and, 261
Soft skills, 59
Solvability factors, 13, 80, 81
Sparrow, M. K., 92
Spiral pattern, 30, 30 (figure)
Spohn, C., 253
Spree killing, 138
Stalking, 172, 173–175

Stalking Resource Center (SRC), 173, 174
Stand-your-ground laws, 150, 257
Stanford Research Institute (SRI), 80
State attorney. *See* Prosecutors
Statements, 108–110, 112
    voluntariness of, 109, 218
    See also Fifth Amendment
Stippling, 130
Stockholm Syndrome, 171
Stop and frisk, 102, 104
Strain theory, 54
Strangulation, 131–132, 172
Strip pattern, 30, 30 (figure)
Strong arm robbery, 151
Subcultural perspective, 54
Substance abuse. *See* Drug crimes; Drugs, illegal
Substantial capacity, 258
Sudden Infant Death Syndrome (SIDS), 134, 180
Sudden Unexplained Infant Death (SUID), 134, 180
Suicide, 123–124, 133–136
    in mass shootings, 139
    methods of, 133 (table)
Suicide bombings, 235–236
Supervisors, 83–84. *See also* Investigation, managing
Supplemental Homicide Report (SHR), 121
Supplemental Victimization Survey, 174
Surveillance, hard, 58
Surveillance, soft, 59
Survivors, 136
Suspects
    false identification of, 60
    field identification of, 57
    identifying, 56–57
    managing, in death investigations, 125
    See also Accused, rights of; Lineups; *Individual amendments*
SWAT, 152
System 1 thinking, 86
System 2 thinking, 86

Task forces, 15–16, 82
Tattersall, A., 58, 85
Tattooing, 130
Tatum, K. M., 136
Team policing, 82
Technology, 19, 25
    adapting to, 92
    arson/explosives and, 35
    cold cases and, 137
    development of, 28–29
    drug crimes and, 210
    ethics in, 32
    gang activity and, 210
    in interviews/interrogations, 67–68
    See also Case management software (CMS); Forensic evidence

Tenth Amendment, 76
Terrorism
    background, 231
    compared to hate crimes, 232
    defining, 230–231
    domestic, 230–231, 232–234
    fusion centers, 240–241
    homegrown violent extremists (HVE), 233
    homeland security, 239–241
    international, 230, 234–235
    investigating, 235, 236, 240–241
    local law enforcement and, 229
    lone wolf terrorists, 234, 236–237, 238 (table)
    media and, 237, 239
    militia groups, 234
    motivations for, 229, 240
    suicide bombings, 235–236
Terry v. Ohio, 104
Texas Rangers, 4–6
Theft, 190, 192–194
    employee theft, 194–195
    motor vehicle theft, 195–196
    organized retail crime (ORC), 194
    preliminary investigation of, 192–193
    rates of, 188
Thornton v. United States, 107
Tool marks, 34, 35, 190
Torquemada, Tomás de, 67
Totality of circumstances, 100
Trespass, 190
Trial
    right to, 112
    See also Courtroom trials
Triangulation, 43
Tunnel vision, 86–87, 88
Twenty-First Amendment, 210

Undercover operations, 215–216
Uniform Crime Report (UCR), 121, 135, 147, 150, 154, 155, 187, 189, 192, 195, 196, 213
United Nations Office on Drugs and Crime (UNODC), 216
United States Marshal Service (USMS), 9
United States v. Leon, 106
United States v. Martinez-Fuerte, 105
University Research Corporation, 78
Unreasonable searches and seizures, 99–100, 104. *See also* Searches; Seizures
Urbanization, 4
USA.gov, 199

Vandalism, 190
Vecchi, G., 253
Vehicles, 102
Vehicle theft, 195–196
Venire, 250

Verdict, 251, 261
Verification bias, 87–88
Vice crimes, 8
Victim advocates, 55–56
Victim blaming, 150
Victim impact statements, 53, 251
Victimology, 54, 154
Victim-precipitated homicide, 150
Victims, 52–56
   in assault, 147, 148, 149
   in burglary, 189
   children, 55
   in cybercrime, 201
   elderly, 55, 181–182, 197–198
   in IPV, 55, 171, 172, 176
   LGBTQ, 176, 261
   National Crime Victimization Survey (NCVS), 154, 155, 168, 169, 174
   Office for Victims of Crime Training and Technical Assistance Center, 16
   in rape/sexual assault, 155, 156, 157–159
   in robbery, 152
   socioeconomic status of, verdict and, 261
   special, 55
   in stalking, 175
   treatment of, 53
   as witness, 60
Victims' rights, 55
Vidocq, François, 27
Violent crimes, 121, 146–147. *See also* Assault; Battery; Rape; Robbery; Sex crimes; Sexual battery/rape
Violent Criminal Apprehension Program (ViCAP), 15–16, 39
Voice stress analyzer (VSA), 68
Voir dire, 113, 250
Vollmer, August, 28

Voluntary consent, 103
Voyeurism, 160

Walsh, J. P., 91
War on Drugs, 210
Weeks v. United States, 106
Wellford, C., 127
Westera, N. J., 58
Wheatcroft, J. M., 255
White-collar crime, 187, 196–199
Wilson v. Arkansas, 100
Witnesses, 60–62
   accuracy/reliability of, 61–62
   in assault, 148, 149
   cold case investigations and, 79
   controlling testimony of, 256
   in death investigations, 126
   defense, 253
   direct examination of, 256
   expert, 35–36
   identifying, 12
   interviewing, 61, 62–64
   preparing, 253, 255
   prosecution, 253
   in robbery, 152, 153
   testimony of, 253, 256
   in trials, 255
   See also Eyewitnesses
Wolfe, S. E., 253
Wolfgang, Marvin, 150
Womack, C., 93
Workplace violence, 151
Wounds, 131

Young, T. J., 34